AF443675

FEATURE INTERACTIONS IN TELECOMMUNICATIONS AND SOFTWARE SYSTEMS VII

Organised and hosted by

University of Ottawa, Canada

School of Information
Technology and Engineering,
University of Ottawa

In Co-operation with

Université du Québec en
Outaouais, Canada

Ottawa Centre for Research
and Innovation

Feature Interactions in Telecommunications and Software Systems VII

Edited by

Daniel Amyot

School of Information Technology and Engineering, University of Ottawa,
Ottawa, Canada

and

Luigi Logrippo

Département d'informatique et d'ingénierie,
Université du Québec en Outaouais, Gatineau, Canada

IOS
Press

Ohmsha

Amsterdam • Berlin • Oxford • Tokyo • Washington, DC

ISBN 1 58603 348 4 (IOS Press)
ISBN 4 274 90594 2 C3055 (Ohmsha)
Library of Congress Control Number: 2003104683

Publisher
IOS Press
Nieuwe Hemweg 6B
1013 BG Amsterdam
The Netherlands
fax: +31 20 620 3419
e-mail: order@iospress.nl

Distributor in the UK and Ireland
IOS Press/Lavis Marketing
73 Lime Walk
Headington
Oxford OX3 7AD
England
fax: +44 1865 75 0079

Distributor in the USA and Canada
IOS Press, Inc.
5795-G Burke Centre Parkway
Burke, VA 22015
USA
fax: +1 703 323 3668
e-mail: iosbooks@iospress.com

Distributor in Germany, Austria and Switzerland
IOS Press/LSL.de
Gerichtsweg 28
D-04103 Leipzig
Germany
fax: +49 341 995 4255

Distributor in Japan
Ohmsha, Ltd.
3-1 Kanda Nishiki-cho
Chiyoda-ku, Tokyo 101-8460
Japan
fax: +81 3 3233 2426

LEGAL NOTICE
The publisher is not responsible for the use which might be made of the following information.

PRINTED IN THE NETHERLANDS

Introduction

This book is the Proceedings of the 7th International Workshop on Feature Interactions in Telecommunications and Software Systems, held in Ottawa, Canada, June 11–13, 2003, in the School of Information Technology and Engineering of the University of Ottawa.

Communications systems offer services and features to their subscribers. Features are often among the main selling points of such systems, and they can be implemented and provided independently by specialized operators under request of users, or by users directly. Features can build on each other, but can also mutually interact in subtle and unexpected ways. A feature can modify or subvert the behavior of another one, possibly leading to system malfunctions. The topic of feature interaction deals with preventing, detecting and resolving such interactions. This phenomenon is not unique to the domain of telecommunications systems: it can occur in any software system that is subject to changes (not to mention areas of medicine, engineering and law, which are not directly covered in this workshop but may be taken into consideration in order to exploit useful analogies).

Although interactions among classical telephony features are now fairly well understood, the feature interaction problem presents new challenges in emerging types of systems based on policies, dynamic (Web) services, mobility, or new architectures such as Parlay, 3G, .NET, or grid and active networks. The proliferation of players and software/service engineering techniques coupled with the constant pressure for the rapid introduction of new services and features can lead to undesirable interactions that jeopardize the quality of the products delivered as well as the satisfaction of the users. Note that, ideally, each subscriber should be able to create or customize his or her own features, dynamically as the need arises, but different subscribers (who may or may not become connected to him/her) may have conflicting goals and objectives. Techniques successfully applied to conventional telecommunications systems are still useful in many cases, yet they may no longer be able to cope with the complexity of emerging systems.

This workshop series was born over 10 years ago, at the initiative of a group from Bellcore (now Telcordia). The following table summarizes its history:

Year	Location	Chairs
FIW I (1992)	St. Petersburg, Florida, USA	N. Griffeth and Y.-J. Lin
FIW II (1994)	Amsterdam, The Netherlands	L.G. Bouma and H. Velthuijsen
FIW III (1995)	Kyoto, Japan	K.E. Cheng and T. Ohta
FIW IV (1997)	Montréal, Canada	P. Dini, R. Boutaba and L. Logrippo
FIW V (1998)	Lund, Sweden	K. Kimbler and L.G. Bouma
FIW VI (2000)	Glasgow, Scotland	M. Calder and E. Magill

The tradition in this workshop has always been to establish a meeting point between industrial and academic views. In this workshop, 17 program committee members are from universities and 9 are from industry or are independent consultants. 14 accepted papers come from university groups, 4 papers from industrial groups, 2 from mixed university and industrial groups, and 1 from a government research group. All invited speakers were from industry or independent consultants. There are 7 papers from the UK (of which one with a Belgian collaborator), 5 from Canada, 3 from Germany and 3 from Japan, while China, France and the USA have one each.

There are 8 short papers (8 pages maximum) and 13 regular papers (18 pages maximum), but both types of papers were refereed by four experts.

The papers are grouped in the following categories, which roughly correspond to sessions (the number at the beginning indicates the number of papers):
- (4) Architecture and Design Methods
- (5) Emerging Application Domains
- (2) Human Factors
- (2) Detection and Resolution Methods I
- (4) Emerging Architectures
- (2) Foundations
- (2) Detection and Resolution Methods II

We must thank the Program Committee members, who helped us in many important decisions (the most important one being, of course, which papers to accept), as well as the reviewers, a list of whom is given on page viii; the University of Ottawa, who made available the premises (in the new SITE building) and helped us in the organization; the Ottawa Centre for Research and Innovation, who provided logistic support; and IOS Press, who helped prepare this book, printed and distributed it. We also thank all the authors for sharing their wisdom and experience, and the three invited speakers for providing enlightening talks: Bill Buxton, Debbie Pinard and Pamela Zave. The paper submission and review software used was Cyberchair (http://www.cyberchair.org). We thank Richard van de Stadt of the University of Twente for having readily made available such a good tool. We are indebted towards Jacques Sincennes, who provided invaluable technical support with the submission and review process. Merci Jacques!

But above all, we must thank the Feature Interaction community for their continuing support of this workshop series.

The results of this workshop show that the Feature Interaction problem, several times said to have been "solved" by various techniques and in various contexts, is very much alive and is acquiring new dimensions as the concept of feature evolves.

We would like to take this occasion to commemorate the passing away of L.G. (Wiet) Bouma of KPN Research. Wiet was present at the very first workshop, co-chaired FIW II and V, and had a role in almost all workshops, as an author or PC member. Sadly, he died of cancer on October 26, 2001. The Best Paper Award in this workshop will carry his name.

Daniel Amyot and Luigi Logrippo, Workshop co-chairs
University of Ottawa and Université du Québec en Outaouais
March 2003

Committees

Workshop Co-Chairs

Daniel Amyot	University of Ottawa, Canada
Luigi Logrippo	Université du Québec en Outaouais, Canada

Programme Committee

Joanne Atlee	University of Waterloo, Canada
Lynne Blair	University of Lancaster, England
Muffy Calder	University of Glasgow, Scotland
Pierre Combes	France Télécom R&D, France
Bernard Cohen	City University, England
Petre Dini	Cisco Systems / Concordia University, USA
J. Paul Gibson	NUI Maynooth, Ireland
Tom Gray	GRconsultants, Canada
Jean-Charles Grégoire	INRS-Télécommunications, Canada
Robert J. Hall	AT&T Labs Research, USA
Bengt Jonsson	Uppsala University, Sweden
Ferhat Khendek	Concordia University, Canada
Ahmed Khoumsi	Université de Sherbrooke, Canada
David Lee	Bell Labs Research, China
Yow-Jian Lin	SUNY Stony Brook, USA
Evan Magill	University of Stirling, Scotland
Dave Marples	Global Inventures, USA
Masahide Nakamura	Nara Institute of Science and Technology, Japan
Tadashi Ohta	Soka University, Tokyo, Japan
Farid Ouabdesselam	LSR-IMAG, Grenoble, France
Mark Ryan	University of Birmingham, England
Henning Schulzrinne	Columbia University, USA
Simon Tsang	Telcordia Technologies, USA
Ken Turner	University of Stirling, Scotland
Greg Utas	Sonim Technologies, USA
Pamela Zave	AT&T, USA

Chair, Most Novel Domain Award Committee

Mark Ryan	University of Birmingham, England

Reviewers

Joanne Atlee
Lynne Blair
Muffy Calder
Zohair Chentouf
Soumaya Cherkaoui
Caixia Chi
Pierre Combes
Bernard Cohen
Alessandro De Marco
J. Paul Gibson
Tom Gray
Jean-Charles Grégoire
Dimitar Guelev
Robert J. Hall
Geoff Hamilton
Ruibing Hao
Hannah Harris
Bengt Jonsson
Ferhat Khendek
Ahmed Khoumsi
David Lee

Yow-Jian Lin
Christine Liu
Evan Magill
Dave Marples
Dominique Méry
Alice Miller
Masahide Nakamura
Tadashi Ohta
Farid Ouabdesselam
Jianxiong Pang
Stephan Reiff-Marganiec
Jean-Luc Richier
Mark Ryan
Henning Schulzrinne
Simon Tsang
Kenneth J. Turner
Greg Utas
Dong Wang
Pamela Zave
Nan Zhang

Contents

Invited Papers

*Feature Interactions in Telecommunications
and Software Systems VII*
D. Amyot and L. Logrippo (Eds.)
IOS Press, 2003

3

Feature Disambiguation

Pamela ZAVE

AT&T Laboratories—Research, Florham Park, New Jersey, USA
`pamela@research.att.com`

Abstract. This essay proposes the related goals of discovering user-level telecommunication requirements and making telecommunication features unambiguous with respect to user-level concepts such as roles and purposes. It also offers a preliminary list of techniques for feature disambiguation.

1   Toward telecommunication requirements

Within software engineering, the subfield of *requirements engineering* is concerned with the nature and development of requirements [4]. A true requirement is a desirable property of the system's environment, the portion of the real world that affects and is affected by the system. It is a property that might not be true of the real world in the absence of the system, but that should be true of the real world in the presence of the system [5].

For example, consider two people miles apart, both standing by idle telephones. In the absence of a telecommunication system, no matter what each does with his telephone, they will not be able to hear each other speak. It is a requirement on a telecommunication system that if both telephones are connected to the system, and if each person performs the correct actions, they will be able to hear each other speak.

As we know, only an extremely simple system would satisfy such a requirement. The telecommunication systems we are concerned with have hundreds of features, many of which could prevent a connection between the two people under various circumstances. To be satisfiable, a connection requirement would have to include exceptions for all these features.

Currently there are no user-level requirements for telecommunication systems. We simply do not know of any properties that are widely agreed to be desirable to users, can be stated precisely, and are satisfiable (let alone satisfied) by systems with normal levels of complexity and customizability.

This situation is entirely understandable, being due to the following factors:

- The long history of telecommunications, including the technological limitations of the past.

- The incremental nature of feature development.

- The geographical and administrative distribution of telecommunication networks.

- The conflicting goals of subscribers.

- The fact that telecommunication services are so entwined in our lives that we hope they will do exactly what we want in every situation. No matter how hard we try, we will never keep our software systems up-to-date with the complexity of real life [1]. And if we did, people would not use them because there would be too much administration to do.

Despite these past and present obstacles, we have an obligation to do better. People today are demanding privacy, security, predictability, and reliability from the computer systems they rely on so heavily. We cannot expect them to exempt telecommunications from these expectations. When they start asking what guarantees we can provide, we do not want to have to admit that we understand nothing about the global behavior of our systems.

2   Requirements and architecture

Most of the research on feature interaction being done now assumes a telecommunication architecture. Features exist within the architecture as components or other recognizable entities. The architecture is an infrastructure that composes the behavior or properties of the various features to generate the behavior of a system as a whole.

Ideally, telecommunication requirements would be stated without reference to the architecture of the telecommunication system, because architecture is internal and requirements are supposed to be external. Requirements stated in this way would be valid constraints on any architecture. That does not seem possible now, however, and I do not know if it will ever be possible in the future.

For example, it would be good to have a responsiveness requirement, stating that a system responds to each user request within a bounded amount of time. But there are so many possible responses, and so many different reasons for lack of response, that any precise *external* statement of responsiveness falls apart immediately—it has too many legitimate exceptions.

Taking an architectural view, on the other hand, we might formulate one aspect of responsiveness in terms of a chain of requests for communication. The chain begins at a source device interface, passes through feature modules, and possibly ends at a target device interface. The requirement states that if each feature module that receives a request continues the chain with another request, then the chain reaches a target interface module within a bounded number of steps. This requirement prevents forwarding loops. Yet it allows screening features, because a screening feature module would not continue a request chain if it is blocking calls from this caller.

I assume that, for the forseeable future, each attempt to study telecommunication requirements will have a specific architecture as its formal framework. This has advantages as well as disadvantages. It will lead to a valuable diversity of approaches. The attempt to relate requirements and architectures should illuminate both, leading to a faster convergence on the most effective architecture.

Understanding requirements will necessitate going beyond the mechanisms of current architectures, to study the semantics of specific features or classes of features. For example, an architecture might define how concurrent events at different priority levels are mediated. This is necessary, but it is not sufficient to explain anything about the external behavior of the system unless we know which events have which priorities and why.

In this context, another advantage of relating requirements to architecture is that an architecture can provide a framework for formalizing semantic information, and effective places to attach it. In other words, the architecture will acquire layers of semantic information, relating it more closely to the environment and to requirements. Subsequent sections will illustrate this concept.

3 Feature ambiguity

Consider two addresses *t1* and *t2*, each of which subscribes to some features, such as voice mail, that are failure treatments. A particular call to *t1* is forwarded by some feature of *t1* to *t2*. Then the attempt to reach *t2* fails. Should the requirements specify application of the failure treatments of *t1* or *t2*?

This scenario can occur within a large number of contexts, each of which might justify a different answer, or at least provide a different justification for one of a small set of answers. Here are two such contexts:

- *t1* is the address of a sales group, and *t2* is the address of a sales representative in the group.

- *t1* and *t2* are the work addresses of coworkers. The worker with address *t1* is on a leave of absence.

In the first context, the features of the group address *t1* should handle the failure. They can find another sales representative who is available, or record voice mail that is accessible by anyone in the group.

In the second context, the features of the delegate address *t2* should handle the failure, because the delegate is working and will answer messages promptly. Voice mail for *t1* may not be received for a very long time.

This example illustrates that most of today's features and feature interactions are laden with ambiguity about what is being accomplished, why it is being accomplished, and on behalf of whom. This ambiguity makes it impossible to determine desirable system behavior, because any decision we might make will be wrong in many circumstances. The only way to rectify this situation is to reduce the ambiguity of features until there is consensus on every case. The information that reduces ambiguity is semantic information of the kind mentioned in the previous section.

The remainder of this essay explores some possible means of feature disambiguation. They are discussed in order, from those that are straightforward and easy to apply, to those that are mysterious and challenging. We have much to learn on the subject of feature disambiguation; my present purpose is to stimulate discussion and further exploration of it.

4 No open-ended "user preferences"

Often papers on feature interaction fall back on an open-ended notion of "user preferences." These preferences are supposed to determine the detailed behavior of some feature.

There is nothing wrong with user customization of features, but the user must always be making choices from a predetermined set of options. ¿From a formal viewpoint the behavior of the feature is a nondeterministic choice from that set, which is perfectly well-defined.

It is obvious that enumerating user options is the only way to obtain a complete formal description of the behavior of a feature. There are also deeper reasons, however, for constraining user preferences.

The behavior of a feature affects how it interacts with other features, which in turn affects the global behavior of the system of which it is a part. Users cannot be expected to understand the global consequences of a local decision. Global consequences must be understood ahead of time by system designers, and user choices must be limited to those whose global consequences are acceptable.

Even further, the interests and goals of different users often conflict. Users cannot be allowed complete control over the behavior of their features, even if they do understand all its consequences, because then there is no way to balance conflicting interests. Finding this balance is one of the most important tasks facing requirements engineering for telecommunication systems. It will be the basis of any privacy or security guarantees we can make.

5   Associate features with purposes, not mechanisms

Telecommunication systems use a relatively small number of basic mechanisms, combined in many different ways, to create their range of functions. For example, the mechanism of changing the target address of a request chain before it has reached an actual target can be used for the following purposes:

- To reach a targeted person at a device near his expected location.

- To delegate responsibilities of one worker to a coworker.

- To reach a representative of a group, who is interchangeable with all other representatives with respect to the group.

- To conceal from a caller, behind an anonymous alias, the true identity of the callee.

- To assemble a group of co-located devices into a single virtual device having more media capabilities than any of the devices alone.

- To connect a caller with a resource such as a voice-mail system.

Clearly there is no enlightenment to be found in lumping all of these together under the name of "call forwarding"! "Call forwarding" is a mechanism that plays a part in the implementation of many features.

It makes much more sense to associate features with specific purposes. Then a feature is a coherent semantic entity that can be understood in relation to the semantics of other features. This is easy to do, as there is little cost to having multiple feature names that map to similar or even identical code.

6   Know what addresses identify

Just as a mechanism can be used for many different purposes in a telecommunication system, so can an address. In general, an address identifies an entity that someone is allowed to call. A callable entity might be a physical device, virtual device, resource, person, group of people, institution, anonymous alias, or other role.

Categorizing addresses is just as important as distinguishing purposes. In fact, the two often go hand-in-hand. If the purpose of a forwarding feature is to reach a representative of a group, it is only being used correctly if the subscribing address identifies a group and the forwarding address represents a person.

Address categories can be ordered by abstractness. In an architecture with request chains, this ordering can be used to constrain feature behavior in a way that guarantees certain boundedness, monotonicity, privacy, authenticity, and reversibility properties [6]. These properties are proposed as telecommunication requirements, thus illustrating the potential relationship among architecture, feature disamiguation, and requirements.

Address disambiguation will be more difficult to apply and enforce in the real world than feature disambiguation, because addresses are necessarily global, while features can be local. It will undoubtedly be necessary to qualify categories and properties with information about administrative domains.

7　Protocol disambiguation

Anyone who has ever done any development of telecommunication features knows that telecommunication protocols are full of ambiguity. A protocol must provide a standard, universal encoding of information, or it is not a protocol in any useful sense. At the same time it must handle an open-ended set of situations, which is the source of the ambiguity.

One approach to protocol disambiguation is taken in the SIP standard [3]. The standard lists 44 possible header fields and 50 response codes. There is an illustration of a header field with 9 parameters. The complexity of the entire SIP message set is combinatorial. Clearly this is an attempt to include in the protocol an encoding for almost any conceivably relevant information.

This encyclopedic approach places a tremendous burden on implementors of features and infrastructure alike. It is simply too complex to lead to the kind of conceptual understanding needed for intelligible user requirements.

It seems obvious that a smaller protocol would be better, but how can a protocol be made smaller without increasing its ambiguity? The only possible solution to this problem is to find the right abstraction, so that the information that cannot be encoded in the protocol is generated and used in the same location, and the information that is encoded in the protocol is globally unambiguous.

For example, the default assumption in SIP is that a server (feature node) has no necessary persistence in a signaling path. Because of this assumption, SIP messages commonly carry information that enables one server to pass responsibility to another server. For example, SIP has formal syntax to indicate "the intended user cannot be found at address a, but can instead be found at address b." Because such information is never enough, SIP syntax also makes it possible to indicate whether the move is permanent or temporary and, if temporary, how long it will last.

In DFC [2], on the other hand, once a feature node becomes part of a signaling path, it stays there until the entire path is destroyed. This minimizes the need for passing responsibility, and therefore also minimizes the need for transmitting descriptive information through the protocol.

Thus, in DFC, it is expected that the feature module associated with address a will be part of any signaling path attempting to reach a. This module always knows how to handle calls to a, and is the *only* module that needs to know how to handle calls to a. Because all information about how to handle calls to a is centralized in this module, there is no need for this information to be encoded in messages of the protocol and passed around.

It will be a long time before these issues are understood completely. In the meantime, we must accept incremental progress. Here is another example of protocol disambiguation in DFC.

Although the telecommunications infrastructure is evolving rapidly, there is still a huge installed base of PSTN telephones, and tremendous pressure to make anything new interoperate with the old. As a result, it is still extremely common for various telecommunication services to use a voice channel for signaling purposes (through tones, announcements, and DTMF tones).

When a call is being placed and the offered voice channel is opened, is the far end of the voice channel a person who is ready to talk or a machine that is ready to use the voice channel for signaling? Many features need to know the difference. In DFC the signaling for media channels is separated from call-level signaling. The call-level success signal indicates true end-to-end success—usually success in reaching a person—so that its presence or absence disambiguates opening of the voice channel.

For an operational example, consider a call that is first handled by an automated routing service. It opens the voice channel to the caller and uses it to recite a menu of possible call destinations. The caller then chooses one over the voice channel by entering a DTMF symbol. Throughout this phase no success or failure signal has been generated.

Next the routing service directs the call to the chosen destination. If the device there is busy, a call-level failure signal is generated. If the destination device alerts and is answered, then a call-level success signal is generated. ¿From the perspective of the caller and the caller's features, the voice channel is opened before the success signal; it is used for signaling before the success signal and for talking after the success signal.

8 Conclusion

This essay proposes the related goals of discovering user-level telecommunication requirements and making telecommunication features unambiguous with respect to user-level concepts such as roles and purposes. The features must act and interact in such a way that the requirements are satisfied. Telecommunication requirements need not describe system behavior completely—it is enough if they document the most critical aspects of privacy, security, predictability, *etc.*

These goals will not be achievable without making significant alterations in existing infrastructures and services. Although such change is always difficult, now is a better time for it than most. The telecommunication industry is in the midst of a conversion to IP-based networks. Many other technological innovations, for example in devices, media processing, and Web integration, are being absorbed. The inevitable confusion creates a window of opportunity for other changes.

Software engineers are currently very interested in the co-evolution of requirements and architectures. We are in a good position to contribute to their discussion. Telecommunication services have rich user-level behavior, which makes their requirements interesting. For a long time, our awareness of feature interaction has led us to explore the relationship between architecture and behavior. Telecommunication experts may have a deeper appreciation of the issues than experts in other application domains.

References

[1] Frederick P. Brooks, Jr. No silver bullet: Essence and accidents of software engineering. *Computer* XX(4):10-19, April 1987.

[2] Michael Jackson and Pamela Zave. The DFC Manual. AT&T Research Technical Report, February 2003, in `http://www.research.att.com/info/pamela`.

[3] J. Rosenberg, H. Schulzrinne, G. Camarillo, A. Johnston, J. Peterson, R. Sparks, M. Handley, and E. Schooler. SIP: Session Initiation Protocol. IETF Network Working Group RFC 3261, June 2002.

[4] Pamela Zave. Classification of research efforts in requirements engineering. *ACM Computing Surveys* XXIX(4):315-321, December 1997.

[5] Pamela Zave and Michael Jackson. Four dark corners of requirements engineering. *ACM Transaction on Software Engineering* XXII(7):508-528, July 1996.

[6] Pamela Zave. Ideal address translation: Principles, properties, and applications. In this volume.

The Periphery, Context and Intent: Architectural Considerations in the Design of Interactive Appliances

Bill BUXTON
Principal, Buxton Design
888 Queen St. East
Toronto, Ontario, Canada, M4M 1J3
Email: bill@billbuxton.com
http://www.billbuxton.com

Abstract. Much of the focus in the design and human factors of complex, but ubiquitous systems (such as telephones) has been on the usability and learnability of specific features. In the early days, this would have included the industrial design of the dial. More recently, the choice of coding the functions of voicemail would be an example.

While this class of consideration remains important, it is no longer sufficient. In fact, it is increasingly becoming secondary to the success, value usability and usefulness of the product. As we add more and more features, this will become increasingly true, especially if we continue along our current paths and criteria of analysis and evaluation.

Why? The reason lies in a classic forest vs. trees situation. As long as there are few trees, the "tree" (feature) and the "forest" (system) are almost the same thing. However, as we add features, no matter how well each tree is formed, the overall design of the forest may constitute a veritable maze, in which any normal person would become lost.

In this case, the quality of the design of the forest becomes more important to the intended user than any of the trees. The objective of this paper is to address some of the design issues that emerge from this seemingly trivial analysis, and to make a strong argument that the way to do a better job at what we are trying to do is to stop doing what we are doing, and follow a different path. Hopefully, the result will be a reasonable aid in navigating the maze. Chainsaws and landscaping tools will be provided.

*Feature Interactions in Telecommunications
and Software Systems VII*
D. Amyot and L. Logrippo (Eds.)
IOS Press, 2003

13

Reducing the Feature Interaction Problem in Communication Systems Using an Agent-Based Architecture

Debbie PINARD

Aphrodite Telecom Research, a Division of Pika Technologies
20 Cope Dr. Kanata Ontario
debbie.pinard@pikatech.com

Abstract. This paper provides a brief overview of traditional approaches to building communication systems and describes the limitations inherent in today's newer PBX architectures. It offers an in-depth look at a different approach, the Aphrodite agent-based architecture, and underscores its ability to reduce and in some cases eliminate some of the feature interactions.

1. Introduction

Legacy PBXs have formed the cornerstone of most business communications systems for over seventy years. Thanks largely to advancements in processor and bus designs, these systems today do more than just distribute calls originating from the PSTN to a bank of office phones. Unlike their ancestors, today's PBX solutions combine voice, data, and in some cases, video from the PSTN, Internet, Ethernet, and various other data networks and distributes the information over LANs, WANs, wireless or other modes of transport to office phones, faxes, pagers, PDA's, email and cellular phones all over the world. A product that previously handled just two basic functions - call distribution and voice messaging - can now perform many advanced tasks.

Feature interactions are a big problem in traditional PBX's, whose architectures exacerbate the problem. Enhanced PBX functionality has not resulted in substantial architectural improvements. The sheer complexity of software code used to implement core modules means that such systems remain expensive to maintain, hard to understand and extremely difficult to upgrade. The addition of new applications requires substantial investments in both time and skilled personnel. A small handful of key employees may understand some aspects of the complicated volumes of software code. New features in many cases, however, are often simply "barnacled" to existing base PBX software. This results in more and more feature interaction problems, increased inflexibility and additional support and maintenance costs. Traditional PBX equipment is also built on a uni-dimensional view of business communication requirements. Such systems are based on the assumption that businesses primarily need to simply connect two end points in order to satisfy their communication needs. This linear approach collapses the systems handling of business resources - such as computing platforms, workers, employee groups and network nodes - into functional software silos, limiting possibilities for code reuse, modularity and overall system simplicity.

Businesses however, also require the flexibility to easily configure who is contacted, at what time, in what location, for what reason and in what manner. A new architectural approach modeled on a business's own rules rather than a linear connection between telephones is therefore needed.

2. Second Generation Architectures, Just a Repackaging of the Same Old Thing

First generation PC-PBX software delivered basic applications such as voice-mail using client/sever based architectures. This provides rudimentary voice and data convergence via a complicated series of line interfaces. The separation of call control from applications within this architecture, of course, results in serious feature and application interaction problems and a tangled mass of spaghetti code to cope with messaging complexities as well as issues such as glares. Glares occur when two contradictory messages 'cross paths'. For example, a message saying a line is busy is sent at the same time as one asking to seize the line. Much like legacy PBX systems, the addition of more and more applications means increased rigidity, further integration challenges and additional costs.

Newer PC-based computer telephony solutions hold out the promise of open systems and commercially available components to deliver products with the features and inherent flexibility to meet next-generation business communications needs. Such software is designed to allow Original Equipment Manufacturers to focus more of their efforts on developing next generation applications, rather than worrying about integrating closed system architectures from various manufacturers. These new products are variously called UnPBXs, IP-PBXs, LAN-PBXs and Communications Servers. Most are based on standard components, such as PC hardware and OSs, and support IP as well as other common standards such as TAPI and H.323. Performance, cost and scalability are crucial to these systems. Most systems have strong integrated messaging features, including voice mail, automated attendant, desktop dialing software, fax, and screen pops. They are usually based on a Windows or Linux server. In many respects, the architecture of these second generation products is the same as that of earlier systems, as applications remain separated from call control functionality, producing the same feature and application interaction problems as first generation architectures (see Figure 1).

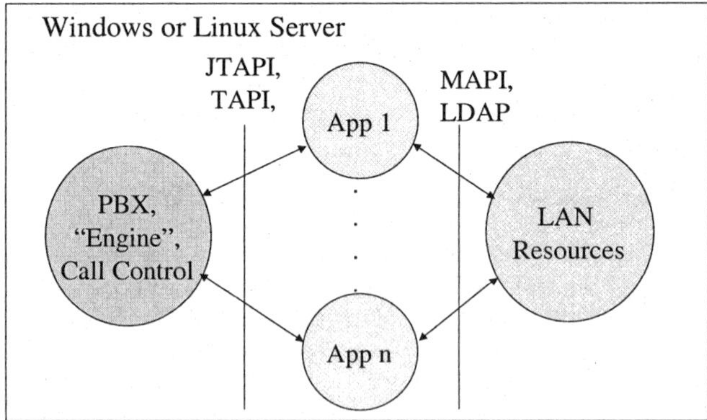

Figure 1: 1990's Architecture Combining PBX and Application Functionality

While somewhat more feature rich, this architecture also suffers from a number of fundamental design flaws. Call control in this scenario remains simply a "slave" to application demands, processing each on a "first come, first served" basis. This architecture has failed to resolve the glare problem, and has just added more complexity to the feature interaction problem, by adding in applications that have no knowledge of each other, or means of resolving conflicts. This ad-hoc approach limits the ability to tailor communication systems according to changing business processes, priorities, or needs, and will eventually produce a system that mirrors many of the earlier rigid, complex and expensive products it was designed to replace.

Today's preoccupation with Soft Switches has led many to conclude that call control can and should be separated from application delivery in any type of networking scenario without any negative consequences. This scenario really only works well in cases where each application is assigned its own port, or is triggered by a different event on the same connection as another application.

3. Aphrodite: An Agent-Based Architecture

The Aphrodite architecture takes a new approach [2]. Built from the ground up to model how a business operates, the framework implements previously external applications as fully integrated PBX features. This gets rid of the problem of Call Control being a slave to applications, and the glare situations that are inherent to that approach. *Control* over *who* gets access to *which* resources and *when* is defined by a data driven policy engine. In this environment, applications can be built to mimic any businesses communication process and can change according to new priorities, and allows for applications and features to compete fairly. Aphrodite's architecture enables the dynamic reconfiguration of system behaviors such as: whom employees are allowed to contact, at what time of day, from which location, to which device, and for what reason. These feature-rich capabilities are derived from both a data-driven approach to overall system design as well as an agent-oriented design methodology. Aphrodite's design results in a base-level system or agency analogous in many ways to a chameleon in its ability to quickly change "color" or function according to shifts in user requirements, organizational design or business priorities. Software developers use agent-based systems to help restore the lost dimension of individual perspective to the content-rich, context-poor world.

3.1. Why Agent-Oriented?

Aphrodite is built from the ground up to seamlessly overlay any business communication process and manage each resource within logical or physical groups according to shifting corporate priorities and needs. The first cornerstone of this architecture therefore involves modeling each business resource as a software entity that can represent that resource's functionality, constraints and information. Software agents deliver the most robust methodology to implement this approach [1]. Aphrodite's agent-oriented approach allows construction of a communications system or agency, based on organizational design.

Agents are capable of performing self-contained tasks, operating semi-autonomously and facilitating communication between the user and system resources. From a software engineering perspective, Aphrodite's agents are objects, each coded with its own set of "mental components" in the form of data that is acted upon by a state machine.

For numerous years, PBX designers have set out to build systems by representing individual ports as the principal business resource to monitor, manage and configure. This rigid and linear view fails to take into account that other resources such as roles, workgroups, individual employees and networked devices also interact and have value in their own right.

In the Aphrodite architecture, each entity, device and application service is represented as an agent. Relationships between the layers or between individual agents can be easily managed through policies, in much the same way that policy engines can direct, prioritize and control the flow of data packets in a policy-based data networking scenario. Agent relationships can be created at initialization. They can also be created and destroyed at run time, for the period of one communication session or until a user action destroys the relationship.

Aphrodite's agents adhere to policies set by the system administrator acting on behalf of the organization. Policies are basically data that affect when a feature is activated or what path a call takes.

Agents correspond to the organizational layers within a business. Functional agents (IDNumbers and Nodes) are represented by logical addresses assigned by the user, like name or phone number. Resource agents (Devices) are represented by physical addresses assigned by the system, like channel or port number. A many-to-many relationship, defined by policies, exists between the three layers (see Figure 2).

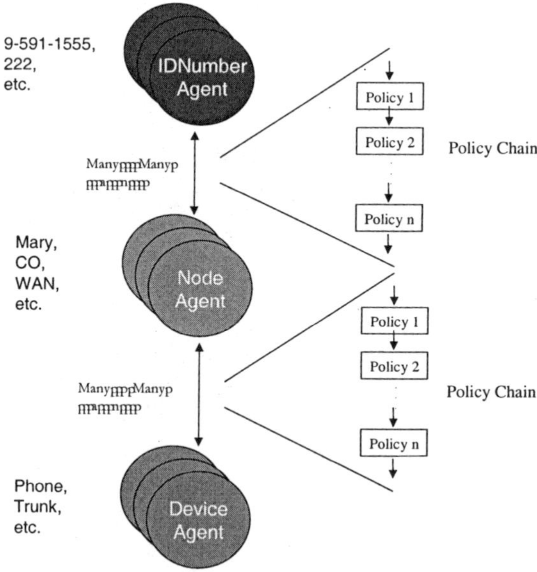

Figure 2: Agent Relationships

Policies can be chained, for example you can have a Day of Week Policy pointing from 1 to 7 different Time of Day Policies, etc. (see Figure 3). Other examples of policies are Calling Line ID (CLID), Calendar Status, Group, and Closed User Group.

The many-to-many relationships at each level mean that devices do not need to be tied to a particular phone number or person. For example, calls to a particular ID Number representing a help desk could be routed to different groups of people, based on the day of the week, or the time of day. As well, each person in the group can route the help desk call to one or more devices dependent on their policy settings. The IDNumber that the call is coming from can direct the policy path that is taken from the person to the device(s). The help desk represents a "role" rather than physical device or employee, and can be handled by any number of specified employees from any type of platform according to who has been assigned to that role.

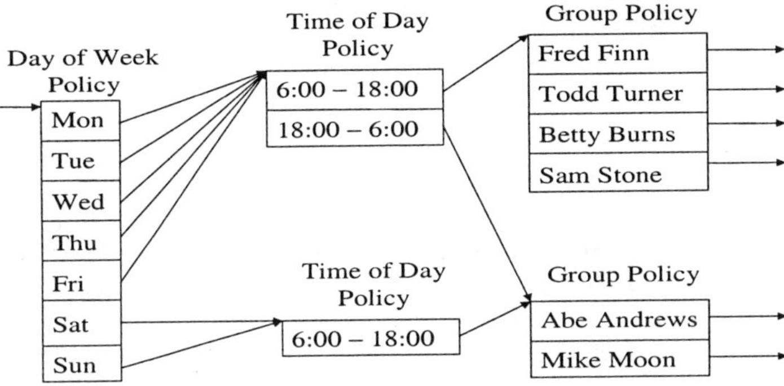

Figure 3: Example of a Policy Chain

This shows that the Aphrodite architecture allows the system administrator to use policies to set up the system based on their particular companies set of communication rules.

3.2.   *Why a Data-Driven Design?*

The second cornerstone of Aphrodite's architecture is its data-driven design approach. This helps reduce the complexity of software code and means that changes and enhancements to each component of the program can be made with minimal disruption to others. Overall system maintenance and use are therefore greatly simplified. Data driven design, in general, relies on the expansion of relationships among data to transform system requirements into programs. While process-driven design relies on the decomposition of processes to transform requirements into programs, expanding relationships among data entities reveals the Aphrodite program requirements. With policies, it is possible to design and program parts of an Aphrodite based application at a higher level of abstraction, with less concern about flow of control.

A record in a table represents an agent. Linked policy records represent the relationships between agents. Tailoring each system to meet individual business requirements can be accomplished by simply updating records within a database.

The use of a data-driven programming construct simplifies the difficult task of maintaining overall model integrity for complex applications in two important ways. First, it enables invariants and constraints to be stated explicitly in a single place, rather than having several scattered in multiple places. This makes code easier to understand and modify. Second, because it is data driven, the invariant/constraint is reevaluated automatically when relevant changes are made to object memory. This relieves the programmer of some of the burden of explicit procedural control. In particular, Aphrodite's procedural logic is no longer cluttered with code for maintaining model integrity. This architecture actually *eliminates* one form of feature interaction, since the type of call is inherent in the path it is following based on the policy rules, and does not need to be checked for or compared with other types of calls before invoking a feature. For example, in the old implementation of call forwarding, a *phone* has call forwarding set. If anyone calls, then call forwarding takes place, unless it is a hunt group call, then in some cases, it doesn't. These rules are all hard-coded. In the Aphrodite case, whether or not call forwarding is done is based on the path the call takes, which is based on data programmed by an administrator, so no checking has to be done on whether or not it is a hunt group call.

If it gets to the call forwarding code that means it was programmed to get there, and it should occur.

4. Features

Aphrodite has two types of features: Standard features, found in any PBX (e.g. camp on, call forwarding, do not disturb, etc.) and Services. Services can include auto attendant, In-queue IVR, voice mail and directory information. Each uses a trigger table that can follow policy chains before invoking a feature. A feature within this paradigm can be defined as a sequence of events that temporarily takes over from the basic state machine.

4.1. Standard Features

Aphrodite agents manage features based on policies and priorities. Agents contain a mainline routine that handles the state machine for the entity they represent, and call appropriate routines to handle events based on the state. These routines can call utility routines to handle generic situations. Handling of the message can include calling high-level API routines, or sending new messages to other agents.

The main state machine only handles basic call processing. Handling of features is done in the feature manager. The feature manager uses a state/event trigger table, followed by a policy chain, to determine if a feature has been invoked. A data-driven design means that the order of features such as camp-on or call forward can be changed within Aphrodite by simply altering records in a specific agent's trigger table. A designer can know simply by looking in one location - the trigger table - that a certain set of features will interact based on the fact there is more than one trigger for example "on-hook" in "connected" state.

As a result of the data driven trigger table approach, traditional features have been broken down into smaller sets of 'mini' features represented by their own objects. Each object can define its own behavior, and can be changed without jeopardizing the code integrity as long as interfaces remain the same. For example, 'transfer' is no longer a feature, but is made up of three different smaller features: 'invoke transfer' (the transferor), 'try transfer' (the held party), and 'offer transfer' (the called party). Each one can have a policy that governs whether or not it can be done. As well, once a feature has been triggered at one level, while it is active, each level of agent (IDNumber, Node and Device) has its own set of code in the feature object to handle messages related to the feature. So for example, all three levels of agent receive the 'offer transfer' message, and each has the ability to accept or reject it. If the 'invoke transfer' feature fails, then next in line in the device trigger table is the 'invoke recall' feature. This brings up an interesting point; there are some features that are tied to other features and have to be part of a basic feature package.

Because each agent maintains its own state machine, call control functionality is not only distributed throughout the system, but also executed within the context of each agent object. This means, for example, that there is no longer any need to run through every line of code for every incoming call for every user, in order to select features that apply within the various states as they occur. Every agent within the Aphrodite framework has all the logic to handle calls based on its present context and forward communications to the next ID Number, Node or Device agent according to policy table rules. This means *significant* performance improvements and reductions in code size.

Services designed today bring with them implicit assumptions about their operation. These assumptions become invalid with the addition of new services and new technologies.

Techniques must be used to make these assumptions explicit and to protect existing applications as much as possible.

This can most clearly be seen in the reliance of voice systems on the concepts of "busy" or "engaged" to trigger features. A common feature is Call Forward Busy in which a user can specify that if his or her telephone is busy, any incoming call should be forwarded to another number. Everything changes in the case of a PC telephone capable of simultaneously handling several calls and other applications such as word processing. It is not at all clear how "busy" will be defined in this case. Someone could be considered "busy" if they were using a word processor rather than talking on the telephone. In this case, a device agent can have knowledge about the specific "busy" assumptions needed for any number of different clients that are each capable of terminating a telephone call. As well, an agency can determine which application has priority on a 'busy' condition, or give the caller the choice as to what application they want to invoke. In current PBX's, voice mail overrides all other possible busy scenarios like camp on, or callback.

4.2.　Repatriating "External" Applications

The treatment of applications within this model represents a complete break from traditional PBX mindsets. From the initial advent of voice-mail systems, newer feature - rich types of programs have been considered something external to basic call control functionality.

In most traditional PBX systems, applications such as Interactive Voice Response (IVR) are required to understand a battery of complex line signaling, line flash, and connection functions in addition to providing what amounts to intelligent dial-tone. In this scenario, when a call arrives at a specific port on a PBX, it is forwarded over another trunk line to the IVR application itself, which must then reroute the call back to the PBX by generating a flash on that line, getting dial-tone, dialing, and disconnecting, "hoping" that everything is transferred correctly.

In contrast, Aphrodite implements these applications as simple internal features with all the capabilities of much larger systems, but without the complexities inherent in coding message logic.

The Aphrodite model implements functionality such as IVR as an integrated service. Services are a special type of feature that are made up of a linked list of basic building blocks, like play and record or get an e-mail message. Once a building block object is created, it is available to any service. Services are triggered by feature access codes, or by the system responding to a user request with a specific tone and reason. Handling of services is done in the service manager. A data driven tone/reason trigger table is used. For example, when busy tone is given to a user because the called party was busy, then this could trigger a service that asks the caller if they would like to camp on or leave a message. If no entry existed in the trigger table, then busy tone would be given.

4.3.　Single Threading

Having a single threaded call control cuts down on feature interactions and glare situations enormously. It means that in a half call context, only one feature can be active at a time, and a feature runs to completion. Multi-threading is used where it makes sense, in the lower layers which interface to the hardware, and doing timing. If the middleware is done properly, no timing whatsoever needs to be done inside call control, as it simply needs to keep the data that defines the timing, for example, how long to give busy tone.

5. Aphrodite vs. (A)IN

There are some similarities between Aphrodite and the Intelligent Network, but there are marked differences. The concepts of feature triggers happening at a point in a call, and temporarily taking over from the basic call are the same. However the path that is followed to get to that point is much different in Aphrodite.

(A)IN takes a very linear and point-to-point approach to calls. It assumes that the *call* is the main entity, and that features are triggered at certain points in the *call*, do their thing, and then return. There is no distinction made between the number, the person and the device being used. For example, Call Screening is based on the calling number and the called number, not the people or devices involved in the call.

In Aphrodite, it is not the *call* that is important, but the *entities* that the call is passing through at a particular time that is given the emphasis. The *data* in each agent or policy automatically has an affect on the path of the call. Features can be triggered, just like in (A)IN, but the context is on *why* and *when* and *if* they are triggered as well. For example Call Screening is triggered at a similar point when a call is offered, but screening can follow a policy chain before being triggered, and once triggered, the feature can make the distinction of screening on calling line id, person or device. As well, each agent in the call path (IDNumber, Node, Device) can have its own call screening feature. So if Joe is not allowed to call Sally, then it does not matter what phone Joe uses, or what number Sally is related to, Joe will not be allowed to call Sally if Joe is on Sally's screening list at the Node level.

Another concept covered better by Aphrodite is that of groups. It allows for the grouping of people and devices. It allows for many people and many devices to be offered the same call at the same time. It allows for each entity to invoke their own set of features and policies at the same time for the same call.

Aphrodite has no call object with a state machine that gets acted upon. Aphrodite has the concept of a *session* per call leg. It has a *call record* that contains transient data. Each agent has a state machine that acts upon messages. Each agent also has its own trigger table for invoking features. You could say that each agent is a mini-(A)IN architecture.

6. Aphrodite Implementation

A working prototype of Aphrodite combines a large percentage of base PBX functionality as well as many integrated applications such as voice mail and in-queue IVR. The system is made up of an NT Server containing specialized hardware and software. Telecom resource cards from Pika Technologies Inc. provide the hardware for circuit switching, voice resources and telecom connectivity. The boards come with corresponding drivers, which are software modules that control the hardware and report events. *Figure 4* is a screenshot of the system administrator's programming interface. It is a drag and drop interface, which allows the system administrator to visualize the relationships between the different levels of agents, and the policy chains connecting them. Simply by double-clicking on a box, opens up a screen for programming the data associated with the agent or policy. It also allows for programming of outpulse plans, service creation and trigger tables, carrier plans, busy lamp filed groups, closed user groups, etc.

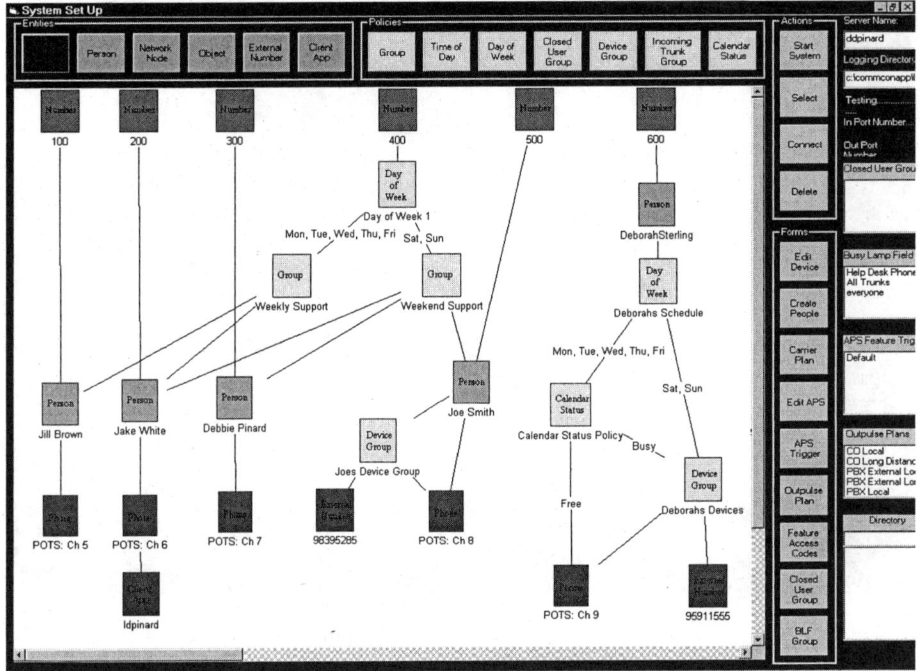

Figure 4: System Administrator Interface

Figure 5 shows the interface for setting up an integrated service. Each service is given a name, and if necessary an access code. Each device agent has a programmable trigger table that equates a tone given for a particular reason to the name of the service to be invoked. It is also a drag and drop interface which allows the system administrator to create services and then assign them to be triggered when a particular tone occurs for a reason. So if instead of giving reorder tone when a user dials a wrong number, the administrator can set up a service which plays a message saying "That is an invalid number, please try again", and then give the user dial tone. He then assigns the service trigger table pair for reorder-tone/invalid-number this new service. Many services such as retrieve voice mail, leave voice mail, auto attendant, and in-queue IVR can be set up this way. In the case of an auto attendant, a trunk would be hot-lined to the access code assigned to the auto attendant service. In the case of leaving voice mail, in would be triggered on a busy-tone/called-party-busy pair, or a reorder-tone/called-party-no-answer pair. In this case, the service could be programmed so that the user could also be given the option of camping on or setting a callback as well as leaving voice mail. The queue-tone/called-party-busy pair can trigger in-queue IVR. What is significant is not the services themselves, but the opportunity the architecture provides for simply adding new services in as new conditions or requirements come up, without having to change any API's or going through the trouble of creating a whole new application that interfaces to call control through a messy messaging interface.

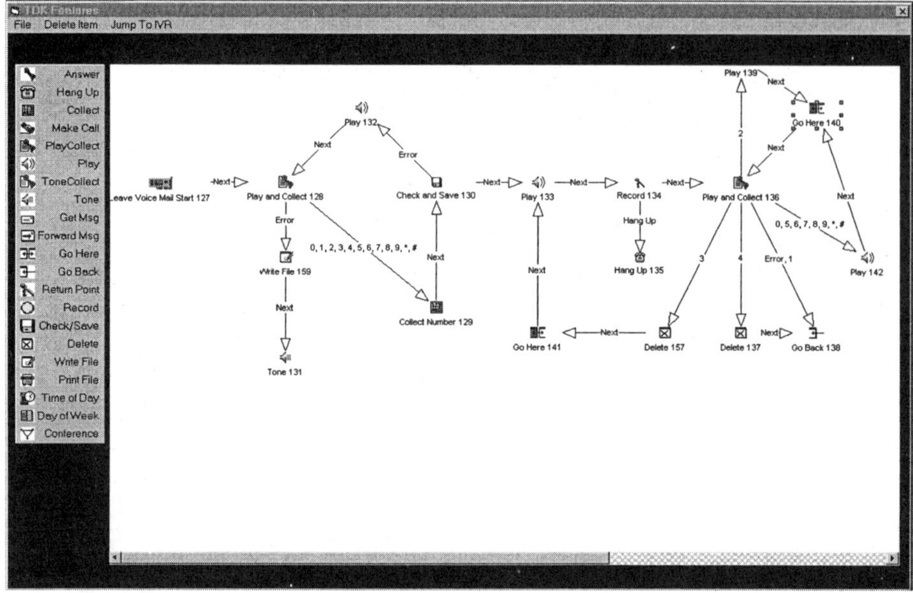

Figure 5: Setting Up an Integrated Service

7. Summary

We are entering a period of dramatic change in the way people communicate. Driven by the rapid growth of new markets and business models, the need to stay in touch from anywhere and at any time is quickly becoming a corporate priority. Most employees now use numerous devices simultaneously to keep in touch.

Next-generation communication systems will provide a distributed, enterprise-wide multimedia environment for information delivery and communications. They will also be an integral part of the business process. Such systems must facilitate and enable overall system customization and incremental evolution in order to accommodate the changing needs of the enterprise, its different work groups, and individual users. Despite the increased complexity of such systems, developers will also be faced with ever increasing demands for greater ease of use and fast turn around times for new applications with no degradation of service.

All these changes are just adding more and more complexity to the feature interaction problem. Aphrodite represents a significant advancement in overall PBX design to meet all these evolving needs. It leverages decades of expertise in the development of traditional systems to offer a solution capable of satisfying the most demanding communication requirements both today and tomorrow within hosted application or customer premise environments.

References

[1] D. Pinard, T. Gray, S. Mankovskii, and M. Weiss (1997) Issues in Using and Agent Framework for Converged Voice and Data Applications. *Proceedings of PAAM'97*, London, UK.

[2] D. Pinard, *Enabling Next-Generation Solutions for Advanced Business Communications.* http://www.aphroditetelecom.com

Architecture and Design Methods

*Feature Interactions in Telecommunications
and Software Systems VII*
D. Amyot and L. Logrippo (Eds.)
IOS Press, 2003

An Environment for Interactive Service Specification

Karim BERKANI[†‡], Rémy CAVE[†*], Sophie COUDERT[‡], Francis KLAY[†],
Pascale Le GALL[‡], Farid OUABDESSELAM[*], and Jean-Luc RICHIER[*]

[†]*France Télécom R&D*
2 avenue Pierre Marzin
22307 Lannion Cedex, France
Francis.Klay@francetelecom.com

[‡]*LaMI, Université d'Evry*
Cours Monseigneur Romero
91025 Evry Cedex, France

[*]*LSR-IMAG*
BP 72
38402 Saint Martin d'Hères
Cedex, France

Abstract. In many practical cases, deciding whether an interaction is desired or harmful is a subjective choice depending on the current state of the specification, and which results from a trial and error process. Thus, we propose an environment for service creation, at the logical level, which takes advantage of both the specifier's expertise and formal methods.

Service integration is performed by the expert following a methodology which preserves the specification formal semantics. A specification is a couple composed of a behavioral description and a set of properties. During the integration phase, the latter are ordered into three categories: the properties which are desired, those which are to be rejected, and the ones which have not yet been classified. In order to help the expert to carry out his selection, the environment offers different automatic or semi automatic services: a static expert-assisted verification for the specification consistence based on heuristics, a controlled animation of the executable behavioral description based on some guides. The animation guides are generated by the tool from interaction patterns (independent from the current specification) provided by the expert. A pattern is a high-level description which corresponds to execution sequences potentially leading to situations where a harmful interaction may appear; a pattern synthesizes the expert knowledge. [*]

1   Introduction

The intelligent network (IN) is an architectural concept which has been implemented to support the rapid and cost efficient introduction of new services in a telecommunication network. However, the service interaction problem continues to impair the safe deployment of services, since the malfunction it causes leads to customers' protest and loss of clients, in addition to the fact that this problem requires significant modifications of the conflicting services. Thus, the interaction problem prevents telecommunication operators from taking full advantage of the IN.

---

[*]This work is being supported by the "Réseau National de Recherche en Télécommunications", which is funded by the French ministries of Research and Industry.

A very usual measure to tackle the interaction problem is to work on off-line interaction detection. Various formal methods, together with their associated validation techniques (based on proving, model-checking or testing) have been proposed. They are all time-consuming, and relatively expensive: they require an important amount of specification and validation work and the number of possible service combinations which are potential sources of interactions is very large. Therefore, and even if their detection capability is interesting, using them could be in direct contradiction with rapid service provisioning.

Actually, some methods [6, 13] have been proposed to screen out the irrelevant cases of service combinations which do not lead to undesired interactions. The concept of filtering [7] and its various implementations provide a way of satisfying the IN goals.

ValiServ is a project whose objective is to facilitate the service designers' work by offering them a simple specification language which helps them to concentrate their effort on interaction avoidance and detection. The focus is on the integration process rather than on the detection operations. Our goal is to cope with the interaction problem after its complexity has been reduced. On the contrary, classical approaches consist of providing a complete specification of services and, afterwards, to check the absence of interactions. To our mind, this leaves too many problems to be detected. We argue that classifying an interaction as desirable is a subjective decision in the very first place. Therefore, interaction detection cannot rely solely on verification tools, and it is worth to take advantage of the service designer's expertise from the beginning. Our thesis distinguishes itself by a service integration activity which embeds some form of interaction detection and a testing phase which complements interaction detection by some systematic analysis of interaction-prone situations. In this approach, the interactions are signalled by the violation of the properties that the expert has written out. Not all these violations lead to dangerous situations. ValiServ aims at providing assistance to the user to classify these properties.

The approach deliberately does not give any special status to the Basic Call Service, additional services or features. All are considered as sets of formulas, and there is no structuring means to organise the specification in terms of services. The notion of service is certainly useful from a software engineering point of view. But, as far as interaction detection is concerned, we consider that the notion of service is not the right level of granularity. Indeed, interactions may occur in one single service and between several services, as well.

Service specifications are provided in the form of state transition rules at the user interface level, very much as in [10, 11]. The original aspects come from the definition of a synchronous interpretation of these rules which allows the expert to reason at the subscribers' interface level, and from the definition of a service integration methodology which reinforces the specification semantics.

Interaction detection is carried out at two levels. During service design and composition through some static expert-assisted analyses, and afterwards when the whole service model is executed and tested using Lutess [4] and some specific testing guides. The purpose of those guides is to exercise the service combinations which are more likely to exhibit interactions.

The paper is structured as follows. Section 2 presents the ValiServ methodology, and the motivations for the two phases it includes (design/composition, animation/testing). Section 3 details the semantics of the specification language and its synchronous interpretation. Section 4 explains what the integration process is all about. In section 5, the principles of the scenario generation process are introduced.

2　The ValiServ project

2.1　General purposes

The ValiServ project is a specific implementation of a methodology for service integration. It advocates an environment whose main purpose is to integrate service specification in an assisted but not totally secure (heuristic) way. As illustrated in figure 1, this environment offers facilities to enter system and feature specifications which are subsequently integrated, and to save them. While a specification is being entered, some basic syntax and type checking are performed. The first system to be inserted in the base is the famous "Plain Old Telephone Service" ($POTS$); new systems are specified by composing existing service specifications and new service specifications. The service specifications are partial; they just provide the modifications which the service leads to with respect to the system they are supposed to be plugged on.

With this environment, the user can select a system specification SYS and a service specification F, and compose them in order to obtain a new system specification. During this composition, the expert is informed of any interaction-prone situation his specifications may introduce. As a consequence, he has to rewrite some parts of these specifications. The new statements correspond to some design choices. Thus, from a formal point of view, this composition is parameterised by the expert's choices, i.e., it is abstractly denoted by $SYS +_{choices} F$. It generally leads to some modifications of SYS; thus, we do not easily get the addition of two services together to a system ($SYS +_c \{F_1, F_2\}$) from the addition of each of them, $SYS +_{c1} F_1$ and $SYS +_{c2} F_2$. Indeed, it would suppose not only to confront the specifications F_1 and F_2, but also to re-examine c_1 and c_2 because c_1 was thought on SYS and not on SYS modified by c_2, and conversely. Thus our approach is to define the integration of one service on a system, resulting in a new system, and to integrate the services one by one. From $POTS$ and some services $F_1, \ldots, F_n$, we want to build an integration $(\ldots (POTS +_{c_1} F_{i_1}) + \cdots +_{c_n} F_{i_n})$, where the order $i_1, \cdots, i_n$ is significant with respect to the choices $c_1, \cdots, c_n$. The integration order is symbolised in figure 1 by the service indexes.

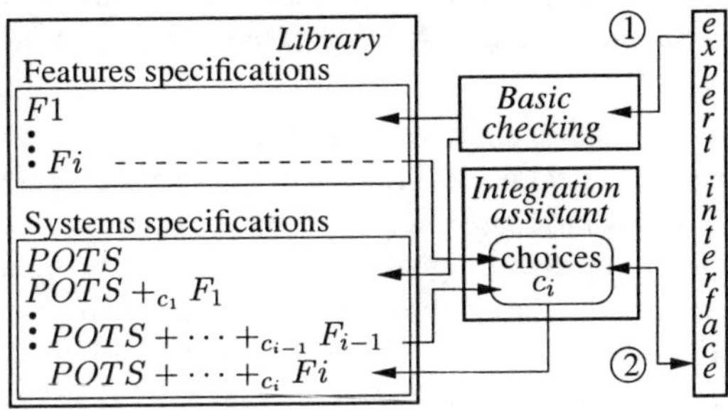

Figure 1

Each integration is carried out in two stages (or phases). The first one is a static composition with some basic properties verification (no deadlock, no livelock, no non-determinism). It is followed by a phase of specification animation in order to validate and improve the composition decisions. This validation is carried out using a set of specific scenarios which are devoted to interaction detection.

We now give the methodological ideas which underlie the ValiServ integration assistant, before presenting the implementation decisions.

2.2   An overview of the methodology

Theoretically, the methodology is based on formal specifications (written in a given language) and a correctness criterion associated with them. Practically, and from a general point of view, the methodology allows the correctness criterion to be approximated by the observance of some specific methodological steps to integrate services. More precisely:

- The basic principle we adopt is that a system is correct if it runs (i.e. no deadlock, no livelock, no non-determinism) and if it satisfies some associated requirements. To lay down an intrinsic correctness criterion, we consider specifications as couples $\langle Sys, Prop \rangle$ where Sys describes the behavior of the system and $Prop$ describes its requirements, i.e. properties which must be preserved by all possible behaviors allowed by Sys. Consequently, to be correct, a system must meet three conditions: each of the Sys and $Prop$ descriptions are consistent, and Sys satisfies $Prop$. The two Sys and $Prop$ consistency relations, as well as the satisfaction relation between Sys and $Prop$ are theoretical criteria. The fact that a compound specification $\langle Sys, Prop \rangle$ is theoretically correct will be expressed by the notation $\langle Sys \models Prop \rangle$.

- Actually, since our work is based on heuristic checks and test [5], we don't ensure the theoretical correctness. We will use the notation $\langle Sys \approx Prop \rangle$ to mean that the specifications have been well integrated with respect to the methodology, i.e. the different required steps whose goal is to corroborate the theoretical correctness have been successfully performed. These steps take place in an integration process which comprises two major phases (see figure 2). The first one, the specification composition, provides a way of building a new system $\langle Sys, Prop \rangle$ from a reference system $\langle Sys_S, Prop_S \rangle$ and the specification $\langle Sys_F, Prop_F \rangle$ of a service to be plugged on it. During this phase, the expert works with static methods which operate on textual specifications. The second phase is intended to be a dynamic one, in which the resulting specification is animated and tested.

- Note that all specifications (of systems and services as well) have two components, since it is natural to provide $Prop$ requirements for any service. However, the correctness of such a service specification alone (which is a partial specification) is not a reasonable objective. Firstly, it would probably not run. Secondly, specifying a service requires the provision of not only its specific requirements but also some other behaviors which are a consequence of this service introduction; these latter behaviors represent alterations or complements of the system on

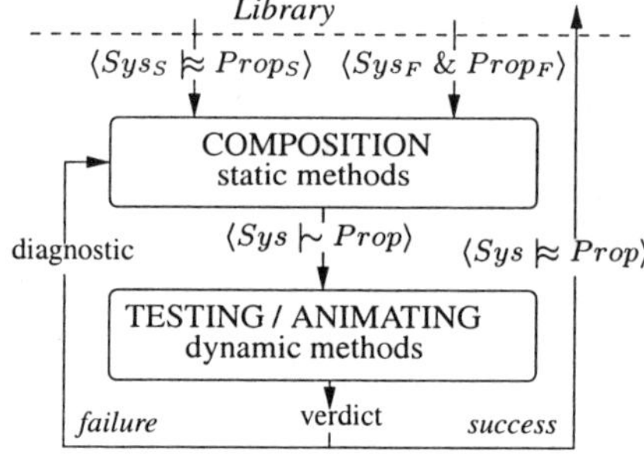

Figure 2

which the service should be plugged. So, the satisfaction of the $Prop$ requirements can depend on hypotheses on the underlying system. Therefore, the strong theoretical correctness only concerns system specifications. But, service specifications can have their own consistency criteria, and check tools can also exist for this purpose. Thus, services in the library are supposed to be well specified, which is denoted by $\langle Sys \,\&\, Prop \rangle$.

- The system resulting from the composition phase has not been tested yet and is not considered to be reliable enough, which is denoted by $\langle Sys \mathbin{\mid\!\sim} Prop \rangle$. If the testing phase is successful, it yields $\langle Sys \approx Prop \rangle$ and the new integrated system can be saved in the library; otherwise a diagnosis can help the expert to revise the design. Working over a design again can modify either components: the $\langle Sys \rangle$ or the $\langle Prop \rangle$ one, since there is no reason for any to be complete from the beginning.

- Basically, both phases can be automated or interactive. We have opted for an interactive approach which relies on the expert knowledge. This expertise is used "on line" when the expert writes and fixes his specifications, while composing a new system. Furthermore, there is a lot of "off line" expertise implicitly in the scenarios which guide the testing process, since these scenarios are automatically generated from interaction patterns which are built in the specification animator. The interaction patterns model and synthesize the experts' knowledge on the main interaction classes which have been derived from existing benchmarks.

2.3 Technical choices

Another aspect of the decomposition in two phases is the opportunity to make use of two technologies. The design phase is likely to be based on high-level formalisms in order to reach concision, to use high-level tools (proof, refinement, variables ... [1, 5, 14, 12, 3]) and to facilitate the expert's design work. The test phase can benefit from more low-level tools, such as model-checking, test, propositional logic, animation ... In addition, the two levels can be complementary, i.e. they do not exactly address the same classes of problems.

Of course, this approach needs a good interfacing of these two aspects, and compatible tools. A difficulty of importance is to translate data from one phase to the other. In the following, we give some motivations for our choices, and sketch out how they can fit in our generic process.

The design phase is based on finite-state machines described by means of formulas of two kinds: state transition rules (STR) and state invariants (I). This is a rather classical approach in the field of specifications of service-oriented telecommunication systems [12, 14]. A STR sentence is defined as a triple (*precondition, event, postcondition*). The *precondition* and *postcondition* terms are properties holding on the current and next states with respect to the transition while the *event* term models the user's action triggering the state evolution. In ValiServ, STR and I are universally quantified formulas. An invariant, or I sentence, is a first-order formula over states. Such a formula allows the specifier to restrict the set of reachable states. The advantages of STR/I specification style for telecom-

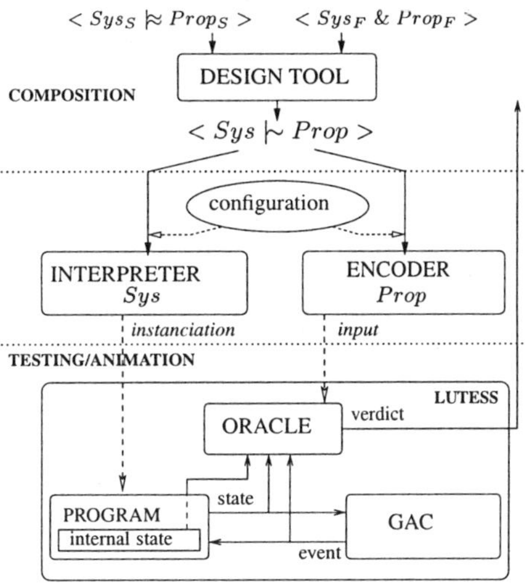

Figure 3

munication systems are essentially their flexibility and their associated set of verification tools devoted to interaction detections. STR and I sentences are quite intuitive expressions; thus, the specifier can easily modify them when necessary. Several previous works [12, 14] have proposed specialized techniques for the interaction detection based on STR-like specifications. Depending on the cases, they study the properties of the reachability graph in terms of non-determinism or deadlock or contradictions raised by the simultaneous application of two

STR sentences ... Roughly speaking, the STR sentences will be the Sys component from the design phase, while the I sentences will be the $Prop$ component (see figure 3).

For the animation phase, we have chosen a specification-based testing method. We use Lutess, an environment dedicated to the test of synchronous reactive software. This tool has already proved its efficiency for detecting interactions (see the first Feature Interaction Detection Contest [4]). A minimal hypothesis to use Lutess is that the system under test is a continuously operating program with instantaneous reaction to its environment evolutions. The properties checked by testing are expressed in the Lustre temporal logic. Schematically (see figure3), Lutess is composed of two main units interacting with the system under test inherited from the Sys component of the design phase. The first unit acts as an oracle and aims at checking whether the system outputs are correct with respect to the previous inputs and the expected properties. Thus, the $Prop$ component provided by the design stage is encoded within the Lutess oracle. The second unit is the input data generator (GAC in figure 3), generating "on-the-fly" input sequences, based on previous outputs and considering preferred scenarios. These scenarios are intended to both exercise each input with a strict positive probability and promote input sequences prone to reveal interactions. In our setting, it simply means a contradiction between the system under test and the properties implemented in the oracle.

The interfacing of the two parts is described by figure 3. The twofold decision of using STR/I-based specifications and Lustre-based specifications is fully convenient to fill in the ValiServ process since not only they have their respective advantages but also they share some underlying technical restrictive hypotheses, as the synchronous one. Thus, a translation from the STR/I level to the Lustre level is clearly easy, up to the choice of a particular configuration of a finite number of users and subscriptions. Since this restriction is common to most verification methods (whether they are based on model-checking or testing), this choice is unavoidable and left to the expert. In the oracle, each invariant is translated into a set of ground sentences by substituting configuration phone numbers for variables. Let us note that ground invariants are of course temporal Lustre formulas. In the same way, the Sys component and the configuration may be simply interpreted as a running system likely to be tested by the Lutess tool if it makes its internal state observable by the oracle.

3 The specification framework

In this section, we show how STR and invariants can be provided with a semantics and a prototype on ground formulas. Then, we introduce STR with variables and give some examples of specifications. Finally, we explicit the link between design and testing through variable substitution, and we briefly explain how deadlock and non-determinism problems are treated by the tool.

3.1 STR and automata

A set of STR sentences characterizes an automaton in the following way:

- Let P be a set of atomic properties and E a set of events. A state is a subset of P and satisfies the properties it contents. A transition is a tuple (s, e, s') where s and s' are states, and e is an event. Then an automaton is a tuple $A = (S, T)$ where S (also denoted by $s(A)$) is a non empty set of states and T (denoted by $t(A)$) is a set of transitions between states of S. A is said to be deterministic if all transitions with the same origin state have different events.

- A STR is a formula $\phi = pre\ \overline{pre} \xrightarrow{e} post\ \overline{post}$, where pre, $\overline{pre}$, $post$ and $\overline{post}$ are subsets of P and e an event. ϕ matches a state s when $pre \subset s$ and $\overline{pre} \cap s = \emptyset$, and then, $\phi(s)$ denotes $((s - pre) \cup post) - \overline{post}$.
  $A = (S, T)$ satisfies ϕ when for each $s \in S$, if ϕ matches s, then $t = (s, e, \phi(s)) \in T$ (t is an *instance* of ϕ). A satisfies a set R of STR when it satisfies all ϕ in R and it is deterministic and each $t \in T$ is an instance of a unique ϕ in R.

- A STR set R and a state set S generate an automaton $\mathcal{G}(S, R) = (S', T)$, where S' and T are the smallest sets verifying $S \subseteq S'$ and $\forall s, \phi \in S' \times R$ (ϕ matches $s \Rightarrow \phi(s) \in S' \wedge (s, e, \phi(s)) \in T$), where e is the event of ϕ.
  S and R are compatible when $\mathcal{G}(S, R)$ satisfies R, and strongly compatible when furthermore $s(\mathcal{G}(S, R)) = S$. R is inconsistent when there isn't any non empty set compatible with it.

With this semantics, for each STR set R, we build a program $Encoder(R)$ implementing state transitions. Ideally, $Encoder(R)$ should be a function which associates a "next state" with any state s and event e. But this "next state" is not always well defined, due to inconsistency problems. The good case is when there exists a unique STR ϕ in R, labelled by e, matching s, giving $\phi(e)$ as "next state". There are two problematic cases for the encoder: no matching STR or more than one applicable STR.

We reject ambiguity between STRs at the semantical level and in the program. Indeed, ambiguity can correspond to non-determinism (when there are two concurrent "next states"), which is not adequate for the behavior of telecommunication systems, in the service perspective. Else, many STRs are equivalent for the considered state (the "next states" are the same), which should be acceptable but too complex to manage in the tool when considering variables in the design step. Moreover, the prototype for the test phase is not equipped to identify these cases (it could probably be done).

The lack of a matching STR for a given state corresponds to a deadlock. To us, this is not an inconsistency case because semantics does not require that any event can occur in any state. This corresponds to reality: one cannot put on her phone when it is hooked on. If the specification is supposed to completely characterize the behavior of the system (which is the case for $\langle Sys \models Prop \rangle$), s must be a state where e never occurs in the real world. Otherwise, a deadlock can simply be due to an unspecified case, which is a common situation in the partial specifications $\langle Sys\ \&\ Prop \rangle$ of services.

Since there isn't anything yet in the high level language to distinguish normal and pathological deadlock cases, the detection of the second ones takes place during the test phase. Indeed, testing produces execution traces starting from an initial state, and thus, it examines only reachable states. Moreover this phase only deals with system specifications Therefore, the problems detected here should be real problems.

Similarly, non-determinism exhibited by testing should concern real situations when we have no effective identification of reachable states at the design level. However an approximation can be done for non-determinism at the high level, allowing some abusive warnings and some missing ones in order to help an expert to correctly integrate. We can propose a free help, allowing this expert to force the input of non-deterministic STR, or a constraining help, refusing it and then asking the specifier to also avoid abusive detection. The good choice depends of the quality of the approximation.

Only considering non exclusive preconditions (which is reasonably decidable, also when considering variables) as a warning for non-determinism leads to a large surapproximation of the problems because it consists in requiring determinism on all theoretically possible states (i.e. the compatibility between R and $\mathcal{P}(P)$ the set of all subsets of atomic properties), which is generally far from the reality. A better solution is to enrich the specifications with an invariant I defining an approximation of the real set of states, and this is our choice, associated with a constraining help. Then the high level check consists in detecting couples of STR with same event such that their preconditions are not exclusive and not exclusive with I. It remains heuristic because I is an approximation and managing variables leads to complexity and decidability problems, and the algorithm itself misses some problematic cases (those implying a lot of terminals). But each detected case is a real case with respect to I and the expert has then two solutions to treat the problem: he can choose to give priority to one of the STR on the problematic state and the tools refine preconditions in order to realize this choice, or he can reinforce I in order to exclude this state. The fact that applying STR does not break I (i.e strong compatibility: $s(\mathcal{G}(I, R) \models I$ where here, I denotes both an invariant and the set of all states satisfying it) could be checked with complementary tools such as the B method. In our framework we only do partial checks between I and the post-conditions at the high level and testing give a check for reachable states. The result of this approach is a tuple (I, R) intended to be strongly compatible where the adequation between I and reality relies on the expert and where checks corroborate compatibility without proving it. Notice that this compatibility is a week approximation for service specification if we allow them to be very partial about invariants. Contrary to [14], there is no elicited knowledge; invariant properties are explicitely provided.

A least anomaly can reside in useless STR, with no instances for I or for reachable states. This is acceptable for service specifications with regard to the reachable states because they are partial and don't really characterize such states. In the other cases, this should correspond to a specification error. For this purpose a check is done, verifying non exclusivity between invariants and STR precondition, with the same restrictions to an heuristic approach than when restricting non-determinism cases with invariants. Moreover, an extension of the language is under study in order to better characterize when the events can occur and thus, allow some deadlock checks at the high level and introduce this specification complement in the design phase where it should naturally takes place. It would be an assistance for producing guidelines for acceptable sequences of events in the test phase which is yet the only one supporting this aspect.

Although this presentation is done without the variables and the unification techniques we use to manage them, the used high level specification language allows generic STR which is more adequate to concisely specify services.

3.2 Specifications with variables

The integration assistant does not distinguish between invariants which are service specific or system intrinsic: it manages both kinds in the same way. All these invariants constitute the $Prop$ part of a specification, while STRs constitute the Sys component. Thus, in the sequel, $\langle$STRs, Invariants$\rangle$ corresponds to the $\langle Sys, Prop \rangle$ of the methodology.

The specifications include variables which are instantiated by terminal numbers for the test phase: each phone is supposed to have the same behavior for the same subscription. Consequently, system states correspond to terminal states described with two kinds of predicates: status predicates for the communication situation (idle, busy, ...) and service subscription predicates. The basic vocabulary also comprises terminal-parametrised events. The atoms defined with these predicates and parametrised events correspond to the basic property set P and the event set E of the previous section. Thus a specification must respect the following shape with some conventions which are given below.

Definition 3.1 — *A signature is a tuple $\Sigma = (S_\Sigma, F_\Sigma, E_\Sigma)$ where S_Σ is a set of status predicates, F_Σ a set of subscription predicates and E_Σ a set of event names. Members α of S_Σ, F_Σ or E_Σ are provided with an arity $n \in \mathbb{N}^*$, which is denoted by $\alpha.n$.*

— Terms on an a set V of symbols provided with an arity in $\mathbb{N}^*$ and a set $\mathcal{X}$ of variables are inductively defined by $T_V(\mathcal{X}) = \{\alpha(k_1, \ldots, k_n) \mid \forall i \in [1 \ldots n],\ k_i \in \mathcal{X}\ \&\ \alpha.n \in V\}$. Members of $T_{S_\Sigma \cup F_\Sigma}(\mathcal{X})$ are atoms, those of $T_{S_\Sigma}(\mathcal{X})$, status, those of $T_{F_\Sigma}(\mathcal{X})$, subscriptions and those of $T_{E_\Sigma}(K)$, events.

— Considering a set $\mathcal{X}$ of variables, a specification is a tuple $SP = (\Sigma, Prop, Sys, Y)$ where Σ is a signature, $Prop$ a set of propositional formulas on $T_{S_\Sigma \cup F_\Sigma}(\mathcal{X})$ called invariants, Sys a set of STRs using $T_{S_\Sigma \cup F_\Sigma}(\mathcal{X})$ as atomic property set and $T_{E_\Sigma}(\mathcal{X})$ as event set, and Y is a set of subsets of S_Σ.

Here, the formulas are implicitly universally quantified. The set Y corresponds to exclusivity between predicates on their first parameter: for example, if the set $\{talking.2, idle.1, \ldots\}$ belongs to Y then in any state, $\forall x\ \forall y\ \neg(idle(x) \wedge talking(x, y))$ and thus, this exclusivity is a specific kind of system intrinsic invariants. Another point is that we do not deal with events which modify subscriptions. Lastly, it is possible to require that two variables are differently instantiated in a formula.

Let us now give some examples with the following conventions to facilitate reading: negative preconditions are overlined and the subscription atoms which are automatically preserved are not repeated in the formula right handsides. Moreover status predicates which are self exclusive are overlined in the Y declaration. $\{\overline{talking.2}, idle.1, \ldots\} \in Y$ means: $\forall x\ \forall y\ \forall z(y \neq z \Rightarrow \neg(talking(x, y) \wedge talking(x, z)))$.

example 1: $POTS$, the plain old telephone service
$\overline{S_{POTS}}$ contains $idle(x)$ (standing for "x is idle"), $dialwait(x)$ ("x is in dial waiting state"), $caller(x, y)$ (resp. $callee(x, y)$) (meaning: "x is in communication with y as the caller (resp. callee) part"), $ringing(x, y)$ ("x is ringing from the caller y"), $hearing(x, y)$ ("x is hearing the tone of the call to y"), $busytone(x)$ ("x is hearing the busy tone").
F_{POTS} and I_{POTS} are empty. E_{POTS} contains $offhook(x) : x$ is hooked off, $onhook(x)$ (x is hooked on), $dial(x, y)$ (x dials y). $Y_{POTS} = \{idle, busytone, dialwait, \overline{hearing}, \overline{ringing}, \overline{caller}, \overline{callee}\}$. R_{POTS} contains:

$\phi_1:\ <\ |\ idle(A)\ \xrightarrow{offhook(A)}\ dialwait(A)\ >$

$\phi_2:\ <\ A \neq B\ |\ dialwait(A), idle(B)\ \xrightarrow{dial(A,B)}\ hearing(A, B), ringing(B, A)\ >$

$\phi_3:\ <\ |\ dialwait(A), \overline{idle(B)}\ \xrightarrow{dial(A,B)}\ busytone(A)\ >$

$\phi_4:\ <\ A \neq B\ |\ hearing(A, B), ringing(B, A)\ \xrightarrow{offhook(B)}\ caller(A, B), callee(B, A)\ >$

$\phi_5:\ <\ A \neq B\ |\ caller(A, B), callee(B, A)\ \xrightarrow{onhook(A)}\ idle(A), busytone(B)\ >$

$\phi_{5'}$: $\ < A \neq B \mid caller(A, B), callee(B, A) \overset{onhook(B)}{\longrightarrow} idle(B), busytone(A) >$

ϕ_6 : $\ < A \neq B \mid hearing(A, B), ringing(B, A) \overset{onhook(A)}{\longrightarrow} idle(A), idle(B) >$

ϕ_7 : $\ < \ \mid busytone(A) \overset{onhook(A)}{\longrightarrow} idle(A) >$

POTS is a system and thus, its specification is supposed to be complete: it characterizes the behavior of a terminal which has just subscribed to the basic telephone service, when communicating with another terminal with the same subscription. ϕ_5, for example, says that when two terminals are communicating, if the call initiator hangs up then her party gets a busy tone. Notice that using two predicates (*caller* and *callee*) for the communication state preserves a useful information about who initiated the call during the communication. More technically, it makes the specification less concise but provides a status to each phone (a predicate provides a status to its first parameter and in a normal state, each phone should have at least one status) and avoids symmetric predicates.

For the sake of completeness, let us note that some formulas are missing. The example shows a "real" deadlock, which is detected during the testing phase, and fixed by adding a rule ϕ_8 (see section 4.2).

example 2: $\ TCS$, terminating call screening (this service screens out incoming calls from terminals belonging to the TCS subscriber's black list).
$S_{TCS} = S_{POTS}$. F_{TCS} contains $Tcs(y, x)$: calls from x to y are forbidden by y.
$E_{TCS} = E_{POTS}$, $Y_{TCS} = Y_{POTS}$, and I_{TCS} contains
ψ_1 : $\ < A \neq B \mid Tcs(A, B) \Rightarrow \neg hearing(B, A) >$
while R_{TCS} contains

ψ_2 : $\ < \ \mid Tcs(B, A), dialwait(A), idle(B) \overset{dial(A,B)}{\longrightarrow} busytone(A) >$

example 3: $\ CFB$, call forward on busy (this service allows a subscriber to forward all incoming calls to a designated terminal, when he is busy).
$S_{CFB} = S_{POTS}$. F_{CFB} contains $Cfb(x, y)$: when x is not idle, forward incoming calls to y.
$E_{TCS} = E_{POTS}$, $Y_{CFB} = Y_{POTS}$, I_{CFB} is empty, and R_{CFB} contains

χ_2 : $\ < B \neq C \mid Cfb(B, C), dialwait(A), \overline{idle(B)}, idle(C) \overset{dial(A,B)}{\longrightarrow}$
$$hearing(A, C), ringing(C, A) >$$

χ_3 : $\ < B \neq C \mid Cfb(B, C), dialwait(A), \overline{idle(B)}, \overline{idle(C)} \overset{dial(A,B)}{\longrightarrow} busytone(A) >$

Examples 2 and 3 specifications present the service specificities and implicitly say "for the rest, refer to the system". TCS specification contains a service invariant characterizing a newly prohibited situation (the subscriber terminal cannot be put in communication with a terminal from its screening list) and a limited behavioral description (what happens when a forbidden terminal attempts to call the subscribing terminal). Although these specifications are basic ones, the POTS specification looks like any system specification (with very little invariants) and can be checked with the high level algorithms before being instantiated in order to be tested.

Interpreter

Our formulas are first order one and supposed to be verified for any system configuration. It leads to undecidability and high complexity, and a need for approximation at the high level. Moreover, problems in too little models (for example with only one telephone) are not real problems. Thus the testing phase offers a complementary check on some particular relevant models. To obtain a model from a consistent specification $\langle Sys, Prop \rangle$ with signature

$\Sigma = (S_\Sigma, F_\Sigma, E_\Sigma)$, it is sufficient to choose a set T and to consider the ground specification $\langle \widehat{Sys}, \widehat{Prop} \rangle$ where $\widehat{Sys}$ and $\widehat{Prop}$ contains all possible instantiations of Sys and $Prop$ by the terminals of T, respecting the inequality constraints of the formulas. Then consistence means that $\widehat{Sys}$ is strongly compatible with $\widehat{Prop}$ and $\mathcal{G}(\widehat{Prop}, \widehat{Sys})$ is the expected model. Thus, with an initial state satisfying $Prop$ and an iterative application of the interpreter we presented in Section 2.3, we obtain a correspondence between the errors of the program and the inconsistency or deadlock problems. Concretely, we recommend to choose separately a communication initial state known to actually exist (typically all terminals idle) and a sub-scription configuration. These choices are done by the expert, from the knowledge of both the nature of the integration system which makes explicit some problematic configurations in the precondition part of the STRs, and the interaction criteria used in the testing phase (see section 5). In practice, it is well know that few phones are needed to build an interesting configuration.

4   The design/composition phase

4.1   Process

For the design phase, the ValiServ project proposes an assisted integration process. When composing a system specification $\langle Sys_S \models Prop_S \rangle$ and a service specification $\langle Sys_F \ \& \ Prop_F \rangle$ in order to obtain a new integrated system specification $\langle Sys \models Prop \rangle$, we would like to inherit behaviors and properties from both parts, thus we look for some kind of union. In fact, a naive union does not lead to the expected result because there can exist inconsistencies, between invariants, between STRs, or between invariants and STRs of both parts: roughly speaking, we generally do not have $\langle Sys_S \cup Sys_F \models Prop_S \cup Prop_F \rangle$. Indeed, in many cases, a service not only adds something to the system, but also modifies it (e.g. TCS service). In other cases, such modifications are only local, and most of the inherited characteristics are preserved; what we aim at obtaining is an arranged union: sets of formulas which only differ from the union on a few members.

The composition process consists in adding the inherited sentences one by one to a target system T which is empty at the beginning of the process and which is the integrated system at the end. Problems are tackled as they are encountered. The main steps are:

- A syntactic operation defines an initial work context: a tuple (SP_T, SP_S, SP_F), where SP_S and SP_F are respectively the specification of the original system and service, and SP_T is[†] $\langle \Sigma_S \cup \Sigma_F, R_T^0, I_T^0, Y_T^0 \rangle$ where R_T^0, I_T^0 and Y_T^0 are empty.

- Each elementary step removes some pieces from SP_S and SP_F, inserts some related ones in SP_T, while the consistency of the target $(\langle Sys_T \models Prop_T \rangle)$ is preserved by heuristic check algorithms. This keeps many formulas unchanged. Some others are replaced by more refined ones: for example an STR precondition can be completed in order to restrict its scope and solve a non-determinism conflict.

- The process terminates when SP_S and SP_F are empty and the result is contained in SP_T.

The order of the steps is significant because the objective is not to revisit the target context in order to guarantee the termination of the process. Thus, if two sentences are in conflict and if

---

[†]Considering the union on each component of the signatures.

one of them has been put in SP_T, then the other one will necessary be modified or abandoned: inserting it in the target would introduce inconsistency. But, if the expert does want to preserve this second sentence, the only solution is to backtrack to the point where the first one has been selected. In fact, the assistant partially constrains this order for a successful integration: invariants are treated before STRs, because it is useful to know them in order to eliminate abusive cases of non-determinism between STRs, as explained in the previous section, and practically, the exclusivity invariants are treated first because they are naturally related to the signature. Some other considerations also constrain this order. Semantic interactions are captured by the invariants and therefore indirectely reviewed when invariants are violated; in [14], these interaction situations are explicitly exhibited.

The general schema of the composition process is then the following:

1. Choosing exclusivities. The default solution is to preserve the inherited exclusivities (when they are compatible on the common part of the signatures) and to ask the expert for "hybrid" tuples (with one predicate in each signature).

2. Choosing invariant formulas and performing some consistency partial checks before putting them in target. Invariants treated one by one.

3. Choosing STRs, performing consistency checks and modifying the STRs if necessary. A detection *a priori* of non-deterministic tuples (with respect to the design phase heuristic algorithms) is best with some assistance: presenting together sentences in conflict prevents the user from first choosing one of them and discovering that he has abusively excluded the other many steps later.

This gives an idea of some of the essential tools the assistant offers. In fact, this is a base on which a lot of improvement could be studied. Indeed, the interest of such an aid resides in the guides which avoid the expert to make too much choices, and also to make choices leading to very probable backtrack. Thus, our current work consists in searching such guides, helping ordering, both in relation with theoretical criteria and with expert experience. Such criteria lead to new integration strategies in the tool, eliminating some random choices and perhaps automating some systematic expert choices. For the moment, the first even simple strategy we have has been satisfying for the examples we studied, especially because specifications are of high level and then sufficiently concise to avoid explosion of expert choices and backtrack. In practice, formulas are simple and there are not a lot of conflicts between them.

Finally, notice that this process can be applied to different integration schemas:

- To directly design reference systems by introducing all formulas in the same manner (as for POTS and the services specifications in section 3) and to test (this meaningful for POTS but meaningless for service specifications); this schema corresponds to the integration carried out in section 4.2. It concerns $POTS + TCS$, $POTS + CFB$ and $(POTS + TCS) + CFB$.

- To integrate two existing systems, for example $(POTS + TCS) + (POTS + CFB)$. This has not never been performed so far in our environment, because it would generate a lot of conflicts due to redundancies.

Even if we didn't Some utilities are provided by the assistant, such as the properties which have been rejected during composition or the identity of the work context parents; for example when integrating $(POTS + TCS) + CFB$, consulting reports on the integration information of $POTS + TCS$ is useful.

4.2   Examples

Here is a summary of the integration on the examples. POTS has been integrated at the design level and the test phase revealed a deadlock when $onhook(A)$ occurrs in a state containing $dialwait(A)$. Adding the rule

$$\phi_8: \quad < \; | \; dialwait(A) \; \xrightarrow{onhook(A)} \; idle(A) >$$

solved the problem. Some non-determinism cases have been avoided thanks to predicate exclusivities of Y_{POTS}. For example ϕ_5 and ϕ_5' could be in conflict when $onhook(A)$ occurs in a state containing $\{caller(A,B), callee(B,A), callee(A,B), caller(B,A)\}$ or $\{caller(A,B), callee(B,A), caller(C,A), caller(A,C)\}$, but such states are excluded because $caller$ and $callee$ are exclusive: $(caller, callee) \in Y_{POTS}$. In the same way, ϕ_1 and ϕ_4 are saved because $(idle, ringing) \in Y_{POTS}$, and ϕ_6 and ϕ_7 because $(busytone, hearing) \in Y_{POTS}$.

Then $POTS+TCS, POTS+CFB$ and $(POTS+TCS)+CFB$ have been integrated, and the following summarizes the experiment:

- Although it is allowed to modify exclusivity during the process, we didn't need it in this example. Therefore
  $$Y_{POTS+TCS} = Y_{POTS}, Y_{POTS+CFB} = Y_{POTS} \text{ and } Y_{(POTS+TCS)+CFB} = Y_{POTS}.$$

- The high level of integration of $POTS + TCS$ preserved the invariant ψ_1 (we have $I_{POTS+TCS} = I_{TCS}$) but a non-determinism was detected between the $POTS$ rule ϕ_2 and the TCS rule ψ_2: in a state where $dialwait(A,B), idle(B)$ and $Tcs(B,A)$ are true, if the $dial(A,B)$ occurs, then the two rules can be applied. Moreover ϕ_2 can lead to a break of ψ_1 (in $I_{POTS+TCS}$). So ϕ_2 has been replaced by ϕ_2' ($R_{POTS+TCS} = ((R_{POTS} - \{\phi_2\}) \cup R_{TCS}) \cup \{\phi_2'\}$), where

  $$\phi_2': \quad < A \neq B \; | \; \overline{Tcs(A,B)}, dialwait(A), idle(B) \; \xrightarrow{dial(A,B)}$$
  $$hearing(A,B), ringing(B,A) >$$

- The integration of $POTS$ with CFB showed a non-determinism between ϕ_3 and χ_2: in a state where $Cfb(B,C), dialwait(A), caller(B,D), idle(C)$ and $callee(D,B)$ are true, if $dial(A,B)$ occurs, then the two rules can be applied. So ϕ_3 has been replaced by ϕ_3' ($Sys_{POTS+CFB} = ((Sys_{POTS} - \{\phi_3\}) \cup Sys_{CFB}) \cup \{\phi_3'\}$), where

  $$\phi_3': \quad < B \neq C \; | \; \overline{Cfb(B,C)}, dialwait(A), \overline{idle(B)} \; \xrightarrow{dial(A,B)} \; busytone(A) >$$

- The integration of $POTS + TCS + CFB$ followed the steps on $POTS + TCS$ and $POTS+CFB$. $Sys_{POTS+TCS+CFB} = (Sys_{POTS+TCS} - \{\phi_2\}) \cup (Sys_{POTS+CFB} - \{\phi_3\})$ and $Prop_{POTS+TCS+CFB} = Prop_{POTS+TCS}$. But the test phase revealed that $Prop_{POTS+TCS+CFB}$ breaks the instance $Tcs(C,A) \Rightarrow \overline{hearing(A,C)}$ of the TCS invariant ψ_1 when $dial(A,B)$ occurs in the state $\{dialwait(A), Cfb(B,C), caller(B,D), Tcs(C,A), idle(C), callee(D,B)\}$ and leads to the state $\{hearing(A,C), Cfb(B,C), caller(B,D), ringing(C,A), Tcs(C,A), callee(D,B)\}$. So χ_2 has been replaced by χ_2' and χ_4 was added ($Sys_{POTS+TCS+CFB} = (((Sys_{POTS+TCS} - \{\phi_2\}) \cup (Sys_{POTS+CFB} - \{\phi_3\})) - \{\chi_2\}) \cup \{\chi_2', \chi_4\}$), where

  $$\chi_2': < B \neq C \; | \; \overline{Tcs(C,A)}, Cfb(B,C), dialwait(A), \overline{idle(B)}, idle(C) \; \xrightarrow{dial(A,B)}$$
  $$hearing(A,C), ringing(C,A) >$$
  $$\chi_4: < A \neq B, B \neq C \; | \; Tcs(C,A), Cfb(B,C), dialwait(A), \overline{idle(B)}, idle(C) \; \xrightarrow{dial(A,B)}$$
  $$busytone(A) >$$

5 The testing/animation phase

The key idea for interaction detection is to focus the execution of the telephone network model (including the services) on a set of specific scenarios that the experts' shared experience recognizes as the most interaction-prone. A scenario is an means of describing an execution trace from an abstarct point of view. A scenario is made out of a sequence of events which only occur in the same order in the execution trace; in particular, there is no reason for a scenario to be a sub-sequence of adjacent events in the execution trace. This sequence of events is used by the model animator (the GAC box in figure 3) to guide the generation of input events which are sent to the service model (the Sys interpreter). Thus, a scenario represents all possible execution traces which comprise the corresponding list of events.

The scenarios themselves are built as instances of interaction patterns (see section 5.2) which result from interaction criteria (detailed in the following section).

5.1 Interaction criteria

An interaction criterion is an interaction-prone situation that the experts know as problematic. The associated pattern is a more concrete description which is composed of actions in a certain order. The actions are abstracted from the common services internal operations. A typical interaction pattern which has been identified for long [2] is the "usage of a data for decision" followed by a "writing/modification" of this data,

We have drawn up the following list of interaction criteria classes and subclasses by analysing the most significant benchmarks [2, 9]. Informally, the identified interactions depend on:

- non-determinism resulting from the concurrent activations of several services; the non-determinism may be local (i.e. attached to services of one single subscriber) or non-local
- absence of reactivity
- bad resource handling
- bad identifier handling.

The criterion "bad resource handling" can be refined into:

- with one single resource
    - the resource being persistent
        * when there are automatic calls
            · when these calls are predictable
            · when these calls are not predictable (1)
        * when all calls are manual
    - the resource not being persistent (2)

- with two dependent resources resulting either from aliases or information loss
- security violation (3)

Several patterns can be associated with each interaction criterion. In rough terms, the simplest pattern which should reveal an interaction of type 2 can be expressed as a sequence of two operations the first one corresponding to an access to a resource and the second one being a modification of this resource; another pattern would recommend to link up an access, a modification and an access, in that order. For a type 1 interaction (typical of the Automatic Call Back), the pattern introduces an idle situation in between two modifications, the last one preceeding an access.

5.2 The action pattern language

The action pattern language has been defined as a means to express problematic situations, i.e. situations which appear during service executions and which could induce or correspond to service interactions. These situations result from a sequence of actions taken by service codes. Our objective is twofold: to express the situations in an as abstract as possible way, and to make the language adequate to describe potential conflicts between any kind of services (telephone services and internet services as well [8]). Therefore, the language components have been designed as an abstract view of the service implementation, and from the broadest perspective, even though they have been inspired by the classical benchmark. The actions are some kind of normalized operations. They are used to label the state transition rules with the following interpretation: when a rule is fired the associated list of actions takes place. For example, to any predicate $Tcs(B, A)$ in a precondition there is a *Test* action associated since, at the service implementation level, some program will test the called number in order to check, for B, whether A belongs to its black list. Likewise, there will be a *Remap* action associated with any $Cfb(B, A)$ predicate in the left-hand side of a pre/post rule, in order to indicate that a telephone number is transformed into another one.

The central notions are that of *resource* and *controller*, the actions being always carried out by a controller on some resource. The notion of resource encompasses what is called elsewhere [2, 6] data and resources: the various numbers handled by the services like, e.g. the Calling, Dialled and Destination Numbers, the list of filtered numbers for OCS or TCS, are examples of data in the literature, whereas the various signals are examples of resources in the literature. We introduce additional elements in the set of resources, such as the communication bits (see below). In brief, a resource is any type of object that one has to refer to when describing the system functioning.

A resource has several other notions attached to it. For the time being, the most relevant ones we have identified are:

- The *owner* is the one who pays for it (it may be a subscriber through the corresponding service code or the system itself). For example, a communication between two subscribers and y is composed of a chain of communication bits; most of them are charged to x but y may have to pay for a communication link which is subsequent to a forward he has asked for.
- The list of *actions* which can be carried out on this resource. For example, testing a resource in order to know its current value or whether it belongs to a specific list is such an action.
- The list of *controllers* includes all subscribers and the system, i.e. all agents which implicitly or explicitly act on this resource. For example, the previous *Test* action definition must make precise the controller which performs it. A controller is not necessarily the owner of the resource.
- The list of *rights* which indicate for each action the controllers which can perform them.

Each controller has a certain level of priviledges: nominal, intermediate, system.

These notions are structured using two types of descriptors, respectively for resources and controllers. Meta-actions are available to modify the contents of the descriptors (e.g. the list of rights).

So, a typical pattern for type 2 interaction could be the sequence of actions:

$Test(x, R), Remap(y, R)$

with $x \neq y$, and where x and y are controllers and R is a resource.

A typical pattern for type 3 interaction could be the sequence of meta-actions and actions: *Remove_authorization*(C, C', A, R) & *Priv*$(C) \leq$*Priv*(C') & *Owner*$(R) = C'$, $A(C', R)$, where C and C' are controllers, *Remove_authorization*(C, C', A, R) means that C removes the right for C' to perform action A on resource R.

This pattern signals a situation to be analyzed since C' still performs action A after its right has been revoked. For example, this could happen with C being the 911 service, and C' being any POTS subscriber calling 911, and A the termination of the call.

5.3 Generation of scenarios

From any pattern, we can compute several scenarios using an algorithm in three steps:

- identification of all pre-post rules containing the last action of the pattern in their list of attached actions;
- from the post-conditions of the previously selected rules, by backward chaining of the pre-post rules, computation of sequences of events.

When these events are issued from the initial state, one obtains a scenario.

For example the scenario $offhook(A)$, $dial(A, B)$, $Tcs(C, A)$, $Cfb(B, C)$ is generated from the pattern *Test*(x, R), *Remap*(y, R), with $x \neq y$. Likewise, the scenario $offhook(A)$, $dial(A, B)$, $911(B)$, $onhook(A)$ corresponds to the second example above.

6 Conclusion

We have presented a methodolody for an interactive service integration. A specification is a couple composed of a behavioral description and a set of properties and is finalized in two phases. The integration phase consists in designing a high level specification by means of partial and heuristic checks and with the help of an expert. The verification phase uses testing tools such that the expert knowledge allows to select input data prone to reveal interactions. The prototype developed in the ValiServ project is a first expert-assisted environment coping with this methodology. The integration phase is based on state transition rules and invariant specifications while the verification phase uses the Lutess tool and some specialised criteria. Until now, the methodology has been validated on small examples (up to 5 services) and deserves to be experimented on larger examples. An advantage of our approach is to allow the expert to classify detected interactions as desired or harmful. But this can also be viewed as a drawback since the expert is continuously solicited to interact with the tool. It explains why at the moment, in the frame of the ValiServ project, we take care of providing the expert with a tool interface easing the data analysis. From a distant point of view, we want to improve our methodology by inforcing the efficiency of the verification both at the integration and verification levels, and by widening the scope of the environment (more expressivity for the properties, new criteria for guiding test selection, pertinent criteria for chosing the interpreter configuration, model-checking verification, etc).

References

[1] Rafael Accorsi, Carlos Areces, Wiet Bouma, and Maarten de Rijke. Features as Constraints. In *Feature Interactions in Telecommunications and Software Systems VI*, pages 210–224. IOS Press, 2000.

[2] E.J. Cameron, N.D. Griffeth, Y.J. Lin, M.E. Nilson, W.K Schnure, and H. Velthuijsen. A Feature Interaction Benchmark for IN and Beyond. In *Feature Interactions in Telecommunications Systems II*, pages 1–23. IOS Press, 1994.

[3] Dominique Cansell and Dominique Mry. Abstraction and refinement of features. In *Language Constructs for Describing Features*, pages 65–84. Springer, 2000.

[4] L. du Bousquet, F. Ouabdesselam, J.-L. Richier, and N. Zuanon. Incremental Feature Validation: a Synchronous Point of View. In *Feature Interactions in Telecommunications and Software Systems V*, pages 262–275. IOS Press, 1998.

[5] Maritta Heisel and Jeanine Sousquires. A heuristic approach to detect feature interactions in requirements. In *Feature Interactions in Telecommunications and Software Systems V*, pages 165–171. IOS Press, 1998.

[6] D.O. Keck. A Tool for the Identification of Interaction-Prone Call Scenarios. In *Feature Interactions in Telecommunications and Software Systems V*, pages 276–290. IOS Press, 1998.

[7] K. Kimbler. Addressing the Interaction Problem at the Enterprise Level. In *Feature Interactions in Telecommunications Networks IV*, pages 13–22. IOS Press, 1997.

[8] K. Kimbler. Service Interaction in Next Generation Networks: Challenges and Opportunities. In *Feature Interactions in Telecommunications and Software Systems VI*, pages 14–20. IOS Press, 2000.

[9] K. Kimbler and H. Velthuijsen. Feature Interation Benchmark. Presented at the Third Feature Interactions in Telecommunications, 1995.

[10] M. Nakamura, Y. Kakuda, and T. Kikuno. Petri-Net Based Detection Method for Non-Deterministic Feature Interations ans its Experimental Evaluation. In *Feature Interactions in Telecommunications Networks IV*, pages 138–152. IOS Press, 1997.

[11] M. Nakamura, T. Kikuno, J. Hassine, and L. Loggrippo. Feature Interaction Filtering with Use Case Maps at Requirements Stage. In *Feature Interactions in Telecommunications and Software Systems VI*, pages 163–178. IOS Press, 2000.

[12] Masahide Nakamura, Yoshiaki Kakuda, and Tohru Kikuno. Feature Interaction Detection Using Permutation Symmetry. In *Feature Interactions in Telecommunications and Software Systems V*, pages 187–201. IOS Press, 1998.

[13] Y. Peng, F. Khendek, P. Grogono, and G. Butler. Feature Interactions Detection Technique based on Feature Assumptions. In *Feature Interactions in Telecommunications and Software Systems V*, pages 291–298. IOS Press, 1998.

[14] T. Ohta T. Yoneda. A Formal Approach for Definition and Detection of Feature Interaction. In *Feature Interactions in Telecommunications and Software Systems V*, pages 202–216. IOS Press, 1998.

*Feature Interactions in Telecommunications
and Software Systems VII*
D. Amyot and L. Logrippo (Eds.)
IOS Press, 2003

Plug-and-Play Composition of Features and Feature Interactions with Statechart Diagrams

Christian PREHOFER

DoCoMo Euro-Labs, Landsbergerstr. 312, 80687 Munich, Germany
prehofer@docomolab-euro.com

Abstract. This paper presents a new approach for modular design of highly-entangled software components by statechart diagrams. We structure the components into features, which represent reusable, self-contained services. These are modeled individually by statechart diagrams. For composition of components from features, we need to consider the interactions between the features. These feature interactions, which are well known in the telecommunications area, typically describe special cases or cooperations which only occur when features are combined. We describe these interactions graphically as well by partial statecharts. The main novelty is that full component descriptions are created automatically in a plug-and-play fashion by combining the statecharts for the required features and their interactions. Furthermore, we develop different classes of statecharts and show the interactions on a case-by-case basis. For composition, we use semantic refinement concepts for statecharts which preserve the original behavior.

Keywords: graphic description techniques, plug-and-play composition, feature interaction, statechart diagrams, semantic refinement, UML

1 Introduction

When talking about a piece of software, we often speak of a "feature" which this software has. For instance, when collecting requirements, features of a software system are common terminology. We propose here a graphic description method for feature-centric software development which supports composition of components from features in a modular way. The abstract behavior of features is modeled by statechart diagrams. However, features are often not independent and do interact in many ways. The main contribution of this paper is to describe features and their interactions graphically and to model their composition by stepwise refinement relations between statecharts.

A key problem is that features often have to cooperate or interact in unforeseen ways. In larger systems with many features, these interactions can lead to highly entangled code which is difficult to maintain. The problem of feature interactions is well established in the telecommunications area [3]. While most of the work in this area focuses on detecting interaction, we focus on software design methods which consider interactions. The problem is that handling interactions in most cases violates modularity. One often has to fix a special case in one feature which only occurs if another particular feature is present. Hence the code implementing these features becomes overloaded and obscures the core functionality of the feature as well as the dependencies.

For our development method, we employ graphic design of features by statechart diagrams (or abstract state machines). A novel point is that we model features and interactions as partial statecharts and transitions, which can be combined automatically when needed. This approach allows one to cross-cut large statechart diagrams into smaller features and their interactions. Based on semantic refinement concepts for statecharts, we present rules to create component models by combining the statecharts of the required features. These graphic refinement rules ensure that the feature behavior does not change during composition.

Our design method aims at graphic plug-and-play feature composition, which is essential in the following application scenarios:
- Many different versions of a software component are needed, each having different, often alternative features.
- Applications where monolithic software is unsuitable, since only the features needed by a customer are delivered. For instance, when downloading software on a limited mobile device, the download time and storage space on a mobile device should be kept minimal.
- Applications where entangled features are added or updated frequently during the live-cycle of a software component.

As an example consider an email server, where [4] has analyzed about 10 common features and discovered about 25 feature interactions. For instance, encryption and auto-responder interact as follows. The auto-responder answers emails automatically by quoting the subject field of the incoming email. If an encrypted email is decrypted first and then processed by the auto-responder, an email with the subject of the email will be returned. This however leaks the originally encrypted subject if the outgoing email is sent in plain. Hence for combining these seemingly independent features one must consider such special cases. We use in the sequel the above email example. The main features (see also [4]) are:
- Encryption and decryption for encrypted mails.
- Filtering particular emails, e.g. for virus protection.
- An auto-reply feature for automatic answers to all incoming emails.

In the following section we present the background on statecharts. Modular construction of statecharts modeling the behavior of features, interactions, and their combination is discussed in Section 3 through Section 5.

2 Statechart Diagrams and Semantic Refinement

Our graphic description method for developing software components is based on UML statechart diagrams [10]. When we model a software component with statecharts, we aim at specifying the behavior of its functions. We label transitions by the functions which trigger these transitions. An external function call triggers a transition labeled with this function depending on the current state of the statechart. Note that the finite number of states of a statechart usually represents an abstraction of the internal states a component can have.

We use the following UML notation for labeling transitions:

called_function() [condition] / action

A transition can be initiated by an external event, here called_function(). It may have a condition and it may have an action it initiates. This action describes the behavior triggered by the function. Note that all three labels may be empty. In case the trigger function is omitted, we have an internal transition without an external event, also called spontaneous transition.

An example is the statechart in Figure 1 describing the basic functionality of an email system in a feature called BasicEmail. We have two states, one called Waiting (for email) and the other called Emailarrived. The initial state is Waiting.

The transition labeled incoming() is triggered if an email has arrived and issues the operation get() to retrieve the email. The transition labeled deliver is the actual processing of the new email which triggers the store() function to save the email. This simple model of an email system only handles one email at a time. Note that we indicate functions by parentheses as in f(), which does not mean that these are parameter-less functions in an implementation. Later we will introduce parallel statechart diagrams and hierarchical diagrams based on UML notation.

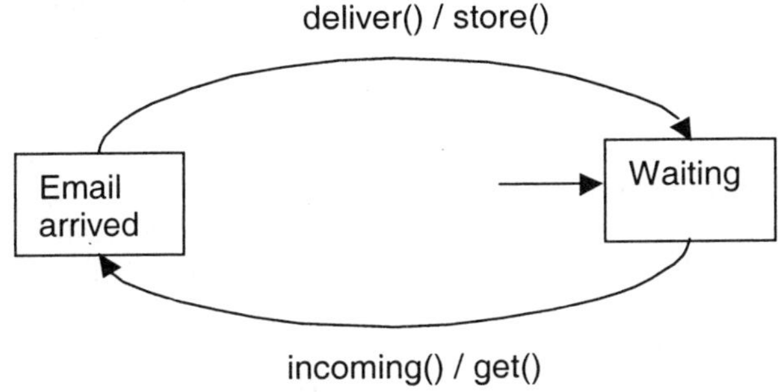

Figure 1: The BasicEmail Feature

2.1　Semantics

Our semantic model employs an external black-box view of the component. It is based on function calls from the outside which trigger transitions. Only the input and output are considered, not the internal states. A possible run can be specified by a trace of the externally called functions and the resulting actions of the statechart.

For instance, consider in this example traces for the statechart in Figure 1 of incoming() and deliver() transitions, triggered by external function calls.

Input sequence:
　　　　incoming(), deliver(), incoming(), deliver(), incoming(), deliver(),
Output sequence:
　　　　get(), store(),get(), store(), get(), store(),

We adopt the loose, "chaos" semantics [9] where the semantics of a component is given as the set of possible traces. The set of traces includes any possible trace by transitions specified in the statechart. In addition, any unspecified function call, e.g. a function call for which no transition is defined in the current state, leaves the statechart in chaos state and any behavior is permitted after that. For instance, the output for the input deliver() in the Waiting state is not defined in the statechart and hence anything is permitted after this. Our semantic model does not specify what happens in case the deliver() function is called in Waiting state. Hence the statechart does not fully specify a single implementation, but permits many possible implementations, which fulfill the specified input/output relation. More formally, we can describe the semantics as a set of deterministic, monotonic functions which are compatible the statechart. Each function represents a possible, deterministic implementation. Since many implementations are possible, we use a set of functions. This set of possible implementations can be reduced, which we view as refinement. Note that our statechart model permits a non-deterministic choice if several transitions are possible in one state, which is just a special case of loose semantics. For a more detailed treatment of the semantic background we refer to [1].

2.2   Refinement

Our notion of semantic refinement of statecharts aims at specifying a component in more detail and hence to reduce under-specification. We speak of semantic refinement of a statechart to another, if the refined one has fewer possible traces. As the loose semantics of a statechart is a set of functions, this can be formally expressed by reducing this set of functions. In this way, it is very suitable for abstract specifications and enables step-wise refinement by adding more specific behavior. For further details on semantic refinement relations for automata models, similar to statecharts, we refer to [9,5,7].

The main benefit of graphic refinement rules is their ease of use. The graphic refinement rules are based on syntactic input and output events. In some cases conditions of transitions have to be considered. Note that the rules do not cover any properties of possible input/output parameter or of internal state variables. To include these, the same semantic models can be used in this case [5,7], but deeper reasoning is required.   In practice, this often means that the refinement is shallow and just implies compatibility with respect to the input and output messages.

The following graphic operations on statecharts are also refinements with respect to our semantics.   These elementary statecharts refinement operations are proven to be semantically correct [5,7]:

- *Add new behavior* which was unspecified before, e.g. add a new state or add a new transition to a state, which did not exist at this state. Since this transition was not specified before, there is less chaotic behavior and the set of traces is decreased.
- *Eliminate alternative transitions*, if alternative transitions exit. This reduces non-determinism and specifies the behavior more precisely. Note that adding a condition to a transition can be handled in the same way, as it removes possible transitions at run time.
- *Add internal or compatible behavior*, which only adds new or internal behavior and does not change the original output. In this case of new behavior, we can abstract from the additional behavior and the old behavior remains unchanged. Under this abstraction, the original behavior is preserved.
- *Eliminating transitions for exceptional cases*. In this case, refinement only holds if some exceptional case does not occur. This is also called conditional refinement. Note that adding conditions to transitions is viewed as removing transitions.

In the following sections, we use the four refinement rules described above.

3   Modeling Features and Interactions by Statecharts

In this section, we describe graphic techniques to model the behavior of features. A feature, like a class in object-oriented design, offers an interface with functions and encapsulates internal state. We describe both features and interactions by partial statecharts to describe a high-level view of their behavior.

The novel point here is that we describe the features and their interactions modularly as fragments of statecharts. For a concrete feature combination, we show later how to combine these automatically to a statechart with the combined functionality. For the combination, we will make extensive use of hierarchical statecharts and parallel composition of statecharts.

When modeling features with statecharts, we can distinguish the following three kinds:

- Base features with a complete statechart, including an initial state and final states.
- Transition-based features which refine a transition locally, but do not add externally visible input transitions to a statechart. These features typically represent a service or an aspect which can be added to a feature combination with at least one base feature.
- State-extending features which add global states and externally-visible transitions. These features extend the states of some other features and also extend the externally visible interface.

We consider in the following these classes of features in the above order.

3.1　Base Features

Base features are self contained and can be used without any other features. An example of the first kind is the statechart in Figure 1 describing the basic functionality of an email system in a feature called BasicEmail.

The base features typically form the basis of a feature implementation. In contrast to others, they can also be used independently. For combination of base features, we will use parallel statecharts, as shown below.

3.2　Transition-based Features

Transition-based features provide services which can be modelled by internal transitions without persistent or global state. These features typically model auxiliary services and hence do not change the global control flow. They do not add externally visible input transitions to a statechart. The transition-based feature may however produce additional output or trigger transitions, e.g. in other, parallel statecharts as shown later.

We present transition-based features in the examples of the Forwarding and Reply features, each of which offers one function of the same name. In Figure 2, we show the internal states of the forward and reply functions. The feature Forwarding includes also a function call do_forward() which is not detailed here. The reply feature is similar. Note that we use two small circles to denote the start and end states of the transitions, which are determined later when composing features.

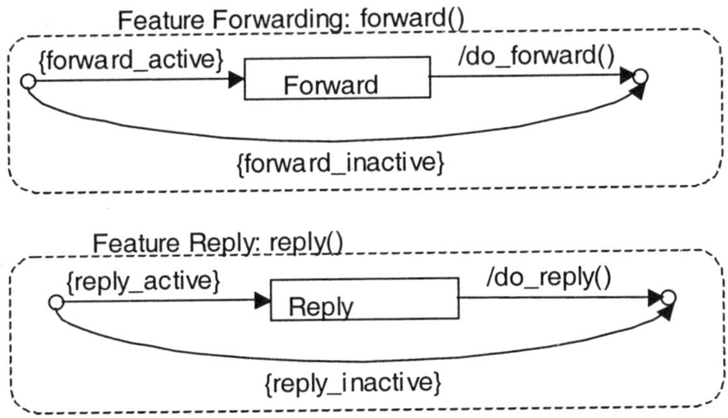

Figure 2: Internal View of Forwarding and Reply Features

Some transition-based features do not have internal states. Examples in the email application are the encryption and decryption features, which consist of a single transition each (Figure 3). An actual implementation of these features may have persistent state, but this is not modeled here.

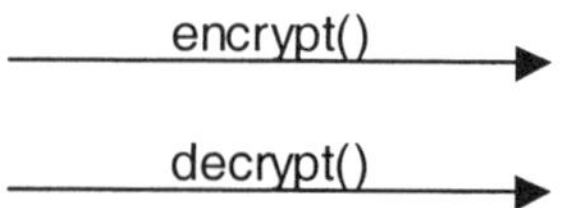

Figure 3: Encryption and Decryption Features

In general, a transition-based feature is modeled by a number of transition functions which we specify without detailing the start and end states. Furthermore, we may use an internal statechart to model the internal behavior of this feature. For refinement, we also make the assumption that the local statechart only consists of internal or spontaneous transitions. A schematic example is shown in Figure 4 below.

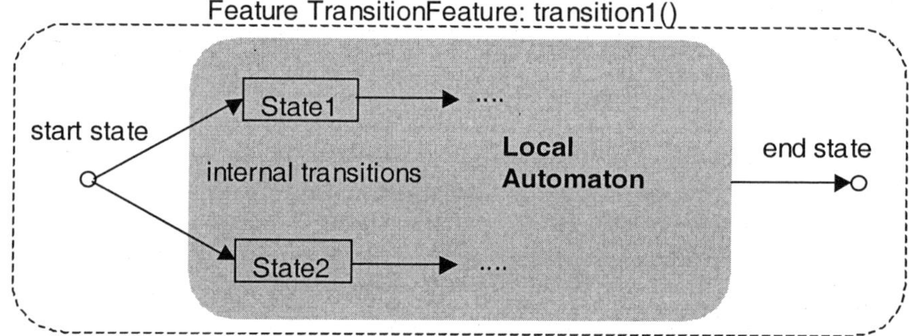

Figure 4: Partial Statechart for Transition-oriented Features

3.3 Global State-extending Features

Another main category of features is that of state-extending features which extend the set of states and add global transitions. In this way, they extend the external interface. For instance, consider a feature with a new state called MaintenanceMode (Figure 5). This partial statechart does not have initial or final states; its states will be reached by transitions from the states of other features. This will be specified separately in the feature interactions. Note that this statechart extends the external interface by new functions.

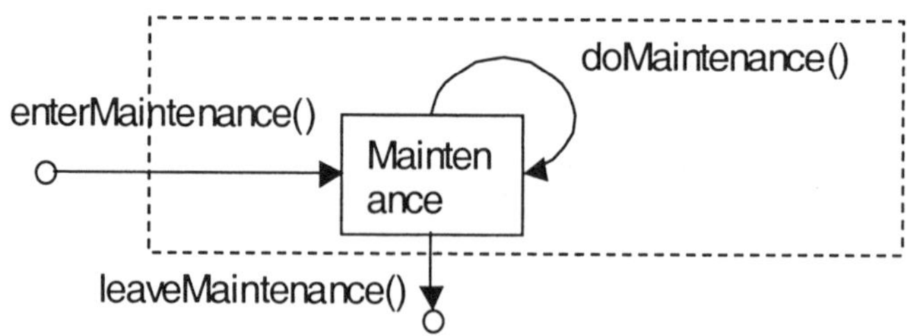

Figure 5: The MaintenanceMode Feature

The definition of a partial statecharts is a statechart with two special transitions for entering and leaving the statechart. These do not have start or end states, respectively. This indicates that the statechart shall not be used in isolation and shall be combined with others. We do not introduce named start/end states, since feature combination in this case would require to rename states which can cause complications, e.g. if other features or interactions refer to this state.

For combination, these start/end transitions will be mapped to specific states. We will also permit that the start function may be triggered from several states, unlike the exit function. Compared to the above transition-oriented features, the new state Maintenance is global, as well as the functions of the transition are visible externally. As we will show in the next section, composition of such features consists of two steps. First we merge this partial statechart with another statechart. In addition, other adaptations may be needed.

4   Resolving Feature Interactions

Interaction handling adapts a feature into the context of another one in order to resolve feature interactions. An adaptor A $\rightarrow$ B defines a refinement of a composed statechart which includes feature A (and possibly others) with feature B. With statecharts, we will use two techniques to refine a feature A to the context of a new feature B:

- The transitions of feature A may be refined. We model this by restricting a transition or by inserting a local statechart, leading to hierarchical statecharts. We will also use this kind of refinement to specify the start or end states of a transition, as shown below.
- New transitions from the states of A to states of B triggered by function calls of B can be added if B is state-oriented. We also refer to [5] for this kind of refinement.

The goal here is to denote this refinement just by describing the necessary refinement steps in isolation, without the context of other features. In addition, it is important to describe this adaptor generically such that it can be used for any combination of features including feature A. This is the essential abstraction which permits to compose features in a flexible way.

4.1   Interaction of Transition-oriented Features

In this section, we describe features which refine transitions by local statecharts. More precisely, this can be seen as an hierarchical statechart. We first focus on adapting base features to transition-oriented features. For instance, Figure 6 shows how the deliver function is adapted to the decryption feature. Note that we do not specify the refinement for individual transitions but for functions labeling the transitions. Hence two transitions with the same label are refined in the same way.

In the example, the function deliver() is expanded by a statechart with two transitions and a newly added state, where decrypt() is a function of the added feature. In this way, we refine the deliver() function to first execute the decrypt() operation and then the original deliver() operation. Note that *BasicEmail.deliver()* is now a special internal operation which is not externally triggered. We denote this by adding the feature name and by italics. It denotes the internal operation to be triggered by deliver() and is implicitly triggered. Externally, it is viewed as a spontaneous transition. Also it is important for further refinement steps, which can refine this transition. This simplifies the technical treatment, because we do not refine local statecharts but only transitions.

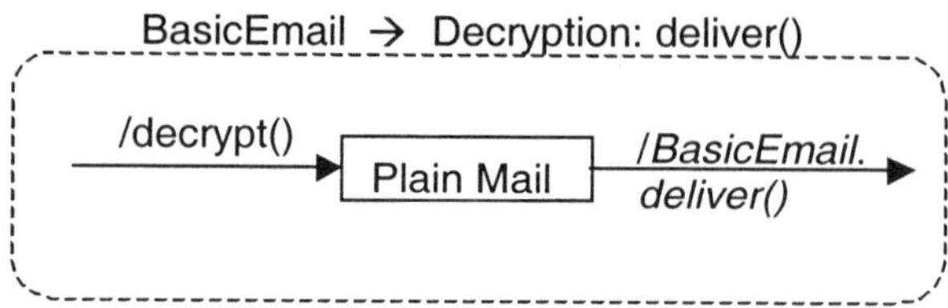

Figure 6: Adapting BasicEmail to Decryption

In a similar way we can adapt BasicEmail to Forwarding, as shown in Figure 7. In this figure, the function forward(), as defined above, is not shown in detail.

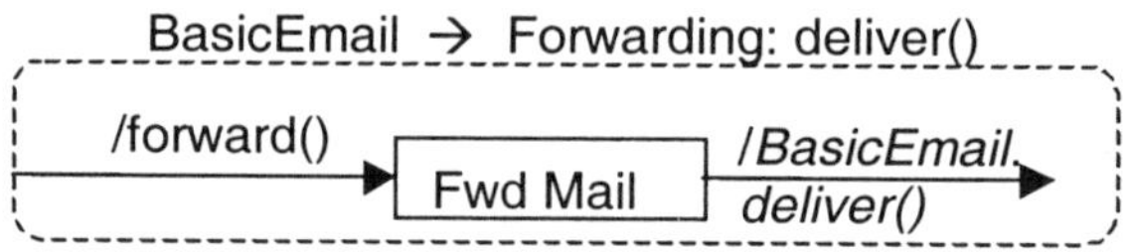

Figure 7: Adapting BasicEmail to Forwarding

Next we consider the case of an adaptor between two transition-oriented features. Interaction between automatic reply and decryption is a typical special case, which can often be modeled by adding or restricting transitions. The interaction here can be handled similar to the above and is shown in Figure 8. Furthermore, we restrict both transitions by adding conditions. As we will see later, do_reply() will occur as an action of a transition which we refine by the statechart in Figure 8.

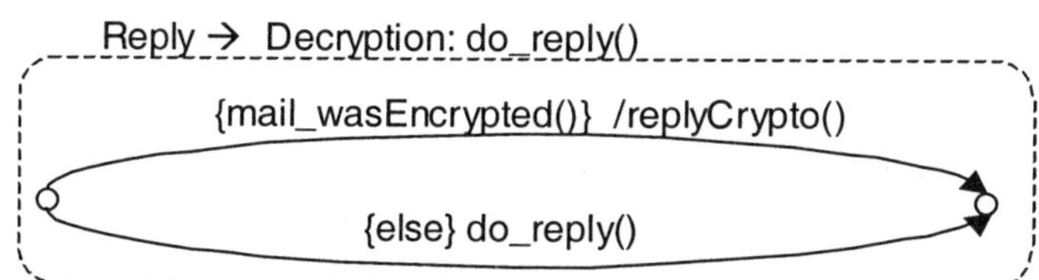

Figure 8: Adapting the Method do_reply of the Feature Reply to the Decryption Feature

In the general case, an adaptor can refine a transition of the adapted feature by extending the transition to a local statechart with internal states and transitions. We do not permit externally triggered functions in this refinement, since this may affect the external semantic behavior. Externally visible transitions in local statecharts would syntactically be possible, but this does not follow our notion of semantic refinement. It would require a different or more complex notion of refinement. (See also [5,1] for a detailed treatment of refinement.) The function which is refined can however be used in the operation associated with an internal transition.

We can express the general case with following schematic adaptor shown in Figure 9. We adapt all transitions labeled with the same function in the same way by one adaptor. In case two transitions (from different states) are triggered with the same function, these may not have different refinements.

The inner "black-box" statechart can use the functions of the feature A, e.g. A.a(), as well as the functions of features B and C. No others are allowed, since the feature adaptors have to be generic to be added to any feature combination which includes the adapted feature. Furthermore, only internal or spontaneous transitions are permitted, as the external input behavior shall not change. Note that we use the notation F.f () to denote a function f of the feature F.

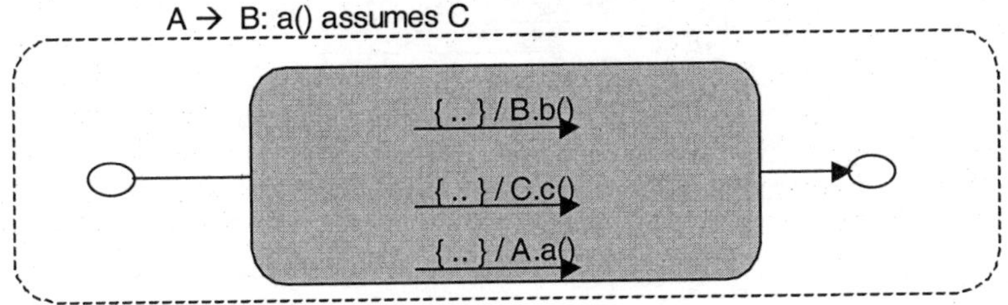

Figure 9: Schematic adaptor for a transition, triggered by a(), of feature A to B

4.2 *Interaction of State-extending Features*

Interaction in this case defines the merging of two statecharts by adding transitions between the states of the features. For this, we need to map the anonymous start/end states of the state-oriented feature to states of another feature. Furthermore, transitions may be refined by conditions or actions. In the following, we consider the possible interactions based on our classification of features.

A typical interaction with a base feature is illustrated by the example in Figure 10 for adding the BasicEmail feature to the MaintenanceMode feature. We only show the relevant states for both features and indicate the MaintenanceMode feature by the gray area. The interaction defines that the latter feature can be reached from the Waiting state of the BasicEmail feature. We view this interaction as a refinement of the MaintenaneMode feature, since the transitions of this feature are refined by specific start/end states.

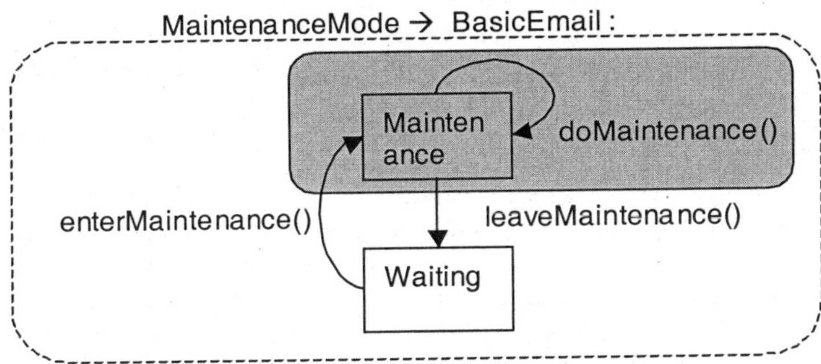

Figure 10: The MaintenanceMode Interaction with BasicEmail Feature

Another example is shown for a feature called ErrorCase with the state EmailError and the two functions error() and resume().

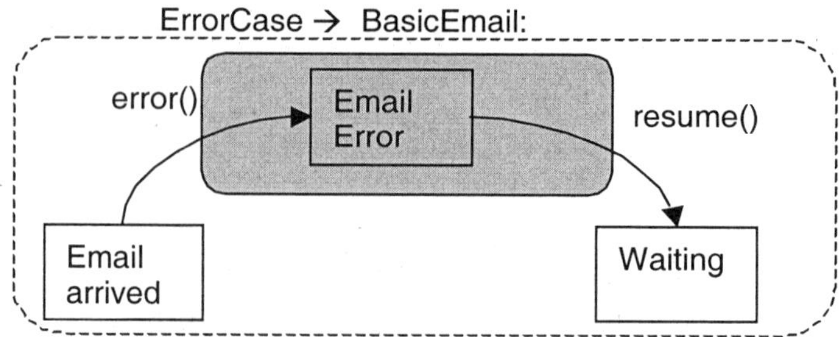

Figure 11: Interaction of Error and BasicEmail features

In the general case, an adaptation A → B, with one base and one state-extending feature means to add new transitions from states of A to states of B, which are labeled by functions of A. For instance, in the above example the other direction for adaptation is also possible. In this alternative, adapting BasicEmail to ErrorCase would allow one to add transitions labeled with deliver() or incoming() of BasicEmail to the EmailError state. Regarding semantic refinement in this variation, we have to make sure that this transition has not been defined before. For instance, it would be legal in this variation to add a transition labeled deliver() from Waiting to EmailError, since this is not defined yet. On the other hand, this is not possible for the Emailarrived state, since this would overlap with an existing transition. Although this might be viewed as a non-deterministic statechart, this leads to semantic refinement problems as discussed in [5].

The case of an adaptation between two state-extending features is similar to the above. The difference is that in combination of several features, only one interaction may define the exit transition of the new statechart. On the other hand, we show that there can be several transitions labeled with the function entering the new feature. An example for this case is the adaptation of MaintenanceMode to ErrorCase, as shown in Figure 12. For this, we add a new transition labeled enterMaintenace() from the EmailError state. In this way, the MaintenanceState is also reachable in an error case, which resolves an important interaction. Note that we do not fix a state for the exit transition from Maintenance, since we assume that this will be done by the base feature, here the BasicEmail feature. As a general rule, only one feature in a combination can define the exit state. In contrast, there can be several transitions for entering the Maintenance state. For instance, there can be transitions from both the Error and BasicEmail features.

In addition to this mapping, we may have to refine the transitions, as for transition-based features. In this example, the transition leaveMaintenance() has been restricted. We refine the leaveMaintenance message to leave the Maintenance state only if no errors exist. In other words, it is possible to enter maintenance from error state, but then the error must be fixed in the maintenance state.

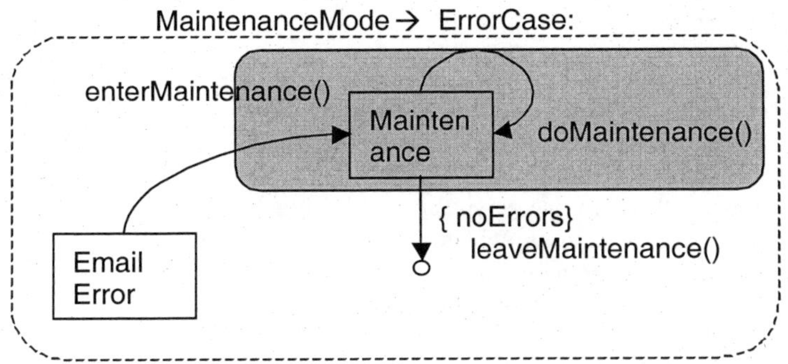

Figure 12: The MaintenanceMode Interaction with BasicEmail Feature

The adaptation of a transition-oriented to a state-extending feature is more restricted, since we may not add transitions from local to global states. Hence only transition refinement by conditions or actions is possible.

We have discussed the combination of state-extending and base features. The remaining case of adaptation of a state-extending to a transition-oriented feature is analogous to the case for base features shown above.

4.3   Interaction of Base Features

For base features, we distinguish several cases. The cases where a base feature interacts with a transition-based or state-extending feature have already been treated above. In case of interactions between two base features, we use parallel statecharts for combination, as shown below. Since these statecharts operate separately, we can only restrict the transitions of the other base feature, as shown in the examples above. Adding transitions between parallel statecharts is not permitted.

5   Combining Features and Interactions as Refinement

We show in the following how to combine features and their interaction handlers in an automatic way. If no adaptor between two features exists, we assume that the features can be combined in any arbitrary order. If feature A must be adapted in the presence of B, A must clearly be added before B. In this way, the interaction handling defines the order of the feature combination. Given a selection of features, a possible sequence for composition can be inferred. Alternatively, a particular order can be given explicitly. Then the statechart for a concrete feature combination can be determined, as we show in this section. We use several forms of statecharts refinement, including transition refinement and parallel statecharts. For a more detailed semantic treatment of refinement we refer to [5,9].

In order to obtain practical composition schemes, we follow several design principles as discussed in [8,6]:

- We limit our model to feature interactions between two features. As we focus on a general purpose software development method, we do not focus on modeling such three-way interactions [12]. We refer to [8] for a more detailed discussion.
- Asymmetric interaction handling: in case of an interaction only one feature is adapted.
- Composition of features in a sequential order is used as it is the simplest and most natural composition technique.

5.1 Transition Refinement

We cover in the following the case of adding transition-based features. In this case, the combined statechart can be obtained by unfolding the interaction handlers in the statechart one after the other. Adding features with global control state is considered in the following section.

We consider in the example a typical case with a base feature, which is externally visible, and add auxiliary features, which do not change the external interface. We first consider adding features to the base feature, and then we look at other kinds of interactions. This combination proceeds sequentially and produces a typical pipeline-architecture, where the input is passed from one feature to another. For instance, the message is first forwarded, and then decrypted, and finally it is delivered to the client. Hence the interaction consists of extending the message delivery function of the basic email feature. For the reverse direction, not shown here, one has to extend the message incoming function in a similar way.

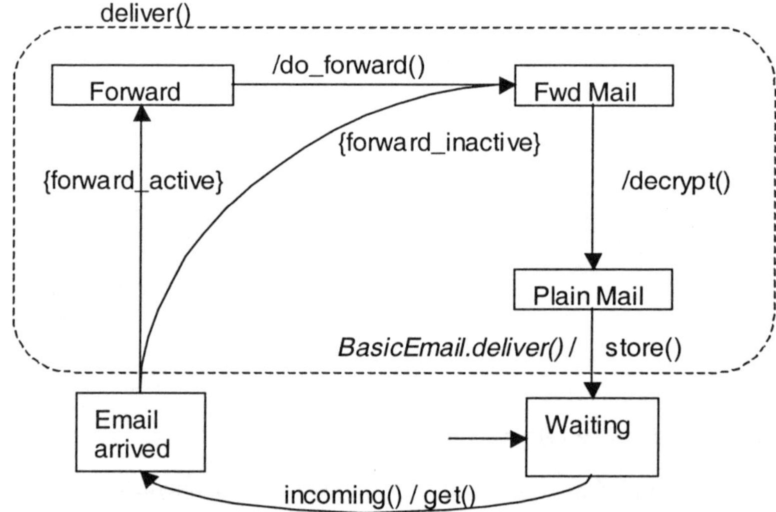

Figure 13: Combination of BasicEmail, Decryption and Forwarding

The hierarchical statechart in Figure 13 shows the combination of three features, extending the basic email feature. Note that the delivery() function has been refined twice by two feature interactions. As only new behavior is added, refinement by abstraction is easy to show.

In similar fashion, we can combine three other features, including the reply and decrypt features. We first define the interaction between the Reply and the BasicEmail feature, as shown in Figure 14. In this figure, the function forward(), as defined above, is not shown in detail.

Figure 14: Adapting BasicEmail to Reply

The combination of the three features in Figure 15 also illustrates the interaction between the reply and decrypt features, which is not shown explicitly. For this interaction, we refine the do_reply() function to account for the interaction and use a function replyCrypto() in case the email was encrypted. This function has to handle the interaction of leaking the previously encrypted header of the email, as detailed above.

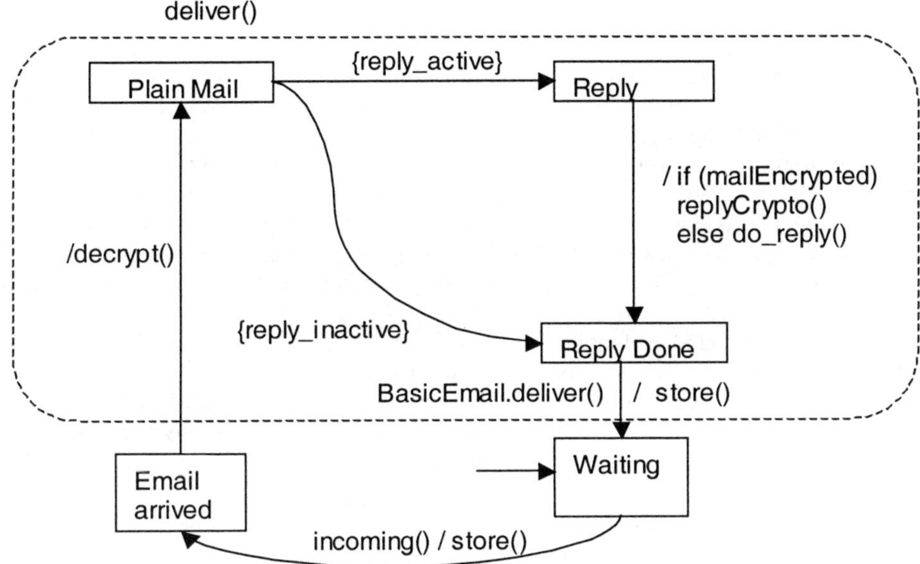

Figure 15: Combination of Basic Email, Auto Reply and Decryption

With respect to the original BasicEmail, the refinement relation is clear; all the new output operations have to be abstracted. For the refinement from the combination of BasicEmail and Auto Reply to this combination of three features, exceptions have to be considered. Since the normal do_reply() function is not used in case of encrypted emails, we have to consider this as an exception. Unless the exception occurs, the refinement holds with the appropriate abstraction of the new output.

5.2 Combination of Base Features

So far, we have expanded transitions for combination. For base features this method is however insufficient. In the case of two base features, we combine features with individual statecharts. We obtain two parallel statecharts with disjoint function labels. This is typically needed if features add external methods and may affect the control state.

For instance, consider a generic locking feature, which disables all function calls to an object which change the state. In the case of the mailer, this may be used to stop receiving/sending, e.g. to configure or inspect the mailer.

The lock feature has two externally visible functions, lock and unlock, and two states. It is shown in Figure 16.

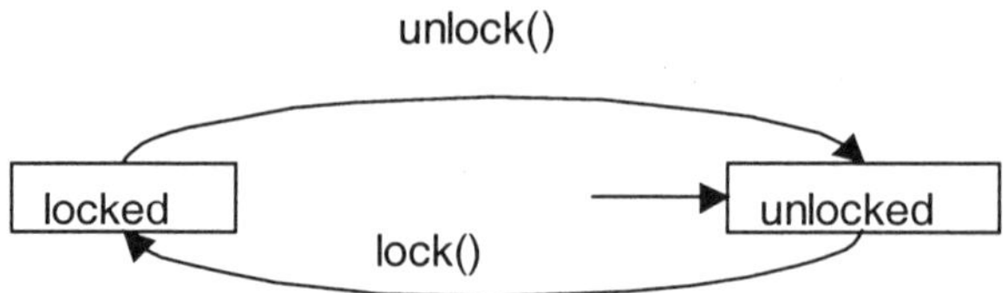

Figure 16: The Lock Feature

When combining the BasicEmail and the lock features, the interaction handling shall block transitions if the statechart is locked. We show the combined statechart in Figure 17; it uses several interaction adaptors which are not shown separately. The first interaction is the blocking of transitions by adding conditions to the transitions, which is a particular form of transition refinement. Adaptors can extend the functions in the parallel statecharts and can add function calls to other statecharts. For instance, an adaptor to lock may invoke the lock() operation of the Lock feature. For illustrating this, we show here a new function of the basic email feature, called reset. When adapting this function to Lock, this transition has to unlock the Lock.

When adding another feature we may have to adapt all the previously adapted functions. With parallel statecharts, only the transitions of one statechart have to be adapted. Furthermore, we need to restrict the transitions of the BasicEmail feature by some conditions which check if the statechart is not locked. These conditions are not detailed here and are specified in an adaptor similar to the above cases.

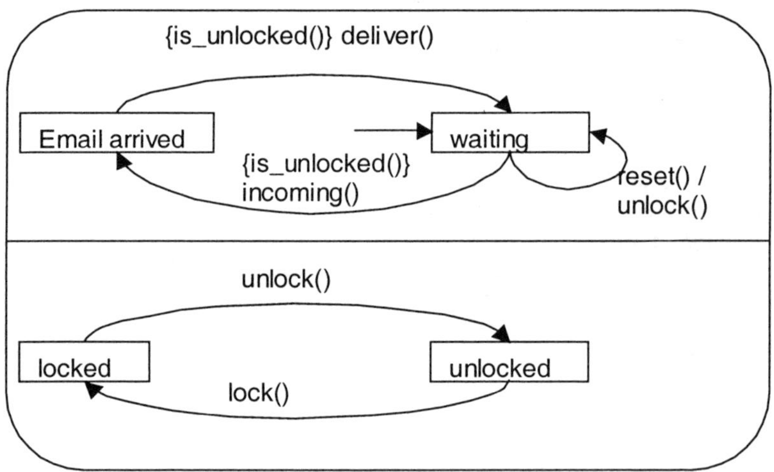

Figure 17: Parallel Statechart with the adapted BasicEmail and Lock Features

An interesting case occurs when a state-oriented feature is added to email which interacts with lock. In this case, only adaptation of transitions is permitted. It is not permitted that interaction handling creates transitions between states of parallel-composed statecharts. This would lead to semantic problems as it contradicts the concept of parallel composition. Since the new statechart only adds new behavior, refinement is easy to show.

5.3   *Features which add Global States*

In the following, we discuss the combination of features which add global states. For this, we have two steps:

- Merge the statecharts of the new feature with an existing statechart with some features.
- Adapt all the previously added features. For this, we add the transitions and transition refinements for each adaptor of all the features added earlier.

For instance, we can combine the Maintenance, Error, and BasicEmail features as shown in the Figure 18 below. Note that we first add the Maintenance feature, which is then first adapted to the Error feature and then to the BasicEmail feature. Each of these adaptations adds one transitions labeled enterMaintenance(). As these transitions and the states are new, it is easy to see that the semantic refinement relation holds.

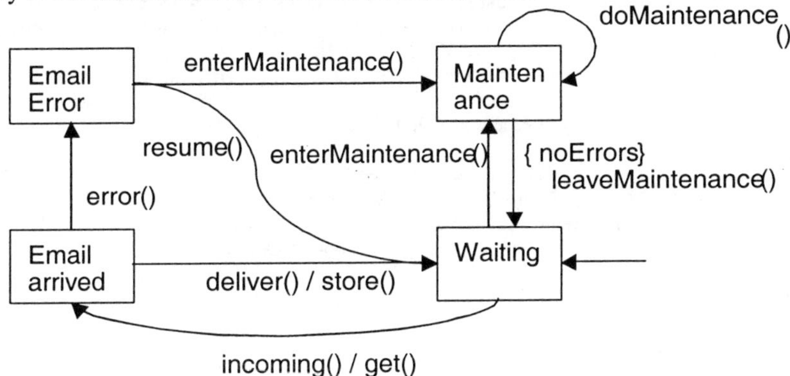

Figure 18: Combination of Maintenance, Error, and BasicEmail features

6   Conclusions

We have presented a set of graphic specification techniques which allow one to structure highly entangled software systems. With our graphic design techniques, we have shown plug-and-play concepts for the construction of complex statecharts, which describe a component with several features. The main contribution is the graphic combination rules of features, which are based on semantic refinement. In this way we can select any combination from a set of features and the complete statechart can be created automatically. As we use each feature only one time and the selected set is ordered implicitly by the feature interaction handlers, we have an exponential number of possible feature combinations. This is based on a quadratic number of interactions.

There exist several interesting extensions in the area of parallel statecharts for further work. For instance, we have focused on sequential composition of statecharts. With parallel statecharts, one may specify features which only affect one of the parallel statecharts or even add the same feature twice in each of the parallel statecharts. Furthermore, we have not considered hiding operations for the external interface. For parallel statecharts, it is possible that one of them is controlled by the other and is not externally visible. In this case, an extension for syntactical interface hiding may be useful.

Our approach was guided by semantic refinement concepts based on a black box component view. We have shown that our combination methods preserve the external interface and only reduce the set of possible execution traces. By nature of our approach, we have only considered syntactic criteria which are compatible with semantic refinement. For a deeper semantic analysis of the behavior of components, formal specification technologies can be used as e.g. considered in [7,1]. The work in[11] is similar as it aims at incremental refinement on different levels of abstraction, but without graphic description.

There is very little work on modularization of statecharts based description. The work in [5] covers incremental development of statecharts, but does not consider features as independent entities and also does not consider interactions. Most of the work on feature interaction aims at detecting interaction, but does not consider systematic development of features. There is other work on modeling telecommunication features using statecharts [12] which however aims at formal verification using model checking.

References

[1]   M. Broy and K. Stolen, Specification and Development of Interactive Systems. Springer-Verlag, 2001

[2]   Tzilla Elrad, R. E. Filman, A. Bader, Aspect-oriented programming: Introduction, Communications of the ACM, Volume 44 , Issue 10 (October 2001)

[3]   Workshops in Feature Interaction 1992-2000, Proceedings published at IOS Press, Netherlands, www.iospress.nl/site/html/tel.html

[4]   R. J. Hall, Feature Interactions in Electronic Mail, IEEE Workshop on Feature Interaction, IOS-Press, 2000

[5]   C. Klein, C. Prehofer und B. Rumpe, Feature Specification and Refinement with State Transition Diagrams, In.: P. Dini (Ed..), Fourth IEEE Workshop on Feature Interactions in Telecommunications Networks and Distributed Systems, IOS-Press, 1997.

[6]   C. Prehofer, Feature-Oriented Programming: A Fresh Look at Objects, ECOOP '97 -- Object-Oriented Programming, Springer LNCS 1241, 1997.

[7]   C. Prehofer, Flexible Construction of Software Components: A Feature-Oriented Approach, Habilitation Thesis, Technical University of Munich, 1999

[8]   C. Prehofer, Feature-Oriented Programming: A New Way of Object Composition, Concurrency and Computation. 2001

[9]   B. Rumpe and C. Klein, Statecharts Describing Object Behavior, In: Specification of Behavioral Semantics in Object-Oriented Information Modeling, Kilov, H. and Harvey, W.(Eds.), Kluwer, 1996

[10] The OMG's Unified Modelling Language, version 1.4, http://www.omg.org/UML/, 2002

[11] Bengt Jonsson, Tiziana Margaria, Gustaf Naeser, Jan Nyström, and Bernhard Steffen. Incremental Requirement Specification for Evolving Systems. *Nordic Journal of Computing*, 8(1):65-87, Spring 2001.

[12] Gregory W. Bond, Franjo Ivancic, Nils Klarlund, Richard Trefler, ECLIPSE Feature Logic Analysis, Internet Telephony Workshop 2001, April 2001, Columbia University, New York City, USA

[13] M. Kolberg, E.H. Magill, D. Marples and S. Reiff. Second Feature Interaction Contest Results. In M. Calder and E. Magill, editors, Feature Interaction in Telecommunications and Software Systems VI, IOS Press, Glasgow, May 2000.

*Feature Interactions in Telecommunications
and Software Systems VII*
D. Amyot and L. Logrippo (Eds.)
IOS Press, 2003

Methods for Designing SIP Services in SDL with Fewer Feature Interactions

Ken Y. CHAN and Gregor v. BOCHMANN
SITE, University of Ottawa
Ottawa, Ontario, Canada, K1N 6N5
{kchan,bochmann}@site.uottawa.ca

Abstract. This paper describes methods for implementing telephony services in SIP with fewer traditional feature interactions. A formal SDL model of SIP and its services has been derived from published SIP specifications for verification and validation. It is known that the SIP RFC describes only the protocol specification. The specifications of SIP services and additional service features are informal and can only be found in various IETF drafts. Nevertheless, the service designers are still faced with new feature interaction problems. These new feature interactions are unique to SIP because SIP has flexible signaling features, such as request forking and dynamic assignment of contact addresses, which have both cooperative and adversarial side effects on each other. This paper also describes an extension to the classical feature interaction taxonomy, which is used to associate the causes, effects/symptoms with the preventive measures of the new and traditional feature interactions. Finally, SIP services can be designed and implemented without certain feature interactions by following certain design rules which are based on the knowledge deduced from the verification.

Introduction

In the telecommunication industry, many people believe that Internet telephony would be the next major market for many carriers. Internet telephony is defined as the provisioning of telephone-like services over the Internet. With many Internet users having already embraced voice chat and webcam conferencing, Internet telephony, which also encompasses convergence of existing Internet and PSTN voice services, is the natural step forward. Although H.323 [14] has been around for a decade or so as the de facto signaling protocol over data networks, SIP (Session Initiation Protocol), which is being standardized under IETF RFC 2543 [3] and backed by powerful industry partners such as Cisco, Microsoft, Nortel, Sonus, and DynamicSoft, is gaining a lot of momentum in establishing itself as the voice over IP (VoIP) signaling protocol of choice. SIP is designed based on the general philosophy of IETF that protocols remain open and support decentralization of control over applications in an Internet environment. SIP is generally better developed than other telephony signaling protocols in areas such as distributed call control between end systems and inter-provider communications. These areas are known to be prone to feature interactions (FI's), thus many researchers believe feature interactions are more pronounced in Internet telephony than PSTN (Public Switched Telephone Network) [1].

This paper is organized as follows: Section 1 presents an enhanced classification of FI's for POTS (Plain Old Telephone Services) and an overview of the SIP protocol. Section 2 gives a detailed description on the SDL specification of SIP services. Section 3 discusses the semi-automated process of verifying traditional FI's that may exist in SIP. Section 4 presents a method for preventing traditional FI's in SIP. Section 5 shows new FI's that are potentially introduced by SIP and their mapping to the extended classification system. Finally, the paper ends with a conclusion and discussion of future work in Section 6.

1   Enhanced Classifications of Feature Interactions and Overview of SIP

Features in a telecommunication system are packages of incrementally added functionality that provide services to subscribers or administrators. In this paper, we do not make any distinction between feature and service; the two terms would be used interchangeably. There are many definitions of *feature interactions (FI's)*. In general, FI's are defined as undesirable side effects caused by interactions between features and/or their environment. The research community of FI's has generally categorized FI's by the nature and the cause of the interaction.

We believe the latter categorization by cause may be useful in classifying FI's, but the classification does not lead explicitly to a general scheme for detecting or resolving FI's. A classification process is most useful if it leads to methods of detecting, preventing, or resolving FI's. While understanding the causes of interactions is helpful in formulating potential resolution policies to prevent FI's, the causes of interactions are conceptually abstract from the actual feature specifications; for example, a resource contention scenario like presenting a voice greeting and call waiting tone to the same user may or may not be considered as a FI. Also how does a service designer know which "resource" to declare for checking resource contention? Thus, the causes cannot be construed as the most efficient means to derive methods for detecting FI's. We propose adding another type of categorization to the taxonomy of FI's, which is categorization by symptom or effect. This new categorization enables FI's to be associated to some of the well-known distributed system properties. As a result, we believe existing verification and validation techniques or tools for these distributed system properties may be used to facilitate detection of FI's, especially in IP Telephony such as SIP.

1.1   The Nature of Feature Interactions

The traditional categorization includes three dimensions: by kind, by the number of users, and by the number of network components. The combinations of these dimensions produce five types of FI's: single-user-single-component (SUSC), single-user-multiple-component (SUMC), multi-user-single-component (MUSC), multi-user-multi-component (MUMC), and customer-system (CUSY) FI's [2]. Detail discussion on this categorization can be found in [2].

1.2   Causes of Feature Interactions

Categorizing FI's by the cause has been suggested in [2]. The suggested causes are violation of feature assumptions, limitations of network support, and intrinsic distributed system problems. Instead of discussing these general causes, we consider below four detailed causes that are also mentioned in [2] and are specializations of the above general causes.

Resource contention (RSC) is definitely a well-known distributed system issue. It is defined as the attempt of two or more nodes to access the same resource. In the context of FI's, an accessing node may be the feature running on a network component. Resource is an abstract term; it needs not be a physical entity. For example, Call Waiting (CW) and Three Way Calling (TWC) in POTS may be considered as contending for the flash hook signal simultaneously.

Violation of Feature Assumptions (VFA) is defined as a set of assumptions that telecommunication features, much like any software features, operate under, and are designed based on. There are many types of assumptions. One example from [2] that is particularly worth examining is assumptions about data availability. Features such as Terminating Call Screening (TCS) cannot function properly without the caller-id of the caller. If the caller-id were made unavailable in the case of Private Call (PC), TCS would permit the call request. In the case of Operating Assisted Call (OAC) and OCS, the hiding of the original caller-id by OAC allows the call to bypass the OCS restriction.

Resource Limitation (RSL) is also mentioned in [2], and is definitely a common cause of FI's in POTS, because most of the traditional POTS end-user devices (e.g. basic phones) have limited user interfaces and computational power. If the classical example of resource limitation were revised (e.g. Call waiting and Three Way Call), the confusion of the flash hook signal can also be attributed to the lack of separate buttons for the two features; thus that feature interaction can be also classified under resource limitation.

Timing and Race conditions (TRC) are also mentioned in [2], and are common, particularly in distributed real-time systems like POTS running on PSTN. A race condition is defined as a condition where two or more nodes have non-mutually exclusive read and modify access to a shared resource. As a result, the value or status of the shared resource may be undefined (different values in the same context) depending on the timing of the accesses between the nodes. A classical example is between Automatic Callback (AC) and Automatic Recall (AR). The AR feature makes a call on behalf of the original caller when the callee becomes idle again. Both AC and AR features depend on the busy signal of the callee. The timing of the busy signal received by the two features would either allow one of the calls to go through on the single line of the POTS phone, or reject both call requests because both ends are calling each other simultaneously.

1.3　Effect of Feature Interactions

This categorization, which is called categorization by effect, focuses on the effects or symptoms of the problems that are more specific and can be used to narrow down the search in detecting FI's. Common distributed system properties such as livelock, deadlock, fairness, and non-determinism are well-defined concepts. Tools such as Telelogic Tau [8] can verify some of these properties in a system. The category called incoherent interactions has also been introduced in the following subsection. The following diagram (Figure 1) illustrates the graphical representation of the extended taxonomy, which is called the "feature interaction tree" (FIT). Since a feature interaction can be associated to just one kind of interaction, but one or more cause-effect category pairs, a feature interaction may be defined as a spanning sub-tree within the following feature interaction tree. Two or more FI's may overlap with each other on the FIT (Figure 1). In addition, each effect-type interaction (e.g. livelocking (LLCK)) is associated to one or more preventive measures (not shown in the diagrams).

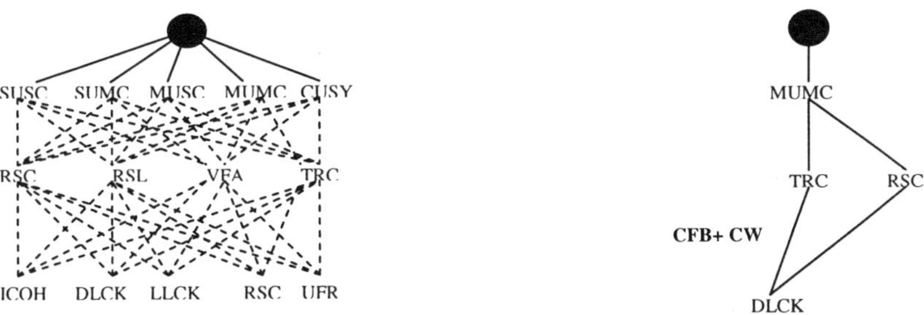

Figure 1: Feature Interaction Tree (FIT) and CFB+CW example

At this moment, the "FIT" is merely a graphical notation of the extended taxonomy to visualize the catalogue of FI's. We have not explored other interesting properties that the "FIT" may offer. For example, if deadlocking interaction could happen in CFB+CW pair and Call Transfer (CT) + CW pair separately, the question would be: could transitive property be applicable to "FIT"? That is, could deadlocking interaction also happen in CFB+CT pair? These questions are beyond the scope of this paper.

1.3.1   Livelocking (LLCK) and deadlocking (DLCK) interactions

Livelock is a well-defined computer science concept that occurs when the affected processes enter state transition loops and make no progress. In the case of Automatic Callback (AC) and Auto Recall (AR), if both features receive the busy signal from the callee and initiate the call simultaneously, both calls would not go through on single-line phones because both ends are busy making calls. Both features are in a livelock situation because they may always get the busy tone when they repeat the process and never complete the call. A livelocking interaction needs not be permanent; in most cases, the relative timing of the two processes leads to an eventual exit from the livelock loop.

Deadlock is another well-known distributed system concept that occurs when two or more processes are in a blocked state because they require exclusive access to a shared resource that belongs to the other, or wait for a message from the other process that will never be sent. An example is described in Figure 2 involving CW and CFB at A communicating with CW and CFB at B.

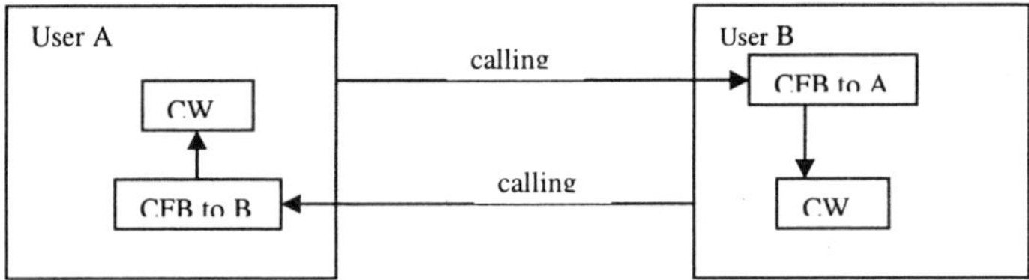

Figure 2: Call waiting and CFB at A versus Call waiting and CFB at B (deadlock)

If A and B call each other simultaneously and are programmed to forward to each other on busy, then both callers would keep hearing the phone ringing at the other end (and perhaps also hear the CW tone). They would be deadlocked at the ringing state and no progress would be made until one of them hangs up. If both ends do not subscribe to CW service, the call would be forwarded to each other on busy (or a call loop). Both users would be presented with a busy tone instead.

1.3.2   Incoherent interactions (ICOH)

An incoherent interaction is a form of a violation of feature assumptions (or properties). This term was first introduced in [9] to describe "the identification of specific incoherence properties" between the affected features. Furthermore, an incoherent interaction can be caused by resource contention and resource limitation. The two properties from the classical case of CFB and OCS is a good example; user A would successfully call user C via CFB even though an OCS entry at A is supposed to forbid any outgoing call to C. The CFB and OCS are programmed with contradictory assumptions.

1.3.3   Unfair interactions (UFR) & Unexpected Non-determinism (NDET)

Fairness is also a well-known distributed system concept in which processes of equal priority should be assigned fair scheduling so that they eventually proceed and have fair access to shared resources. A novel case of unfair feature interaction occurs between Pickup (CP) and Auto Answer (AA) (also known as Call Forward to Voicemail). If one of the CP destinations has subscribed to AA that would unconditionally forward any incoming calls to the voicemail, the call would always be answered by the destination with AA first.

Unexpected non-determinism is introduced here because it is an observable feature interaction. It occurs when a feature with non-deterministic behavior triggers other features that have normally deterministic behaviors, but due to this triggering behave in a non-deterministic fashion. The users are usually confused by the behavior of the affected features because they do not expect such non-deterministic behaviors. This is usually caused by timing and race conditions among the features. A case of this type of feature interaction is between Automatic Call Distribution (ACD) and Call Pickup (CP) (see Figure 3). The ACD feature allows incoming calls to the subscriber be redirected to one of the pre-programmed destinations (e.g. destination A or B). The redirecting policy can be a random selection or a deterministic algorithm. The CP feature allows the subscriber of the service to inform a list of destinations (destination A and B) that an incoming call has been put on hold, and is ready to be picked up by one of the destinations. It is conceivable that the switching element Y sends a call pickup message to all destinations and then another switching element called X with ACD would redirect the call to a particular destination (e.g. B). When destination A decides to pick up the call, the incoming call is no longer available because the call has already been redirected to destination B. In summary, the non-deterministic call redirecting policy of ACD affects the first-come-first-served call pickup policy of CP.

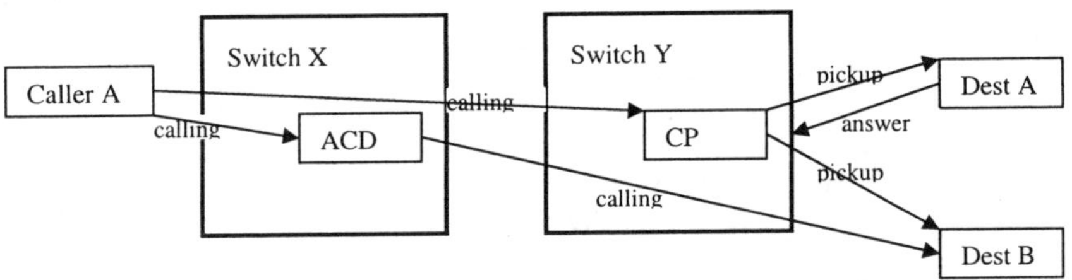

Figure 3: ACD and CP unexpected non-deterministic interaction

1.4   Overview of SIP

SIP (Session Initiation Protocol) is an application-layer multimedia signaling protocol standardized under IETF RFC 2543 [3]. Similar to most World Wide Web protocols, SIP has an ASCII-based syntax that closely resembles HTTP. This protocol can establish, modify, and terminate multimedia sessions that include multimedia conferences, Internet telephony calls, and similar applications. There are additional IETF drafts that describe other important extensions to SIP in efforts to realize VoIP deployment. However, these are beyond the scope of this paper.

In SIP terminology, a call consists of all participants in a conference invited by a common source. A SIP call is identified by a globally unique call-id. Thus, for example, if several people invite a user to the same multicast session, each of these invitations will be a unique call. However, after the multipoint call has been established, the logical connection between two participants is a call leg, which is identified by the combination of "Call-ID", "To", and "From" header fields. The sender of a request message or the receiver of a response message is known as the client, whereas the receiver of a request message or the sender of a response message is known as the server. A user agent (UA) is a logical entity which contains both a user agent client and user agent server, and acts on behalf of an end user for the duration of the call. A proxy is an intermediary system that acts as both a server and a client for the purpose of making requests on behalf of other clients. A proxy may process or interpret requests internally or by passing them on, possibly after translation, to other servers.

What makes SIP interesting and different from other VoIP protocols are the message headers and body. Like HTTP, a SIP message, whether it is a request, response, or acknowledgement, consists of a header and a body. The Request-URI names the current destination of the request. It generally has the same value as the "To" header field, but may be different if the caller has a more direct path to the callee through the "Contact" field. The "From" and "To" header fields indicate the respective registration address of the caller and of the callee. The "Via" header fields are optional and indicate the path that the request has traveled so far. Only a proxy may append or remove its address as a "Via" header value to a request message. The "Record-Route" request and response header fields are optional fields and are added to a request by any proxy that insists on being in the path of subsequent requests for the same call leg. It contains a globally reachable Request-URI that identifies the proxy server. "Call-Id" represents a globally unique identifier for the current call session. The Command Sequence ("CSeq") consists of a unique transaction id and the associated request method. It allows user agents and proxies to trace the request and the corresponding response messages associated to the transaction. The body of a SIP request is usually a SDP [4], which contains the detail descriptions of the session.

The first line of a response message is the status line that includes the response string, code, and the version number. It is important to note that the "Via" header fields are removed from the response message by the corresponding proxies on the return path. When the calling user agent client receives this success response, it would send an "ACK" message back to the callee.

The message header fields that we have included in our SDL model are: "Request-URI", "Method", "Response Code", "From", "To", "Contact", "Via", "Record-Reroute", "Call-Id", and "CSeq". We believe these fields are most important to FI's because they convey the state of all the participants in the session. There are many header fields available in SIP and can become very complex. The associated RFC [3] is recommended for further details on SIP. We will discuss how the features are modeled in the next section.

2   Modeling SIP Services and Traditional Feature Interactions in SDL and MSC

In this section we describe how we modeled SIP services and their FI's. The Specification and Design Language (SDL) [6] was chosen as the modeling language for the following reasons: (1) the SIP protocol and services are state-oriented and map well to the communicating extended finite state machine model of SDL, (2) SDL is well supported by commercial software tool vendors like Telelogic [8] whose tools are used by many telecommunication software developers, and (3) various verification and validation techniques are available in SDL tools. Our approach to model SIP and its services starts with defining the use case and test scenarios using Message Sequence Charts (MSC) [7]. Next, we will describe the structural and behavior definitions of the SDL model. Then we will discuss the verification and validation process. We have described the basic SIP protocol and many additional services like CFB, CW, and OCS in SDL, but only selected diagrams will be presented in this paper.

2.1 Use Case Scenarios and Test scenarios

Since the IETF drafts have provided a call flow diagram or success scenario for each sample service in graphical notation [5], we translate these scenarios into message sequence charts. However, we note that these call flow diagrams only include the sequence of exchanged SIP messages at the protocol level. They do not represent service scenarios in the sense of use cases. Following standard practice of software engineering, we think that it is important to define service usage scenarios at the interface between the user and the system providing the communication service. We have therefore defined an abstract user interface which represents the interactions at the service level. These interactions between the users and the SIP system describe use case scenarios of SIP services. The users are the actors of the use case scenarios and are represented by the environment "env_0" in SDL.

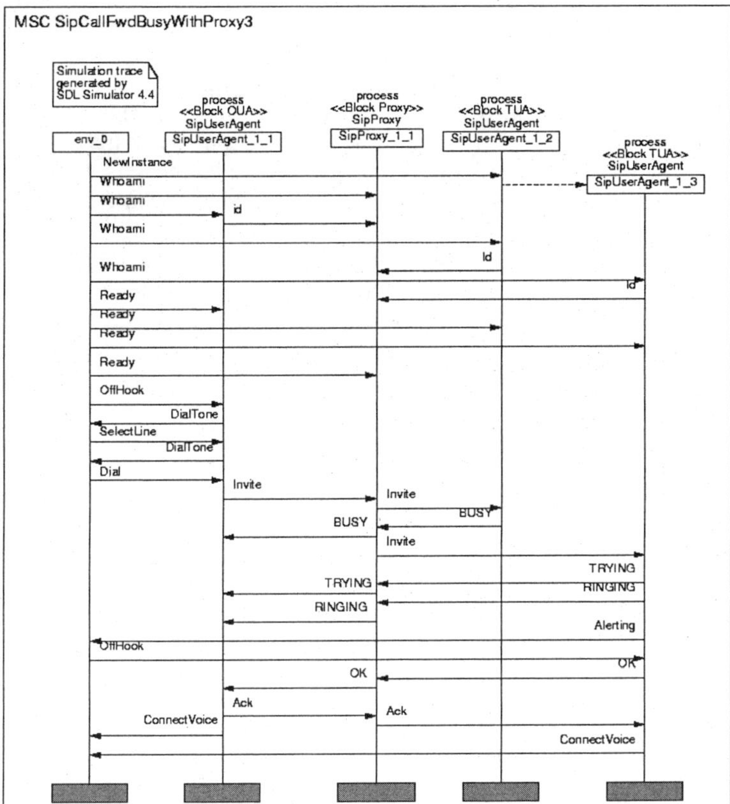

Figure 4: Call Forward Busy "Service and Protocol" Scenario

The sequence chart shown in Figure 4 represents the Call Forward Busy "service and protocol scenario". The sequence chart is the combination of the use case scenario with the corresponding scenario of exchanged SIP messages from [5]. It is written with response codes as message names and without message parameters so that the chart can fit in this paper. The complete MSC can be used as a test case for validating the SDL specification of the SIP protocol, as explained in the next subsection.

As a matter of facts, we have written one or more message sequence chart (service and protocol scenario) as the test case for each service because we use the Telelogic Tau's Validator to verify our SDL model against the combined scenario.

2.2　*Structural Definition*

A system specification in SDL is divided into two parts: the system and its environment. An SDL specification is a formal model that defines the relevant properties of an existing system in the real world. Everything outside the system belongs to the environment. The SDL specification defines how the system reacts to events in the environment that are communicated by messages, called signals, sent to the system. The behavior of a system is the joint behavior of all the process instances contained in the system, which communicate with each other and the environment via signals. The process instances exist in parallel with equal rights. A process instance may create other processes, but once created, these new process instances have equal rights. SDL process instances are extended finite-state machines (FSMs). An extended finite-state machine (EFSM) is based on an FSM with the addition of variables which represent additional state information and signal parameters. The union of the FSM-state and additional state variables represent the complete state space of the process [12].

The relationship between modeled entities, their interfaces, and attributes are considered parts of the structural definition. A SDL system represents static interactions between SIP entities. The channels connected between various block instances specify the signals or SIP messages that are sent between user agents and/or proxies. Block and process types are used to represent SIP entity types such as user agent (SipUserAgentType) and proxy (SipProxyType).

A SIP User Agent contains both, what is called in SIP, a user agent client (UAC) and a user agent server (UAS). Since a user agent can only behave as either a UAC or UAS in a SIP transaction, the user agent is best represented by the inheritance of UAC and UAS interfaces. The inheritance relationship is modeled using separate gates (C2Sgate and S2Cgate) to partition the user agent process and block into two sections: client and server. The "Envgate" gate manages the sending and receiving of "Abstract User" signals between the user agent and the environment (see Figure 5). An instantiation of a block type represents an instance of a SIP entity such as user agent or proxy, and contains a process instance that describes the actual behavior of the entity. The process definition file contains the description of all the state transitions of the features to which the SIP entity has subscribed. In addition, each SIP entity must have a set of permanent and temporary variables for its operations. In the case of a user agent, the permanent variables store the current call processing state values of the call session. The temporary variables store the values of the consumed messages for further processing.

Similar to a user agent, a proxy consists of client and server portion. It tunnels messages between user agents but also intercepts incoming messages, injects new messages, or modifies forwarding messages on behalf of the user agent(s). A proxy is also a favorite entity to which features are deployed. A SIP Proxy has one gate that interacts with the environment (Envgate), and four gates that interact with user agents or proxies: client-to-proxy (C2Pgate), proxy-to-client (P2Cgate), server-to-proxy (S2Pgate), and proxy-to-server (P2Sgate) (see Figure 5).

In our SDL model, we use different SDL systems to represent different structural bindings between SIP entities and to simulate a particular set of call scenarios. The most complex system in our telephony model (see Figure 5) realizes the concept of originating and terminating user endpoints. It contains an originating user agent block, a proxy block, and a terminating user agent block. The originating block contains all the user agent process instances that originate SIP requests while the terminating block contains all the user agent process instances that receive these requests. Upon receiving a request, a terminating user agent would reply with the corresponding response messages. It is important to note that only the originating user agent and proxy instances can send SIP requests (including acknowledgements).

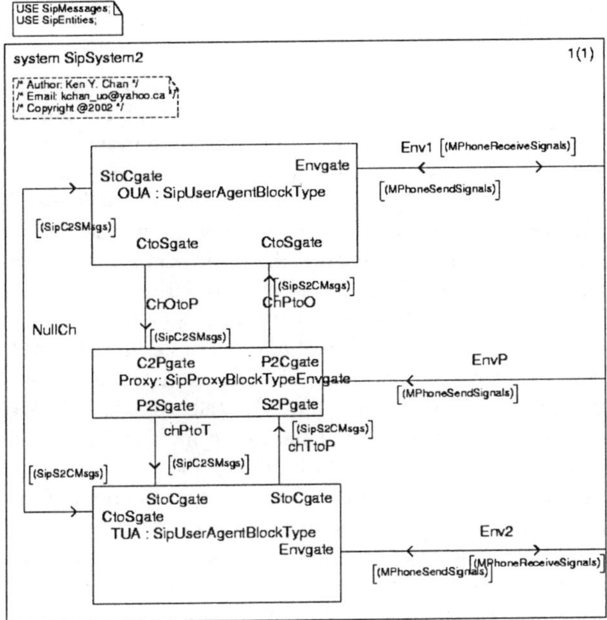

Figure 5: System Diagram of Proxy and User Agents

All blocks are initialized with one process instance. During the simulation, a 'NewInstance' "Abstract User" signal can be sent to a process instance to create a new process instance. Before the first "INVITE" message is sent, the environment must initialize each user agent and proxy instance with a unique Internet address by sending them a message called 'Whoami'. The user agent instances would in turn send an 'Id' message along with its Internet address and process id (PId) to the proxy. Thus, the proxy can establish a routing table for routing signals to the appropriate destinations during simulation. Similar to the user agent, a proxy has a set of permanent and temporary variables for its operations.

SIP messages are defined as SDL signals in the SIPMessage package. The key header fields in a SIP message are represented by the corresponding signal parameters. Since there is no linked list data structure in SDL, the number of variable fields such as 'Via' and 'Contact' must be fixed in our model. Complex data and array types of SDL have been tried for this purpose, but were dropped from the model because they may cause the Tau validation engine to crash. Instead, we used extra parameters to simulate these variable fields. A set of user signals, that is not part of the SIP specifications [3, 5], has also been defined here to facilitate simulation and verification. We call these user signals the "Abstract User interface". They represent the interface between the user and the local IP-telephony equipment in an abstract manner. The modeling of these user-observable behaviors is essential to describe FI's; however, the SIP messages described in the SIP standard do not describe these user interactions, concentrating only on the SIP messages exchanged between the different system components. For example, an INVITE message in SIP serves as more than just a call setup message; it may be used for putting the call on hold in the middle of a call (mid-call features) [5]. The user interactions include such signals as Offhook, Onhook, Dial, SelectLine, Flash (flash hook), CancelKey, RingTone,

AlertTone, TransferKey which relate to the actions that are available on most telephone units on the market.

2.3   Behavior Specification

In general, a SIP feature or service is represented by a set of interactions between users and the user agent processes, and possibly the proxy processes. Each process instance plays a role in a feature instance. A process instance contains state transitions which represent the behaviors that the process instance plays in a feature instance. In our model, we capture the behavior of a SIP entity in an SDL process type (e.g. UserAgentType). The first feature we model is the basic SIP signalling functionality, also known as the basic service. Each process type has state transitions that describe the basic service. In the case of a user agent, the process includes the UAC and UAS behavior. We define a feature role as the behavior of a SIP entity that makes up a feature in a distributed system. A feature role is invoked or triggered by trigger events. The entry states of a feature role in a SIP entity are the states for which the trigger events are defined as inputs. For example, the on-hold feature can be triggered in the 'Connect_Completed' (entry) state by an INVITE message with 'ONHOLD' as one of its parameters. After the invite message is sent, a response timer is immediately set. Then, the user agent is waiting for the on-hold request to be accepted. When the other user agent responds with an OK message, the requesting agent would reset the response timer and inform the device to display the on-hold signal on the screen.

In general, trigger events are expressed as incoming signals; pre-condition, post-conditions, constraints are expressed as enabling conditions or decision. Actions are tasks, procedure calls, or output signals. As a triggering event being consumed by the user agent process, the parameters of the event may be examined along with the pre-conditions of the feature. Then, actions such as sending out a message and modifying the internal variables may be executed. Post-conditions and constraints on the action may also be checked. Finally, the process progresses to the next state.

A state transition occurs when 1) an "Abstract User" signal is received from the environment, 2) a request or response message is received, or 3) a continuous signal is enabled. The so-called Continuous Signal in SDL is used to model the situation in which a transition is initiated when a certain condition is fulfilled. For example if the UAS is busy, the Boolean 'isBusy' would be true in the 'Server_Ring' state. The UAS would immediately send the BUSY response to the caller. This way, we would not have to worry about the timer expiration because we do not need to send a busy toggling signal to simulate busy during a simulation. An asterisk '*' can be used in a state and signal symbol to denote any state or any signal; its semantics is equivalent to a wildcard. A state symbol with a dash '-' means the current state. Error handling, such as response timer expiration, can easily be modeled with SDL timers and a combination of '*' and '-' state symbols. For example, when the response timer expires, a timeout message would be automatically sent to the input queue of the user agent process. The expiration of the response message is generalized as the situation in which the receiving end does not answer the request in time. Thus, a 'NoAnswer' signal is sent to the environment. Finally, the process can either return to the previous state or go directly to the idle state.

Moreover, we can add additional features or services such as CFB, OCS, and other services, to the system. To add behaviors of additional features to a process type, we can subtype a "basic" process type such as UserAgentType. The derived type has the same interfaces and also additional state transitions [12]. We do not want to add new interfaces (signal routes) to the process types because we do not want to change the interfaces of the block types. If a feature requires new SIP methods and response codes, we would not need to change the interfaces because method names and response codes are simply signals parameters in our model. Thus, we avoid the need to add new interfaces whenever a new feature is defined.

2.4 Verifying MSC against SDL Model

The SDL specification that has been discussed in the previous subsection was constructed using the Telelogic TAU tool version 4.3 and 4.4 [8]. TAU offers many verification or reachability analysis features: bit-state, exhaustive, random walk bit state exploration and verification of MSC. Bit-state exploration is particularly useful and efficient [10] in checking for deadlocking interactions because it checks for various reactive system properties without constructing the whole state space like the exhaustive approach does.

We verified our SDL model of SIP mainly by checking whether the model would be able to realize specific interaction scenarios which were described in the form of Message Sequence Charts (MSCs). In fact, we used the scenarios described informally in [3, 5], and rewrote them in the form of MSCs. Then we used the TAU tool to check that our SDL model was able to generate the given MSC. An MSC is verified if there exists an execution path in the SDL model such that the scenario described by the MSC can be satisfied. Thus, An MSC in Tau is considered as an existential quantification of the scenario. When "Verify MSC" is selected, Tau may report three types of results: verification of MSC, violation of MSC, and deadlock. Unless Tau is set to perform 100% state space coverage, partial state space coverage is generally performed and reported in most cases. Verification of an MSC in TAU is apparently achieved by using the MSC to guide the simulation of the SDL model. As a result, the state space of the verification has become manageable [11].

3 Detecting Feature Interactions

One of our objectives in this research is to explore the feasibility of using Tau to detect known and new FI's. Although SIP is fundamentally different from traditional PSTN signaling protocols, most of the traditional POTS FI's still exist in SIP, if the SIP services are designed and implemented with the mindset of POTS. (Note: Alternate design approaches to SIP services are discussed in the next section). We do not believe it is either feasible or practical to come up with an automated feature interaction detection scheme in Tau environment because Tau has the limited validation features in Tau and Tau's Validator frequently crashes. Instead, we write test cases in the forms of either MSC or Tau's Observer Process Assertions to verify whether traditional POTS FI's that were discovered by other feature interaction's researchers still exist or not [2,9].

In the previous sections, many FI's were identified as violations of distributed system properties: deadlock (a form of violation of safety) and livelock (a form of violation of liveness). The Tau tool offers various automated reachability analysis of SDL models and reports any deadlocks found during the analysis. Let us revisit the previous deadlock example on CFB and CW (see Section 1.3). If we have a SDL system with the corresponding SDL blocks or SIP entities which has a user agent process running with CFB and CW behaviors, we can select the "bit state exploration" option to ensure no deadlock would be found. Tau's Validator does not seem to offer reports on the liveness of the analyzed system in any of its reachability analysis functions. Thus, we use Tau's Observer Process features to detect live-locking FI's, as explained in Section 3.2.

Furthermore, Incoherent interactions are not easy to check with the Tau tool because certain interactions that involve contradictory properties cannot be expressed in a straightforward manner. There are two ways to approach this problem: (1) Use an MSC to specify incoherent properties, or (2) use an observer process offered by Tau to specify assertions. By examining the validation report (which shows the simulation trace in the form of MSC), we could trace back to the locations where the problem occurs; thus we could identify the interacting features. Both approaches have their advantages and disadvantages. However, the observer approach is the only practical approach in the current Tau release.

3.1 Specify Incoherencies as MSC

Many service designers like to use MSCs to capture important scenarios of a feature. If the sets of message sequence charts describing two features can be compared/verified against each other, then certain obvious incoherent FI's may be detected in this informal requirement specification stage. However, capturing key behaviors of a feature in MSC is not always possible. For example, the success scenario of OCS cannot be expressed directly as an MSC in the case of OCS and CFB interaction. How can we express in an MSC that user A cannot call user C? The concept of something that can "never happen" is usually expressed using universal quantification. Since verification of MSCs in Tau is based on an existential quantification of the scenario, a property that something should "never" happen can be expressed by negation. If m is a scenario that should never happen, then we could use the Tau tool to check whether the SDL model can satisfy this MSC m. If the result is that m cannot be satisfied by the model, this verifies the property.

If this concept is applied to OCS, "verifying user A can call user C" being false is equivalent to the truth of "user A can never call user C". Thus, the successful verification of the success call scenario from user A to user C concludes the violation of the OCS specification (see OCS MSC diagram). Since features like OCS apply to all calls including mid-call, the pre-enable condition of the MSC must be specified. The pre-enable condition must include all possible situations in which a call can be made to user C. There are very many possibilities; clearly, detecting incoherent interactions using MSC with the current version of Tau is almost impossible.

However, an extension to MSCs, called Live Sequence Chart (LSC) [16], seems to address the above concerns. First of all, an LSC allows the designer to specify messages that cannot be sent in a scenario. This is useful for specifying OCS type of services. Secondly, an LSC can be specified with either the universal or existential quantification of a scenario, which offers great flexibility in verifying scenarios that are contradictory to each other. Last but not least, a universal LSC allows an associated pre-enable condition which defines the scope of the scenario. This is essential to modeling services because such pre-enable conditions can be used as the triggering condition of a feature. Unfortunately, Tau does not support LSC but we believe LSC is a promising tool in this context. Details on LSC are beyond the scope of this paper and will be considered in future work.

3.2 Specify Incoherencies as Assertions in an Observer Process

Observer Processes with powerful access to signals and variables of various processes are provided by Tau as a tool which is useful for feature interaction detection. One or more observer processes can be included in An SDL system to observe the internal state of other processes during validation [8]. When the validation engine is invoked to perform state exploration, the observer processes remain idle until all the observed processes have made their transitions. Then, each observer process would make one transition. All observer processes have access to the internal states of all the observed processes via the Access operators. Therefore, the conditions (e.g. the continuous signals, decisions and enabling condition) in which the observer process makes a transition may be considered for defining assertions for the observed processes. If the observer process finds that the assertion is violated, it will generate a report in the middle of the validation. The validation process would be stopped immediately and the report would be presented to the user.

Furthermore, we have experimented with assertions that verify certain liveness properties of the system. For example, if we have an integer counter for each user agent to keep track of the number of times each user agent has been in 'Ringing' state, we can monitor whether a user agent has been looping at 'Ringing' state or not. More specifically, an assertion, which verifies user agent UA1 and UA2 are not simultaneously in 'Ringing State' for more than three times ensures a certain level of liveness in the system. We have yet to experiment with converting well-known liveness detection algorithms into SDL assertions. However, since we are not aware that Tau supports temporal logic, we believe specifying liveness property without temporal logic would be non-trivial.

In general, we use our intuition to come up with useful assertions for each feature interaction category. Although this is not an automated process, we believe it is a practical approach to a subjective problem such as detecting "undesirable side-effects" between features (FI's) in Tau's environment. We believe the current features offered in Tau are very limited for detecting new FI's, but Tau meets our key objective; we could verify whether the well known FI's still exist in SIP or not. Thus, we could apply preventive measures to make SIP services more robust, as explained in the next section.

4   Preventing Feature Interactions

After a feature interaction is detected, the next natural step is to change the design to prevent it. In this section, we discuss corrective measures that can be incorporated in the design and implementation of a system to prevent FI's. These corrective measures map directly to the causes and effects of FI's. Preventing FI's may be done at design-time or at run-time. Run-time prevention is more difficult to exercise because the features may be running on different nodes and it is not easy to coordinate the actions that should be performed when a feature interaction is detected. Prevention strategies for different feature interaction categories are presented in the following subsections. In addition, the feature interaction examples described in Section 1 will be revisited again. The goal of this discussion is to provide a catalog of FI's to service designers, such that a designer can associate a feature with a set of prevention schemes that would reduce potential interactions with any unknown features. We will also see that it is easier to prevent FI's in SIP than in POTS, because SIP has extra call information (e.g. 'Via', 'Record-Route', 'Contact') in the message headers.

4.1   *Preventing Resource Contention and Limitation – (CW and TWC)*

The natural prevention strategy for resource limitation is to increase the number of available resources. IP Telephony services, particularly SIP, should face fewer resource limitation problems than POTS because SIP end-user devices tend to be more powerful: fast processor, a lot of memory, and flexible user interfaces and displays. For example, the semantics ambiguity of flash hook at the POTS user interface, which is the cause of both resource contention and resource limitation in the case of CW and TWC in the POTS service, should never happen with SIP phones. SIP phones should be able to display the choice of two actions, namely putting the current call on hold and switching to the incoming call, or conferencing in the third party. The phone may assign one or more soft buttons for this purpose. However, if interactive user intervention is not possible, priority schemes (e.g. fuzzy policies [13]) may be used to resolve resource contention. Clearly, these strategies can be applied in both design-time and run-time feature interaction prevention.

4.2   Preventing Incoherent interactions – (OCS and CFB)

Incoherent interactions can either be direct or transitive [9]. Direct incoherencies are present when two features are associated with the same trigger signal but lead to different or contradictory results. A transitive incoherent interaction occurs when one feature triggers another feature, and the latter has results that are contradictory to the former feature. The word 'transitive' refers to the transitivity with respect to features that may lead to loops. Gorse has suggested a scheme to detect this type of interaction [9]. However, no prevention scheme has been presented. Since incoherencies lead to contradictory results, allowing only one of the features to be triggered would prevent the incoherence. However, transitive incoherencies such as OCS and CFB are difficult to prevent at run-time because the interaction may involve indirect triggering of a remote feature. How could we know that the incoming call has triggered CFB running on server Y which contradicts the assumptions of the OCS feature running on server X? Obviously, if the triggering condition of the first feature were made available to the other feature and vice versa, the two features could take advantage of the extra information and negotiate a settlement. Figure 6 illustrates an application of this scheme.

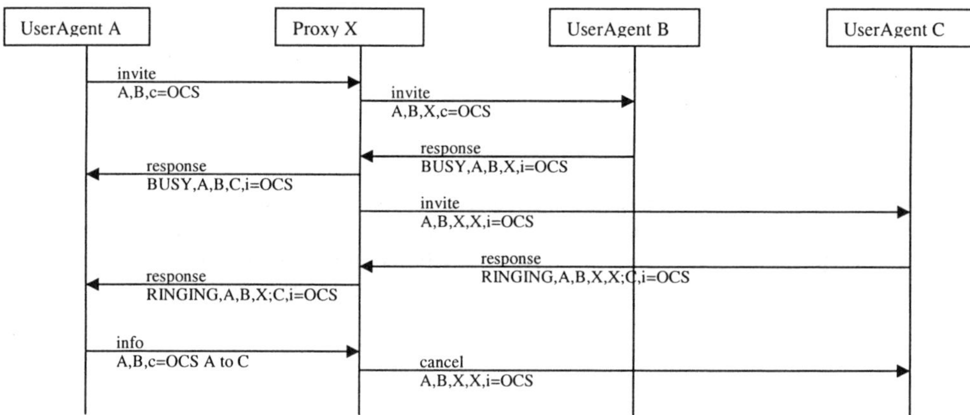

Figure 6: Preventing CFB and OCS incoherent interactions in SIP

In SIP, the 'Via' header indicates the path the request has traveled and can be used to prevent loops. The 'Record-Route' header indicates the proxy path that subsequent requests must follow. The formal specification of OCS and CFB can take advantage of header values in {From, To, Via, Contact, Record-Route} to determine whether a request has been forwarded to restricted destinations or not. For example, if OCS is activated at user agent A, the user agent may add an extra tag to some field (e.g. i=OCS) in the SDP of both the responses to A and all outgoing invitations, indicating that OCS is present at A (see Figure 6). When a downstream proxy or back-to-back user agent that is running CFB receives the invitation, it would add the 'Record-Route' header field with its proxy address and the final destination as the tag to the response header. Upon the reception of the response from either the proxy or the destination, the OCS feature at the user agent A may deduce the call forward path and could determine whether a violation of its screening list occurs or not. If the answer is yes, then the OCS may send an "INFO" message [3] with some negotiation parameters to determine the appropriate action to be taken between CFB and OCS. If OCS was to have a higher priority than the CFB feature, the OCS user agent could send a cancel request instead of the final acknowledgement message, in order to cancel the original invitation. We believe this negotiation approach can be generalized but it is beyond the scope of this paper.

4.3 Preventing Unexpected Non-determinism & Unfair Interactions – (ACD & CP) and (CP & AA)

Unexpected non-deterministic interactions must be caused by at least one feature that has some non-deterministic behavior. The non-deterministic action of one feature typically triggers the other feature. Therefore, if we were to remove the triggering dependency of the second feature on the first feature, the problem would be solved. In the case of ACD and CP, an incoming call may trigger the non-deterministic invitation of ACD and the pickup invitation of CP to the same or different destination(s). If concurrent triggering of ACD and CP was detected and only one feature was activated, the problem would be solved.

An unfair interaction represents the violation of a fairness property. There might be tools that can perform reachability analysis on the system and verify that all processes are treated fairly. Such properties can be specified in temporal logic. We are not aware that Tau has the ability to deals with such properties. An unfair interaction can be avoided by relative priorities between the two features or by introducing randomization in the selection process.

4.4 Preventing Deadlocking & Livelocking Interactions - (CW & CFB) and (ACB & AR)

Since a deadlock between two features is caused by a mutual dependency on each other's resources, removing the cyclic dependency would typically resolve this problem. The dependency in the case of CW and CFB is on the Busy signal. The busy signal is intercepted by CW, therefore CFB at both ends would continuously ring each other. Loop detection can prevent the problem. Loop detection and prevention is a mandatory feature of the core SIP protocol. Thus, CFB features can examine the {From, To, Contact, Via, Record-Route} headers to determine whether a loop would occur or not. If the answer is yes, the subsequent forward would not be allowed and a response message with a response code of LOOP_DETECTED would be returned.

Livelock interactions, like deadlock interactions, also imply cyclic dependency on each other's resources. However, the result is that the system would fail to proceed. To break the lock or the loop, a randomized timer can be introduced to the triggering of the affected features. For example, ACB and AR may be stuck in a livelock if ACB and AR would initiate the callback and the recall simultaneously on single-line phones. If the triggering of one of these features was delayed, one of the calls would go through. However, this livelocking interaction is unlikely to happen because SIP phones usually have more than one line. Instead, ACB and AR would result in an incoherent interaction because both users would be presented with two calls between the same endpoints. Since two different SIP user agents at both ends handle the calls, loop detection in a user agent would normally not detect this interaction at run-time. However, the user agents that are running in the same terminal may be designed such that they would share their route information with all their local user agents. As a result, this incoherent interaction could be prevented.

5 New Feature Interactions in SIP

Lennox introduced two feature interaction categories for IP telephony services called cooperative and adversarial interactions [1]. They are incidentally similar to, but not the same as, the concepts called cooperative and interfering interactions defined by Gorse [9]. In the following subsections, the SIP FI's described by Lennox are associated to our proposed taxonomy. The corresponding preventive measures will also be presented. Furthermore, new FI's that are unique to SIP will also be described.

5.1 Cooperative Feature Interactions

Cooperative FI's are multi-component interactions (SUMC or MUMC) where all components share a common goal, but have a different and uncoordinated way of achieving it [1]. Specific examples would be livelock, deadlock, unfairness, and unexpected non-deterministic multi-component interactions. The fact that features contend for a single resource or for each other's resources in order to establish call connections, resulting in violations of safeness and liveness properties, is a form of cooperative interactions. Let us examine the following SIP feature interaction proposed by Lennox.

Request Forking (RF) and Auto-answer (voicemail) is a potential cooperative FI. Request forking is unique to SIP; it allows a SIP proxy server P to attempt to locate a user by forwarding a request to multiple destinations in parallel [1]. The first destination to accept the request will be connected, and the call attempts to the others will be cancelled. The Auto-Answer (AA) or Call Forward to Voice Mail is designed to accept all incoming calls while the user is unavailable. Both features are intended to ensure that the user does not miss the call. However, since AA would answer the call request almost immediately, RF would always forward the call request to the same AA destination. Therefore, the user who may be closer to the other destinations would never be able to answer the call, and hence the intention of RF is ignored. To resolve this violation of fairness, one may introduce a delay timer to AA (e.g. answer after 4 rings) such that all destinations would be given a fair chance to answer the call.

5.2 Adversarial Feature Interactions

Adversarial FI's, by contrast, are multi-component interactions (SUMC, MUMC) where all components disagree about something to be done with the call (e.g. violation of feature assumptions) [1]. They are a form of incoherent interactions and are difficult to detect. The following new interactions are specific to SIP and have not been mentioned in [1].

Timed ACD and Timed Terminating Call Screening (TCS) is a potential adversarial FI. With IP telephony, service designers can rapidly create innovative features; for example, time or calendar-based forwarding and screening features. In this example, TCS restricts incoming calls that are originating from certain callers at the destination. ACD may be programmed to distribute different kinds of incoming calls to different sets of destinations depending on the time of the day (e.g. calls from A or B to destinations S or T in the daytime, and calls from C or D to destinations X or Y in the nighttime). However, destinations S and T are programmed to screen calls from A or B in daytime and X and Y to screen calls from C and D at nighttime. Although ACD is intended to select the best route for incoming calls, TCS at various destinations is programmed to screen calls from these destinations at given time periods. As a result, their policies contradict with each other and no calls between these callers and callees can ever be completed in the conflicting time period. When time is involved in incoherent interactions, the solution to such conflicting policies is not trivial.

Call Screening and Register is another form of multi-component interaction unique to SIP involves dynamic address changing in SIP. A user agent or proxy may add, delete, or modify the current contact address of a user stored in the registrar. The address can be changed at any time by sending a "Register" request message to the registrar. Since a user cannot possibly stay current with the latest address change of another user, the call screening features would be rendered useless. It is a form of incoherent interaction. This interaction involves a debate of privacy for the call screener and the caller. These security issues are beyond the scope of this paper.

Dynamic Addressing and User Mobility and Anonymity is another adversarial FI that centers on the controversial issue on the balance between the lack of address scarcity in the Internet [1] and the correct programming of features that depend on reliable addresses. As it becomes the norm that a user owns more than one intelligent wireless and/or wired personal communication device, user mobility is a serious issue in designing reliable services. Furthermore, anonymous email accounts and email spamming are big problems in Internet. Could anonymous accounts and telemarketing spam calls emerge in the SIP world when SIP becomes popular? If SIP addresses were just as easy to obtain as email addresses, the call screening and forwarding feature in SIP would need to be more sophisticated than in POTS.

6   Conclusions and future work

The main contribution of this research described here is to provide a formal model of SIP services using the SDL language. The SDL model covers some of the important characteristics of SIP, such as caller-id, command sequence and, to a certain extent, most of the mandatory addressing header fields. The benefits and disadvantages of using a modeling and verification tool like Telelogic Tau are discussed. Our experience with verifying and validating our SDL model is also described. The second most important contribution is the discussion of new SIP FI's and some details on how to prevent the traditional FI's, and potentially new FI's, in SIP.

Another significant contribution is the discussion on how to detect FI's with the Tau tool. Although Tau has a reliable editor and simulation engine, we have run into many difficulties with the Tau validation engine. It frequently crashes with General Application Errors, which are unrecoverable in the Windows environment. We are not certain of the root cause of this problem but we suspect that the validation engine may not be able to handle very big state spaces or certain complex data structures. Hopefully, these issues would be addressed in future versions of the tool. We have used MSCs to characterize user interaction scenarios and used these scenarios to verify the SDL model of SIP. However, MSCs have limitations in terms of expressing quantification of instances and their behaviors. Live Sequence Chart (LSC) [16], which has not been discussed much in this paper, appear to be a promising extension to MSCs in this context. We have also used the observer processes provided by Tau to detect FI's. This seems to be a practical semi-automated approach, particularly to detect incoherent interactions and to verify liveness properties.

Finally, an extension to the classical feature interaction classification is presented to show the importance of categorizing FI's and the corresponding prevention strategies. As part of our future works, we would like to investigate potentially useful properties of FIT. In addition, we would like to modify our model to support the new SIP standard [15]. We intended to start the model with the latest standard but unfortunately the new RFC came to our attention only in the later stage of our project. We have some ideas on how to extend our model to support the new standard but it is beyond the scope of this paper.

Acknowledgements

This project would not be possible without the support of Communications and Information Technology Ontario (CITO) and Natural Science and Engineering Research Council (NSERC), Canada. The tools and the support we have received from Telelogic are appreciated. Furthermore, we would like to thank our colleagues at SITE, in particular Dr. Luigi Logrippo from Université du Québec en Outaouais and Dr. Daniel Amyot from University of Ottawa, for their insights to the domain of FI's.

References

[1]   J. Lennox, and H. Schulzrinne, "Feature Interaction in Internet Telephony", Sixth Feature Interaction Workshop, IOS Press, May. 2000.

[2]   E.J.Cameron, N.D.Griffeth, Y.-J. Lin, M. E. Nilson, W. K. Shure, and H. Velthuijsen, "A feature interaction benchmark for IN and beyond," Feature Interactions in Telecommunications Systems, IOS Press, pp. 1-23, 1994.

[3]   M. Handley, H. Shulzrinne, E. Schooler, and J. Rosenberg, "SIP: Session Initiation Protocol", Request For Comments (Proposed Standard) 2543, Internet Engineering Task Force, Mar. 1999.

[4]   M. Handley, and v. Jacobson, "SDP: Session Description Protocol", Request For Comments (Proposed Standard) 2327, Internet Engineering Task Force, April 1998.

[5]   A. Johnston, R. Sparks, C. Cunningham, S. Donovan, and K. Summers, "SIP Service Examples", Internet Draft, Internet Engineering Task Force, June 2001, Work in progress.

[6]   International Telecommunication Union, "ITU-T Recommendation Z.100: Specification and Description Language (SDL)", ITU-T, Geneva, Switzerland, 1999.

[7]   International Telecommunication Union, "ITU-T Recommendation Z.120: Message Sequence Chart (MSC)", ITU-T, Geneva, Switzerland, 1996.

[8]   Telelogic Inc., "Telelogic Tau SDL & TTCN Suite", version 4.3 and 4.4, http://www.telelogic.com, accessed on Dec 20, 2002.

[9]   N. Gorse, "The Feature Interaction Problem: Automatic Filtering of Incoherences & Generation of Validation Test Suites at the Design Stage" (Master Thesis), University of Ottawa, Ottawa, Ontario, Canada, 2001.

[10] Holzmann, G.J. (1988) 'An improved protocol reachability analysis technique', Software Practice and Experience, Vol. 18, No. 2, pp. 137-161.

[11] O. Hargen, "MSC Methodology", http://www.informatics.sintef.no/projects/sisu/sluttrapp/publicen.htm, accessed on Dec. 23, 2002, SISU, DES 94, Oslo, Norway, 1994.

[12] J. Ellsberger, D. Hogrefe, and A. Sarma, "SDL - Formal Object-oriented language for Communication Systems", Prentice Hall Europe, ISBN 0-13-621384-7, 1997.

[13] M. Amer, A. Karmouch, T. Gray, and S. Mankovskii, "Feature-Interaction Resolution Using Fuzzy Policies", Feature Interactions in Telecommunications and Software Systems VI, pp. 94-111, IOS Press, 2000.

[14] ITU, "Packet based multimedia communication systems", Recommendation H.323, ITU-T, Geneva, Switzerland, Feb. 1998.

[15] J. Rosenberg, H. Shulzrinne, G. Camarillo, A. Johnston, J. Peterson, R. Sparks, M. Handley, and E. Schooler, "SIP: Session Initiation Protocol", Request For Comments (Standards Track) 3261, Internet Engineering Task Force, June 2002.

[16] W. Damm, and D. Harel, "LSCs: Breathing Life into Message Sequence Charts", Formal Methods in System Design, Kluwer Academic Publishers, pp. 19,45-80, 2001.

*Feature Interactions in Telecommunications
and Software Systems VII*
D. Amyot and L. Logrippo (Eds.)
IOS Press, 2003

Phase Automaton for Requirements Scenarios

Bill MITCHELL[1], Robert THOMSON[2], Clive JERVIS[2]
[1]*Department of Computing, University of Surrey, GU2 7XH, UK*
[2]*Motorola UK Research Lab, Jays Close, Basingstoke, RG22 4PD, UK*
{b.mitchell, brt007, clive.jervis}@motorola.com

1 Introduction

Requirements specifications for wireless telecommunications systems are often only partially defined, usually consisting of a set of normative scenarios together with scenarios for some of the more important exceptional behaviours. A major challenge is to find effective means of analysing such specifications formally in order to detect errors as early in the lifecycle as possible. The challenge arises because existing techniques generate too many bogus reports due to the incomplete nature of the specifications, so masking genuine errors, and generate too large a state space for effective analysis.

Motorola UK Research Lab has been collaborating with testing, architecture and design groups throughout Motorola to investigate this area over the last few years. We have found that standard analysis techniques are inappropriate for scenario based incomplete specifications. For example, an internal study of TETRA [12] MSC [10] engineering scenarios used Spin and algorithms similar to those found in [1], [2] and [9] to analyse the specifications in a standard manner. The final model suffered from the usual state explosion problem and generated many error reports none of which were genuine. Two other Motorola Research Labs used independent methods to construct a complete model of a POTS system incorporating a switch, written in SDL [11], which were analysed with technology built out of the FDR model checking system. It also suffered state explosion problems, and was only able to find conflicts when the exploration algorithm was directed towards the error states.

The technique reported in this paper address the problems found in these earlier studies. We define a particular semantics for combining scenarios that can be conveniently represented by a new kind of automata, called phase automata. The semantics uses information embedded in the specification to factor the resulting automata into phases, which gives a more compact representation than other approaches, and concurs with engineering use. In this paper we assume scenarios are defined as MSCs, which is indeed standard practice in Motorola and other wireless telecommunications companies, though the results apply more generally than this.

An internal software tool, FATCAT, has been developed by Motorola UK Research Labs that constructs phase automaton representations for each instance from a set of normative MSCs, analyses them and, if any conflicts are found, outputs MSCs that describe the problem scenarios. FATCAT has been used to analyse features being developed for 3G handsets, where it has discovered errors in the specifications that had previously gone undetected and which were subsequently uncovered only during field testing of pre-release models.

FATCAT is built on patented technology ([4], [5]) already developed by Motorola UK Labs for the *ptk* software tool [3]. The *ptk* tool generates a set of conformance tests from a collection of MSCs. *ptk* has been developed over a number of years and is now uscd by engineering groups throughout Motorola.

2 Phase Traces and Phase Automaton

We adopt the following notation. For a message m, we use $!m$ to denote the send event generated by m, and $?m$ to denote the receive event generated by m in accordance with MSC semantics [10]. The paper assumes that a protocol specifies message exchanges whose purpose is to define the transitions between different phases of operation in the system. For example in a system such as TETRA there are phases such as *call setup, call active, call roaming, call queued,* and *ruthless resource pre-emption.*

Let the set of specification events be E, let P be the set of phases that can occur in the specifications. Let S be a set of states, and let ϕ be a function $\phi : S \longrightarrow P$. Define a set of deterministic transitions to be a partial function $\partial : S \times E \longrightarrow S$. The tuple $\mathcal{A} = (S, E, \partial, \phi)$ is defined to be a *phase automaton*. A *phase trace* is an alternating sequence of phases and events S_i, a_i, terminated at both ends by a phase. A phase trace is a *specification phase trace* if the sequence of events a_i in the phase trace is a complete trace of events from one of the specification scenarios, and in that scenario each event a_i occurs during phase S_i, and after the event the phase becomes S_{i+1}. A phase trace is an *execution phase trace* of phase automaton $\mathcal{A}$ if there are states x_i for $0 \leq i \leq n + 1$ such that $\phi(x_i) = S_i$ and $\partial(x_i, a_i) = x_{i+1}$. When a specification phase trace t is also an execution phase trace, we say the phase automaton generates t, and that x_0 is a start state. From now on we will refer to both specification phase traces and execution phase traces as simply phase traces whenever it is clear from the context which prefix should apply. Execution traces that are not specification traces are implied by the specification.

For MSCs we use condition symbols to define the current active phase of a scenario. A phase remains active on an instance until another condition is encountered connected to the same instance.

3 Process Algebra Products for Phase Automaton

This section defines the particular semantics for combining MSC scenarios in the form of a process algebra. MSC scenarios are mapped to terms in the algebra. The delayed product of these terms defines the semantics of the combined MSCs.

Let E be a set of atomic actions. Let $+$ be the usual choice operator over process terms. For a set U define $\sum U$ to denote $\sum_{u \in U} u$. Also let $\cdot$ be the usual composition operator of atomic actions and process terms. The set of terms defined by $+$ and $\cdot$ over E is the set of process terms. Let $\sqsupset$ be a binary reflexive relation over E and let $\eta : E \longrightarrow \mathbb{B}$ be a boolean valued function. $P \mid Q$ is the delayed composition of process algebra terms as defined by the axioms in figure 1.

Define the stack automaton $\mathsf{Sk}(I)$ for instance I in M as follows. Intuitively this automaton defines all the combined phase traces that belong to I, and the states of the automaton track the series of events that have occurred during the current active phase. Let $\mathcal{A}$ be any phase automaton that generates the phase traces of I in M. Constructing $\mathcal{A}$ is straightforward when M does not contain any race conditions. It can be derived from any automaton that generates the event traces of I by annotating the states with suitable phase information. Note

par	$P \mid Q$	$=$	$P \triangleleft Q + P \triangleright Q$
skew1	$a \cdot P \triangleleft b \cdot Q$	$=$	$a \cdot P \triangleleft\!\mid b \cdot Q$ if $a \sqsupset b$
skew2	$a \cdot P \triangleleft b \cdot Q$	$=$	$a \cdot (P \triangleleft b \cdot Q)$ if $a \not\sqsupset b$
skew3	$P \triangleright Q$	$=$	$Q \triangleleft P$
zero1	$0 \triangleleft Q$	$=$	0
left-synch1	$a \cdot P \triangleleft\!\mid b \cdot Q$	$=$	$a \cdot (P \triangleleft\!\mid Q)$ if $a \sqsupset b$ and $\neg\eta(a)$
left-synch2	$a \cdot P \triangleleft\!\mid b \cdot Q$	$=$	$a \cdot (P \parallel Q)$ if $a \sqsupset b$ and $\eta(a)$
branch1	$a \cdot P \triangleleft\!\mid b \cdot Q$	$=$	$a \cdot P + b \cdot Q$ if $a \not\sqsupset b$ and $\eta(a)$
prune	$a \cdot P \triangleleft\!\mid b \cdot Q$	$=$	$a \cdot P$ if $a \not\sqsupset b$ and $\neg\eta(a)$
zero2	$0 \triangleleft\!\mid Q$	$=$	0
synch1	$a \cdot P \parallel b \cdot Q$	$=$	$a \cdot (P \parallel Q)$ if $a \sqsupset b$
synch2	$P \parallel Q$	$=$	$Q \parallel P$
zero3	$0 \parallel Q$	$=$	Q
branch2	$a \cdot P \parallel b \cdot Q$	$=$	$a \cdot P + b \cdot Q$ if $a \not\sqsupset b$ and $b \not\sqsupset a$

Figure 1: Delayed Composition Axioms

when M contains infinite traces and there are race events it is quite possible that $\mathcal{A}$ does not exist.

A state in $\mathsf{Sk}(I)$ is a pair (s, Sk), where s is a state of $\mathcal{A}$, and Sk is a stack containing events from I. Let $\partial(s, a) = s'$ be a a transition in $\mathcal{A}$, where $\phi(s) = c$, and $\phi(s') = c'$. Let Sk' be the result of pushing a onto Sk whenever $c = c'$, and set $Sk' = [\,]$ when $c \neq c'$. Then $\partial((s, Sk), (c, a, c')) = (s', Sk')$ is a transition in $\mathsf{Sk}(I)$. A start state of $\mathsf{Sk}(I)$ is a pair $(s_0, [\,])$, where s_0 is a start state of $\mathcal{A}$. For two stacks sk_1 and sk_2, we write $sk_1 \leq sk_2$ if stack sk_1 is the head of stack sk_2. Define $(c, a, c', Sk) \sqsupset (c_1, a_1, c'_1, Sk')$ when $c = c_1, a = a_1, c = c'_1$ and $Sk' \leq Sk$. Define $\eta(c_1, a_1, c'_1, Sk')$ to be true exactly when $c_1 \neq c'_1$. So that η records when an event causes a phase transition. For a stack automaton $\mathsf{Sk}(I)$ define its process algebra term inductively. Let $P(s, Sk, \mathsf{Sk}(I))$ be

$$\sum \{(c, a, c', Sk) \cdot P(s', Sk', \mathsf{Sk}(I)) \mid (s, Sk) \xrightarrow{(c,a,c')} (s', Sk') \in \mathsf{Sk}(I)\}$$

Then define $P(I, M)$ to be the sum of terms $P(s_0, [\,], \mathsf{Sk}(I))$ where $(s_0, [\,])$ is a start state for $\mathsf{Sk}(I)$. The stack component in the atomic actions of $P(I, M)$ are used solely as a mechanism for controlling the delayed composition of these terms.

For an instance I that occurs in MSCs M_i, where $0 \leq i \leq n$, let

$$P(I) = P(I, M_0) \mid P(I, M_1) \cdots \mid P(I, M_n)$$

We define $P(I)$ to be the *semantic representation* of the phase traces of the instance I for the set of MSC scenarios M_i. $P(I)$ contains the explicit phase traces for I, and also the implied concurrent phase traces not explicitly given by the M_i.

4 Phase semantics discussion

This section motivates the definition of $P(I)$ by describing intuitively what this process is capturing. Common practise treats MSC conditions as a convenient mechanism for incorporating important composite states into the scenario specifications. Effectively this makes the phase traces the focus of the specification scenarios, rather than the event traces.

The problem now becomes one of how to combine a set of phase traces within a single phase automaton. Consider the two MSCs contained in Figure 2.

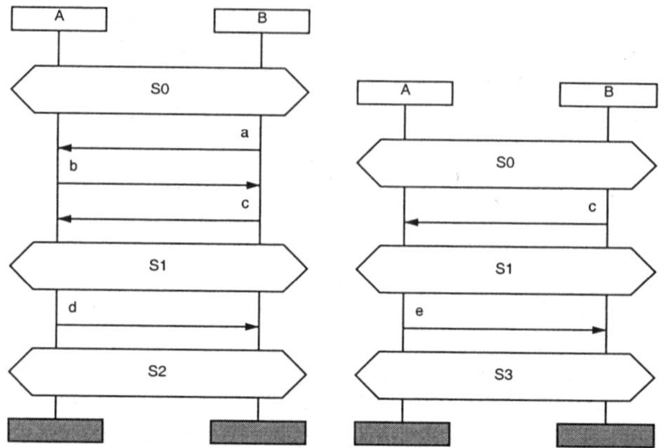

Figure 2: Two MSC scenarios for instances A and B

Instance A has a single phase trace in each MSC of Figure 2:

$$S0, \ ?a, \ S0, \ !b, \ S0, \ ?c, \ S1, \ !d, \ S2$$
$$S0, \ ?c, \ S1, \ !e, \ S3$$

How are these to be combined within a single phase automaton? If we have no further information about how the different phase traces are related to each other, the only way we could incorporate these phase traces within an automaton would be as shown in Figure 3 by the automaton on the left.

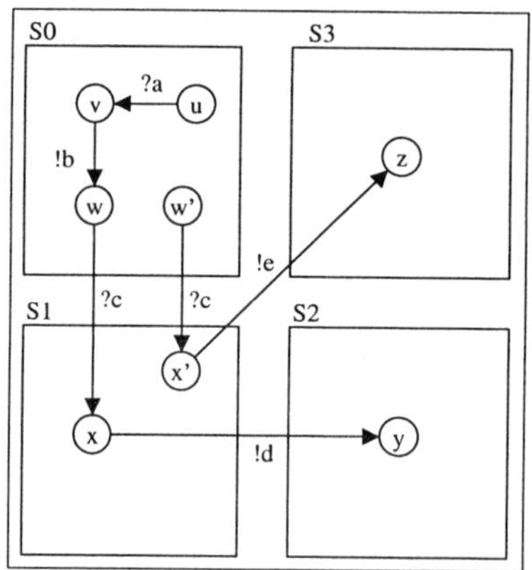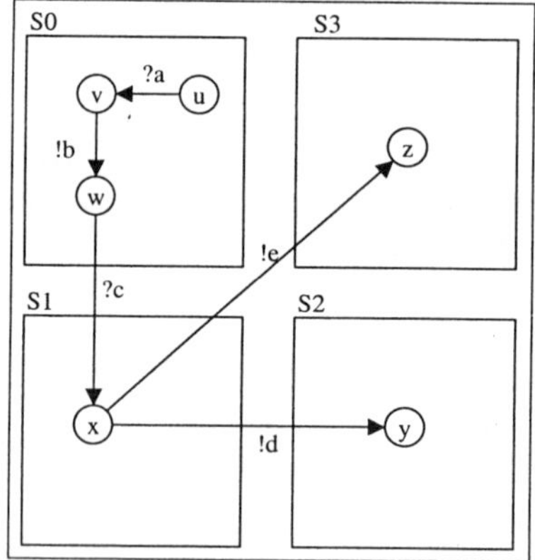

Figure 3: Phase automaton with, and without disjoint states

Here we have depicted a phase as a box that surrounds those states that are part of the phase according to the phase function ϕ. The left automaton does not combine the phase transitions it merely enumerates the phase traces as if they are completely unconnected.

From a number of studies we have formalised phase automata as the way engineers intuitively combine multiple MSC scenarios. That is a phase automaton is the semantic representation of a set of phase traces if:

> during any execution, if it has generated the first part of a specification phase trace, and it has reached a point where a phase transition is possible then the implementation can always generate the rest of the phase trace from that point.

For the phase traces of Figure 2 applying these informal semantics would lead to the phase automaton on the right hand side of 3. This automaton is simulation equivalent (modulo a suitable mapping of action names) to the process term $P(A)$. In general $P(I)$ is always simulation equivalent to the phase automaton that is the semantic representation for instance I.

5　Phase transition errors

Phase automaton can be statically analysed to detect certain simple types of conflict without user input. A phase automaton can also be verified with standard model checking techniques against modal properties, which is ongoing research. Care must be exercised since, for example, searching for unreachable states is not appropriate for partial requirements. Also many safety and liveness errors will not remain valid when the requirements are expanded, that is the errors are not invariant under additions to the requirements. Two types of invariant errors are defined below. Anecdotal evidence suggests that these kinds of errors account for a significant proportion of feature interactions.

Phase inconsistencies A significant static error that can occur is where two phase traces define the same events initially, but disagree with the phase transition that later occurs. Here is an example of two such phase traces.

$$S0, \quad ?u, \quad S0, \quad !a, \quad S1$$
$$S0, \quad !a, \quad S2$$

The phase semantics force there to be two distinct transitions labeled with $!a$ leading to different phases from the same state. In general this conflict leads to a nondeterministic automaton where the next composite state is not uniquely defined. A detailed browser example is given in Section 6.

Structural inconsistencies The simplest form of static analysis is to detect certain kinds of nondeterministic behaviour taken by the instance implemented by the automaton. For example, when the instance has to make an exclusive choice between sending one of two messages. Consider a state x and transitions

$$x \xrightarrow{!a} x_1, \; x \xrightarrow{!b} x_2$$

where a and b are distinct. A system component will not be able to determine if it should initiate the behaviour indicated by send a or the behaviour initiated by send b. Whilst non-deterministic behaviour of a component is acceptable under some circumstances, such as consuming or sending out given messages in any order - as in the case of an MSC coregion, non-deterministic choice of two divergent behaviours usually indicates a specification error.

6 Mobile handset browser example

The MSCs of Figures 5 and 6 describe two hypothetical MSC requirements for a browser like feature for a mobile handset. MSC M_1 describes a high level scenario where the handset browser is active downloading a file when a call is received from another device. The download is suspended while the call is dealt with. After the call is terminated the handset presents a dialogue box to enable the user to resume the file download if they wish.

MSC M_2 describes a different scenario relating to the general operation of the browser like device. This scenario describes how the handset should behave when the browser has been suspended during an active call, and then the END key is pressed. The scenario states that the phone should always return to the browser_active composite state.

Notice that if both of these scenarios are applied to the browser feature then they cause a conflict. When a file download is suspended in order to take a call, and then the END key is pressed, the handset is trapped between the browser_download_dialogue and browser_active composite states. That is the next composite state after the timer expires is not uniquely defined, which is an error.

Figure 4 is the semantic representation given by FATCAT for the phone instances of M_1 and M_2. The states are labelled with integers from 0 to 10. State 9 represents an error state, since from there the next state is not uniquely defined after the timer expires. Note there is a trace from state 1 to state 9. This illustrates how the error state can be reached assuming that state 1 can be reached from some initial configuration. This seems reasonable since state 1 belongs to phase browser_active from Figure 5, which is the initial phase for that scenario. Hence the trace from state 1 to state 9 represents an example of how the conflict can occur in practise and is the one chosen by FATCAT for output in MSC format. This scenario can be represented by M_3 as in Figure 7. The conflict in this MSC is represented by the alternative construct at the end of the MSC, which illustrates that after the timer expires the next composite state for the phone instance is not uniquely defined. Recall an MSC alternative construct describes mutually exclusive possibilities. The dashed line delineates these possibilities.

$$
\begin{aligned}
\partial(9,4) &= \text{'wait for timer to expire'} \\
\partial(9,10) &= \text{'wait for timer to expire'} \\
\partial(8,9) &= \text{?key\_press(END)} \\
\partial(7,8) &= \text{'wait for timer to expire'} \\
\partial(6,7) &= \text{!ack} \\
\partial(5,6) &= \text{'accept call'} \\
\partial(3,4) &= \text{'wait for timer to expire'} \\
\partial(2,5) &= \text{key\_press(right\_soft\_key)} \\
\partial(1,2) &= \text{?incoming\_call\_notification(B)} \\
\partial(0,3) &= \text{?key\_press(END)}
\end{aligned}
$$

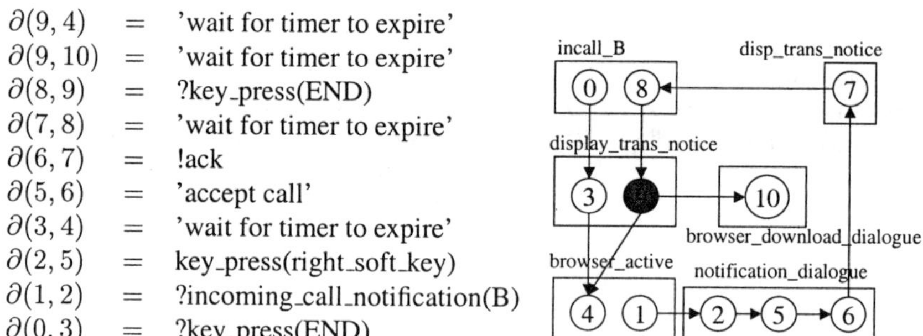

Figure 4: FATCAT phase automaton representing phone instance

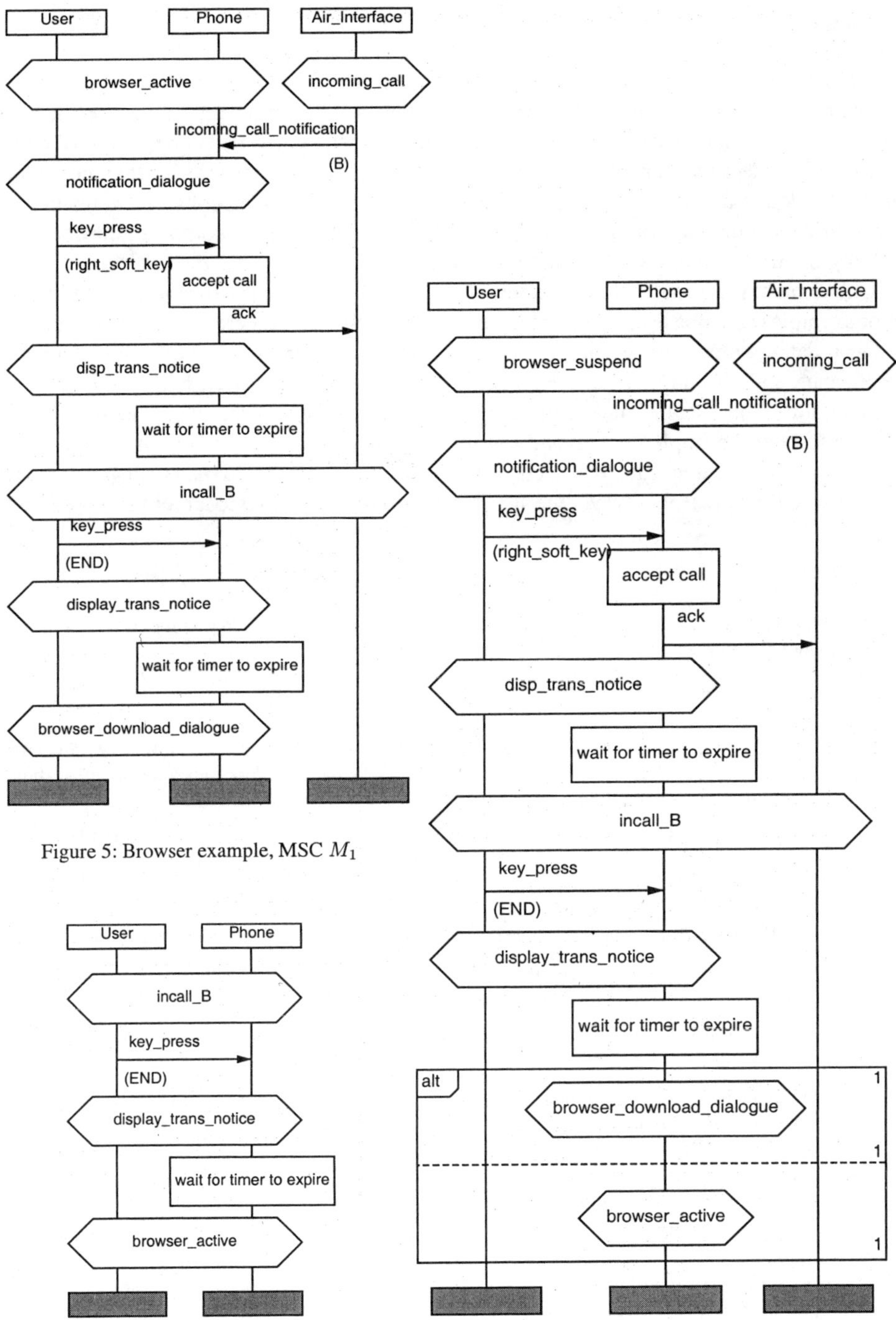

Figure 5: Browser example, MSC M_1

Figure 6: Browser example, MSC M_2

Figure 7: FATCAT output MSC M_3

7 Conclusion

Where the purpose of a protocol specification is to define phase transitions it is possible to construct a phase automaton semantic representation for each process in the specification. These can then be statically analysed to detect phase transition errors. With the right semantic interpretation of phase transitions the phase automaton can be constructed efficiently, and is capable of detecting interesting interactions. Compared to automata constructed using naive compositional semantics, our phase automata approach yields highly compact representations enabling the tractable analysis techniques reported here.

Unlike most work in the area of feature interaction (see [7], [8] or [6] for a comprehensive set of examples) the phase automaton technique is not intended to detect all possible conflicts. Rather it detects phase transition interactions. This class of conflict seems to be of practical significance and simple to detect. Further, as specifications are often highly incomplete it is important to have some means of detecting errors that are inherent in the specification as it is meant to be implemented, and not highlight errors that are purely an artifact of the incompleteness. For this reason the techniques described here are designed to detect persistent errors that will be present however the specifications are extended into a more complete form.

References

[1] R. Alur and M. Yannakakis, Model checking of message sequence charts, Proceedings of the Tenth International Conference on Concurrency Theory, Springer Verlag, 1999

[2] R. Alur, K. Etessami, M. Yannakakis, Inference of Message Sequence Charts, Proceedings 22nd International Conference on Software Engineering, pp 304-313, 2000.

[3] P. Baker, P. Bristow, C. Jervis, D. King, B. Mitchell, Automatic Generation of Conformance Tests From Message Sequence Charts, Proceedings of 3rd SAM (SDL And MSC) Workshop, Telecommunication and Beyond, Aberystwyth 24th-26th June 2002, to appear in LNCS 2003.

[4] P. Baker, C. Jervis, D. King, An optimised algorithm for test script generation, patent GB18137.0, 2000.

[5] P. Baker, C. Jervis, B. Mitchell, Method of Generating Coordinating Messages for Distributed Test Scripts, patent GB18138.8, 2000.

[6] M. Calder, E. Magil, Feature Interaction in Telecommunications and Software Systems VI, IOS, 2000.

[7] N.Griffeth, R. Blumenthal, J-C, Gregorie, T. Ohta, A feature Interaction Benchmark for the first feature interaction detection contest, in journal of Computer Networks, Vol 32, No 4,April 2000

[8] K. Kimbler, L. G. Bouma, Feature Interaction in Telecommunications and Software Systems V, IOS, 1998.

[9] P. Madhusudan, Reasoning about Sequential and Branching Behaviours of Message Sequence Graphs, proceedings of 28th International Colloquium on Automata, Languages and Programming, Crete, Greece 8-12 July 2001, LNCS 2076.

[10] Z.120 (11/99)ITU-T Recommendation - Message Sequence Chart (MSC)

[11] Z.100 (11/99) ITU-T Recommendation - Languages for telecommunications applications - Specification and description language

[12] Annex C, Service Diagrams related to the model of Mobile user, Terrestrial Trunked Radio (TETRA); Voice plus Data (V+D); Designers' guide; Part 2: Radio channels, network protocols and service performance, European Telecommunications Standards Institute 1997.

Emerging Application Domains

*Feature Interactions in Telecommunications
and Software Systems VII*
D. Amyot and L. Logrippo (Eds.)
IOS Press, 2003

Aspect-Oriented Solutions to Feature Interaction Concerns using AspectJ

Lynne BLAIR
Computing Department,
Lancaster University, Bailrigg
Lancaster, LA1 4YR, U.K.
lb@comp.lancs.ac.uk

Jianxiong PANG
Computing Department,
Lancaster University, Bailrigg
Lancaster, LA1 4YR, U.K.
j.pang@lancaster.ac.uk

Abstract. In this paper, we propose a two-level architecture for feature driven software development, consisting of a base layer for a feature's core behaviour and a meta-layer for resolution modules that provide solutions to feature interaction problems. Whilst a standard programming language is used at the base level, e.g. an object-oriented language such as Java, we propose the use of an aspect-oriented programming language for the inherent cross-cutting concerns that exist at the meta-level. We evaluate the use of AspectJ for the implementation of interaction resolution modules at the meta-level. This evaluation is carried out through an in-depth study of an email system. We conclude that aspect-oriented approaches are highly suited for this split-level architecture and that the architecture has many benefits for feature driven software development. Finally, we also highlight a number of problems with AspectJ for our intended use, but discuss how the selection of an alternative aspect-oriented technique would avoid these problems.

Keywords. Feature driven development, aspect-oriented programming, feature interaction and interaction resolution, feature composition.

1. Introduction

Agile methodologies [13][14] are gaining increasing popularity for tackling today's software development challenges. Examples of such methodologies are Extreme Programming, Feature-Driven Development, Adaptive Software Development, and Dynamic Systems Development, ranked in this order of popularity according to the results of a survey reported in [4].

In this paper, we focus on the Feature Driven Development (FDD) approach [7][27]. In this approach the unit of development is a feature: a small piece of client-valued functionality. The granularity of these features is guided by projected development time: if, during decomposition of the software task, a feature looks bigger than two weeks' work then it should be decomposed further.

In addition to being valuable units of development, features can also be seen as valuable units of evolution, for example with respect to system adaptation, re-configuration, versioning and even billing. The telecommunications industry has a tradition of organising development projects, people and marketing by features [32]. Microsoft has also apparently followed this process in their software product line for a number of years [5].

However, as has been witnessed in the telecommunications industry for many years now, whilst the value of feature-oriented approaches is clear, interactions between different parts of a program are an inevitable, and often intended, result of modularisation. Referring back to FDD, unit testing for the features is carefully prescribed as part of the methodology (see table 1), yet across-feature testing is left rather ad-hoc.

This paper proposes a framework in which Feature Driven Development is enhanced with the power of Aspect-Oriented Software Development (AOSD) techniques [20], the enhancement being provided specifically to deal with the handling of the resolution of interactions. In this work, we use the Java programming language to implement the core functionality of a feature (we refer to this as a feature's hard logic), but we do not entangle any code that would be specifically added to this feature to allow it to work with other features. Instead, this is kept separate by utilising aspect-oriented programming (AOP) constructs to implement any composition or interaction concerns (we refer to this as a feature's soft logic). In this paper, we illustrate the implementation of the feature's soft logic by using the notion of the aspects provided by AspectJ [19].

Table 1: FDD Process #5, table taken from [10].

Unit Test	
Each Class Owner tests their code to ensure that all requirements on their classes for the feature(s) in the work package are satisfied. The Chief Programmer determines what, if any, feature team-level unit testing is required - in other words, what testing across the classes developed for the feature(s) is required.	

The structure of the rest of the paper is as follows. We initially give an overview of aspect-oriented software development (section 2), before moving on to describe our two-level architecture (section 3). In section 4, we then present an email system with just three features initially; this is used to illustrate our approach. In section 5, we develop the case study further, to include ten features and a graphical user interface. Rather than describing our approach at this stage, we use this further study to drive our evaluation. This raises a number of significant points about our architecture and the use of AspectJ, and leads us to discussions on other related approaches and plans for our future work. Finally, in section 6, we draw our conclusions.

2. Aspect-oriented software development (AOSD)

Over recent years, AOSD has received a lot of attention as a new paradigm to more effectively achieve the separation of concerns alluded to by Parnas [29] and Dijkstra [8]. Whilst a large degree of success has been achieved through advances such as abstract data types and object-oriented programming, such techniques have failed to elegantly capture concerns that cut across modules. AOSD has become the umbrella term relating to all approaches that achieve this explicit separation of cross-cutting concerns, including aspect-oriented programming [18], subject-oriented programming [12] and reflection (in which AOP has some roots) [17].

AspectJ [19] has been developed by the Xerox PARC team over the last few years and, at the moment, is arguably the most dominant technique within the AOSD community. AspectJ provides new language constructs to deal with cross-cutting concerns, and excellent tool support for the subsequent weaving process: the AspectJ tool acts as a pre-processor that weaves programs together to generate (tangled) Java code. Note that other techniques such as JAC (Java Aspect Components) [30] have been proposed as alternatives to AspectJ should weaving be required dynamically at run-time.

In AspectJ, *aspects* are defined in a similar way to Java classes (with attributes and methods as required). These aspects may also define sets of join points (called *pointcuts*) that represent well-defined points in the execution of the program (thus forming the basis for supporting crosscutting concerns). In addition, two further constructs are provided: an *advice* construct that allows the addition of code before, after or around an existing method (or methods), and an *introduce* construct that allows the introduction of a new attribute or method into an existing class (or classes). For further information we refer the reader to [19] or the associated AspectJ web site.

3. A two-level architecture for features

In our proposed framework, we assume that every feature has a clear specification of its functionality. Although the implementation of that specification varies, it is generally easy to distinguish the *pure feature code*. We call this the *hard logic* of the feature, i.e. the inevitable part of the implementation of the feature's functional specification.

However, in feature driven (or feature-oriented) development, features must clearly be capable of working with other features. Since the feature's hard logic is unable to adapt itself to different execution contexts (different connected features), we also require a corresponding *soft logic* to soften the behaviour, making it flexible enough to adapt to other interacting features. Therefore, a feature's soft logic is responsible for boundary condition checking and taking action to smooth any incompatibilities.

Since a feature designer cannot foresee the future features that will interact with his/ her developed feature, soft logic should be able to be added to the hard feature logic at any stage. To support this kind of addition, the soft logic is ideally raised up to the meta-level so as to provide a separation from the hard logic and facilitate reuse and easier maintenance/ evolution.

It is this soft logic that we believe is ideally suited to aspect oriented software development techniques. The soft logic forms a resolution module that clearly cross-cuts the base-level system representing a feature's hard logic.

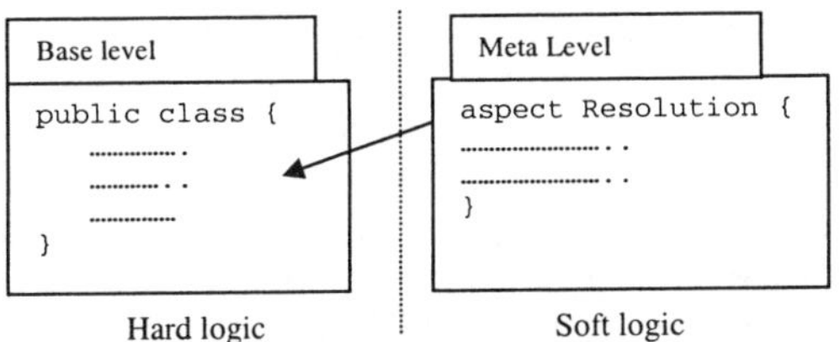

Figure 1. A two-level architecture for features

To achieve this separation of concerns in this piece of work, we will use Java to implement a feature's hard logic and AspectJ to implement the associated soft logic.

4. A case study: an email system

As our case study, we wanted to choose a system beyond the area of traditional telephony, but with known interactions that could be extended to more general distributed systems such as Internet-based systems. This led us to the email system studied in [11], where several different email features were studied and a large number of interactions found.

Our intention in this work is, by using the already identified interactions, to show how AOP techniques can be used to provide resolution modules (soft logic) for the problematic features, allowing them to work together correctly. Crucially, as mentioned above, this solution aims to maintain a clean separation between the core feature code (a feature's hard logic) and the code required to allow it to work correctly with other features (a feature's soft logic). We have chosen to focus initially on three problematic features, all of which have had interactions identified in [11]:

- **RemailMessage:** This feature forwards an incoming message and replaces a sender's true ID with a pseudonym. *RemailMessage* can also handle messages sent in the other direction as well, i.e. delivering a message addressed to a pseudonym to the real receiver (by reversing the address mapping).
- **MailHost:** This feature receives incoming messages for the *MailHost's* users, and delivers the messages to users' accounts.
- **AutoResponder:** This feature automatically replies to each incoming sender with a prescribed message, but only one reply should be returned per incoming sender.

More features will be mentioned later in the paper with respect to evolving this system to include additional functionality (see section 5). Importantly, all of the aspect algorithms in this paper (representing feature resolution modules), along with other features from [11], have been implemented in a simulation of an email system (see also section 5).

In this case study, we will follow the same approach as [11] regarding the architecture and the features, i.e. a user originates a message, conforming to email message format standards [2][23], from an email client program. This message then passes through one or more feature processing components, each termed an *email feature component*, until the message is delivered to the email client of the intended recipient(s). We also borrow terminology from the Distributed Feature Composition (DFC) approach of [16], calling each feature component a *feature box*.

4.1 An interaction resolution for RemailMessage vs MailHost

As a first illustrative example of an undesirable interaction, consider the composition of *RemailMessage* with *MailHost* (both server-side features). A problem arises if a message is sent via *RemailMessage* to a user supposedly associated with the *MailHost*, but who is actually unknown to this feature. In this case, the *MailHost* will create and send a message to the message's originator saying "There is no such a user as ...". As a result, the message body of this reply has leaked the real ID corresponding to the pseudonym, therefore defeating *RemailMessage's* intention of maintaining anonymity.

This feature interaction problem can be resolved by two possible approaches:

- **res1:** allowing the *RemailMessage* feature to query the existence of a recipient before it carries on remailing the message; or alternatively,
- **res2:** forcing any reply message to also be remailed, to ensure the real user name is not leaked (see also section 4.3 on preventing the leaking of user names).

The existence of two possible approaches here highlights an interesting point: when we say something is a resolution of an interaction, this is a subjective judgement since resolutions on the same feature interaction problem may vary from developer to developer. Sometimes the resolution just meets a requirement of the features' users, rather than being a sound rationalisation. In this paper, we assume that any solution that is able to mitigate a feature interaction constitutes a *resolution* of that interaction. Therefore, the simplest resolution is to disable one of the interacting features. However, real world applications are likely to need a more deliberate resolution so as to improve the quality of service.

As an example, an overview of a possible Java implementation of *RemailMessage's* hard logic is shown in figure 2:

```
class Remailer implements Pipe {
  String myID;
  ArrayList pairTable;
  public Remailer(...) {
    ......
  }
  public void send(Message msg) {
    ......
  }
  public void receive(Message msg) {
    if ( !(msg.getReceiver().equals(myID)))
      msg = changeIncomingID(msg); //incoming message
    else
      msg = changeOutGoingID(msg); //outgoing message
    send(msg);
  }
  private Message changeOutGoingID(Message msg) {
    String content= msg.getContent();
    String target = content.substring(0,content.indexOf("\n"));
    content = content.substring(content.indexOf("\n")+
      String alias = findAlias(msg.getSender());
    msg.setReceiver(target);
    msg.setSender(alias);
    msg.setContent(content);
    return msg;
  }
  private Message changeIncomingID(Message msg) {
    String realAddress = findRealID(msg.getReceiver());
    if (realAddress == null) throw new IllegalArgumentException();
    msg.setReceiver(realAddress);
    return msg;
  }
  public String findAlias(String outgoing) {
    ......
  }
  public String findRealID(String incoming) {
    ......
  }
}
```

Figure 2. RemailMessage's hard logic

The hard logic takes care the translation of user address from/ to pseudonym. In order to do this, for an incoming message, it will replace the receiver's address with a real user address; for an outgoing message, it will get the first line of the message body, and put it to the receiver slot, then replace the sender address with a pseudonym. The *Pipe* interface, which contains two methods, `receive(..)` and `send(..)`, must be implemented for the connection of feature boxes.

We can see that the hard logic of a feature is simple, cohesive, and highly consistent to its original specification, and thus easily understood. Typically, these features have two basic parts:

- Some data (structures) such as a forward address, a list of filter addresses or a list of <pseudonym, real name> pairs, or even the prescribed message content.
- Some methods to operate on the data and provide necessary feature logic to implement a service feature.

Under normal circumstances, this hard logic works fine with other features without problems. However, as is well known in the telecommunications domain, some combinations of features lead to undesirable interactions, as with the *RemailMessage* and *MailHost* interaction mentioned above.

With an interaction such as this, one feature (either *RemailMessage* or *MailHost*) is required to incorporate new additional behaviour to allow it to adapt to the countering feature; i.e. interaction resolution code needs to be introduced.

Now suppose we consider a possible resolution to the above and elect for the first resolution option, *res1*, from above (querying the existence of a user).

For the Java code presented above (figure 2), this resolution could take the form of directly modifying the receive method to allow *RemailMessage* to ask *MailHost* if the intended user exists or not. If this returns true, carry on as usual; if false, *RemailMessage* itself should create a "no such user" reply to the sender. An example of this direct modification is shown in figure 3:

```
public void receive(Message msg) {
   if ( !(msg.getReceiver().equals(myID))) //this is an incoming message
      msg = changeIncomingID(msg);
      if (queryExist(MailHost, userID) == false){     // 'tangled' resolution code
        msg = createReply(msg, "no such user");        // ...
        send(msg);                                     // ...
        return;                                        // ...
      }                                                // ...
   else    //this is an outgoing message
      msg = changeOutGoingID(msg);
   send(msg);
}
```

Figure 3. An interaction resolution tangled with the feature code

Alternatively, resolution code could be added to the *MailHost*. For example, the *MailHost* could check if a message is from *RemailMessage*, in which case it should answer to *RemailMessage* instead of directly to the original sender.

As can be seen from figure 3, adding in resolution code for interworking with one other feature is not too drastic concerning the structure or complexity of the feature code. However, considering the number of interactions found in [11], adding resolution code to handle all such interactions is clearly going to very quickly destroy the structure of the features and the original clarity of the hard logic. It is also obviously going to be harmful with respect to the maintenance of the code or future evolution. For this reason, we have proposed the lifting of the resolution code to the meta-level. With the support of AspectJ this can be elegantly implemented as shown in figure 4:

```
aspect ResolveWithMailHost {

  after() returning(Message m):execution(Message Remailer.changeIncomingID(Message)){
    if (queryExist(MailHost, userID) == false) //throw exception if userID not known
      throw new IllegalArgumentException();
  }

  void around(Message msg):execution(void Remailer.receive(Message)) && args(msg) {
    try{
      proceed(msg);  //execute the original receive method
    }
    catch(IllegalArgumentException e) {  //handle new exception that may be thrown
      msg = createReply(msg, "no such user");
      send(msg);
    }
  }
}
```

Figure 4. Separating the interaction resolution (soft logic) from the core feature code

This aspect inserts code after the execution of the *Remailer's* changeIncomingID method, throwing an exception if the userID is false. Having thrown an exception, the receive method now needs exception handling code to be added. This is achieved by wrapping the try and catch constructs *around* the original receive method.

The above example reflects two points about the hard logic and soft logic of a feature:

- the *hard logic* is the relatively stable description of a feature's behaviour, and
- the *soft logic*, representing an interaction resolution, is variable and may change depending on a need to adapt to a context change (such as new deployment of

features and technology updates) or depending on subjective judgement regarding the best mechanism for handling interacting features (as discussed above).

4.2 An interaction resolution for RemailMessage vs AutoResponder

Now consider the deployment of the *AutoResponder* feature. There are two scenarios with the *RemailMessage* being involved, both of which have been documented in [11]:

- **Scenario 1:** Bob has a remailer account and also switches on the *AutoResponder* feature before he goes on vacation. Alice sends a message to Bob's *Remail* account (i.e. to Bob's pseudonym rather than his real ID). When the *AutoResponder* receives the message via *RemailMessage*, it replies automatically and directly to Alice, but using Bob's real ID rather than his pseudonym. Because of the *AutoResponder's* direct reply, Alice can infer the *Remail* account is for Bob, thus defeating the *RemailMessage's* purpose.

- **Scenario 2:** As above, Bob has a *Remail* account and switches on the *AutoResponder* feature before he goes on vacation. Alice sends a message that requires a reply to both Bob's *Remail* account and to his *MailHost* account, since Alice doesn't know that these two accounts are actually for the same person. According to the *AutoResponder's* rule, this feature should reply only once for messages from the same address. For this reason, Alice will only receive a reply from one of Bob's accounts, and never from the other account.

The general problem here is that the *AutoResponder* has no knowledge of whose messages it is answering (e.g. whether the message has been received via a remailer). One possible resolution is for the *AutoResponder* to check whether it is answering a message from the remailer. If so, reply to *RemailMessage* by following the remailing rule, namely set the receiver field as the *RemailMessage* and put the intended recipient's address in the first line of the content. This algorithm can be represented as AspectJ as shown in figure 5 (but note that as discussed above, it is likely that there will be more than one viable resolution – see next section below).

Typical of resolution modules such as this, is the need for them to know information about prior feature boxes. One solution to this is to require each active feature box to add a tag to declare its process status when a message goes through it. Importantly, this simple protocol can also be implemented through AOP techniques as a crosscutting concern. For example, *pointcuts* can be defined that correspond to exit points from the execution of the feature boxes. On exiting a particular feature box, an *after advice* can be declared to manipulate the message and add the required tag.

```
aspect SoftenAutoForRemail {
  before(Message msg):execution(void AutoResponder.send(Message)) && args(msg) {
    //if msg is from Remailer, then respond to Remailer using the remailing rule
    if (isFromRemail(msg)) {
      writeContentFirstLine(msg.getReceiver());
      msg.setReceiver(getRemailAddress(msg));
    }
  }
  boolean isFromRemail(Message msg) {
    //check if msg is from Remailer ...
  }
  void writeContentFirstLine(String str) {
    //write intended recipient's address in the 1st line of message content ...
  }
  String getRemailAddress(Message msg) {
    //get Remail Server's  address from msg ...
  }
}
```

Figure 5. The AutoResponder's soft logic – to handle messages from a Remailer

4.3 A more general interaction resolution for RemailMessage and AutoResponder

To illustrate the possible diversity of resolutions for a given interaction problem, and to further evaluate the appropriateness of AspectJ, we present an alternative resolution for the *RemailMessage* and *AutoResponder* interaction problem. In this second solution, we introduce a mechanism into the *Remailer* (using AspectJ) to enforce the rule:

- For each message sent to *RemailMessage* from a user, any internal reply message must also pass back through *RemailMessage*.

In order to achieve this, *RemailMessage* can replace the sender's userID with a temporary pseudonym, thus ensuring that any return message must also pass through the remailer in order to be mapped back to the sender's real userID. Then, to allow the receiver to know the original sender of the message, the original sender's information should be put in another slot, e.g. the first line of message body. The mechanism for this resolution algorithm is showed in figure 6.

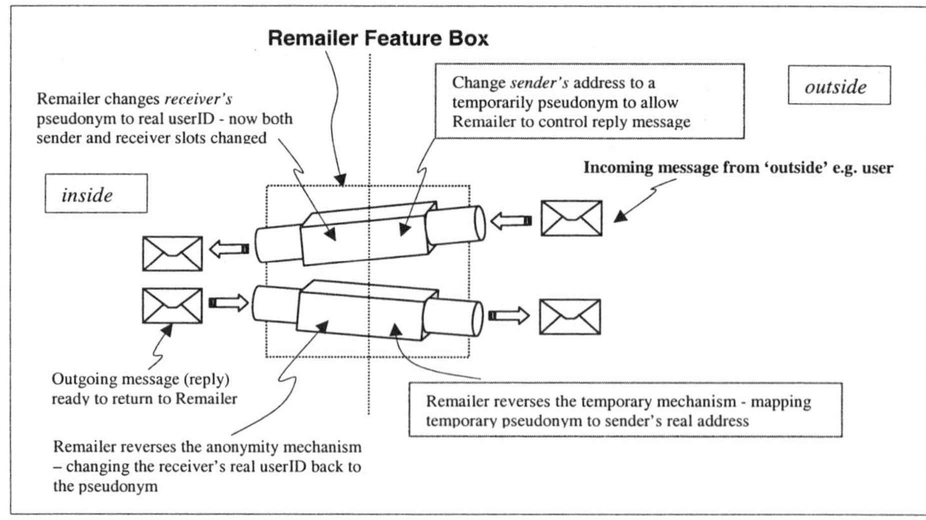

Figure 6. The Remailer feature box and interaction resolution algorithm

To reverse the process, when receiving the inner user's reply (probably from another feature box, such as *AutoResponder*), *RemailMessage* must replace the originating sender's temporary pseudonym with the sender's real address. If required during this translation, *RemailMessage* can also check user names against the email content to make sure there is no further leaking of the user ID (illustrated in figure 7 by the inclusion of a `preventLeaking` method).

To achieve this, *RemailMessage* should keep two lists of <pseudonym, realname> pairs: one for the anonymity of its customers, and one for the enforcement of controlling replies. In this way, the two interaction problems mentioned above are solved readily:

- **Scenario 1:** The *AutoResponder's* reply will be sent back to *RemailMessage* first, rather than directly to the originator. *RemailMessage* can thus anonymise the recipient's real identity.
- **Scenario 2:** In this scenario, two messages arrive at *AutoResponder* with different sender addresses (since the remailer has allocated each message a different temporary pseudonym). Consequently, both messages will be properly replied.

This interaction resolution algorithm can be implemented in Aspect J as shown in figure 7.

```
aspect ResolutionAll {

  public ArrayList Remailer.guestList; //introduce a new dual list

  public String Remailer.createAnAlias(String guest) {
     //introduce new Remailer method to create a pseudonym for the sender
  }
  public String Remailer.findGuestRealID(String alias) {
     //introduce new Remailer method to find the real username given a pseudonym
  }
  public void preventLeaking(Message msg) {
     //check the message content and replace any real userID with pseudonym
  }

  //assign a name to the sender of incoming msg.
  after(Remailer rm) returning(Message msg):
          call(Message Remailer.changeIncomingID(Message)) && target(rm) {
    msg.setSender(rm.createAnAlias(msg.getSender()));
  }
  //make sure outgoing message contains no leaking of real userID
  after(Remailer rm) returning(Message msg):
          call(Message Remailer.changeOutGoingID(Message)) && target(rm) {
    preventLeaking(msg);
  }

  //process the reply message from the inner user (define named pointcut first)
  pointcut messageArrive(Remailer rm):
          target(rm) && call(void Remailer.receive(Message));

//continued ...
```

```
// ... continued
  void around(Message msg,Remailer rm): messageArrive(rm) && args(msg) {
    try {
      proceed(msg,rm);
    }
    catch(IllegalArgumentException e) {
      String realReceiver= rm.findGuestRealID(msg.getReceiver());
      if (realReceiver != null) {
        msg.setReceiver(realReceiver);
        String senderPseudonym = rm.findAlias(msg.getSender());
        if (senderPseudonym != null){
          msg.setSender(senderPseudonym);
          preventLeaking(msg);
          rm.send(msg);
        }
        else throw new IllegalArgumentException();
      }
      else throw new IllegalArgumentException();
    } //end of catch
  } //end of around
} //end of aspect
```

Figure 7. An aspect-oriented implementation of the new interaction resolution

With this more general interaction resolution, it is important to observe that the original interaction with *MailHost* can also be resolved. The reason for this is that the "no such user" reply message will be returned to *RemailMessage* first, rather than directly to the user. As a result, this provides us with an effective mechanism to ensure the userID is always hidden.

4.4 Summary: aspect-oriented programming for interaction resolutions

The examples above have hopefully provided an insight into how resolution modules, implemented in AspectJ, can be used to cleanly separate the interaction resolution code (soft logic) from the feature's core behaviour (hard logic). Benefits can be gained, especially for more complex resolution modules such as the one provided in figure 7, when compared with the more traditional approach of entwining the resolution behaviour with the feature's core behaviour.

In particular, in this example, three new methods and a new attribute (the dual list) have been *introduced* into the remailer's behaviour. In addition, new behaviour (*advice*) has been provided that will be *woven* into the feature code: *after* the feature's two changeID methods and *around* the receive method. Whereas the introductions would be relatively easy to separate structurally in the original feature code (e.g. the appropriate use of comments and grouping of related code), the extra woven behaviour cannot be effectively separated. This leads to difficulties for the maintenance and evolution of the overall system.

Perhaps more importantly, the example in figure 7 also gives us an indication that one resolution module, although using a more complex algorithm, can resolve multiple feature interactions. This is where the power of our aspect-oriented approach brings further benefits. Pointcuts can be defined over any execution points in any of the feature boxes and need not be restricted to one feature box (as has been the case in our examples above). Also, pointcut definitions can use wildcard-matching techniques, thus removing the need for the tight-coupling of aspects with features via actual method names.

With further investigation, we hope that this may lead to the implementation of general interaction resolution *patterns*, although it is inevitable that more concrete (domain-specific) resolution modules will also still be required.

5. Evolution of the email system to further evaluate our approach

The two-level architecture we have proposed consists of:

- **Base level:** contains features' core behaviours, known as *hard logic*; implemented using *Java*
- **Meta-level:** contains interaction resolution modules, known as *soft logic*; implemented using *AspectJ*

To further evaluate the effectiveness of this architecture, we have considered evolving the email system that we have presented so far (with three features), by extending it to all ten features of [11]. In order to make a working system, we have also refactored some of the GUI modules from ICEMail, an email client written in Java and based on the new Java Mail API [15].

We classify our evaluation into five properties: cleanness of separation, re-use, faithfulness of implementation to specification, adaptability to requirement change and support for interaction avoidance, detection and resolution. It should be noted that these properties are, by their nature, more qualitative than quantitative.

A further evaluation criterion we would like to address in the future is that of performance. It can be expected that, in line with other meta-level/ reflective approaches, our approach will incur at least a minor performance overhead. We have not yet investigated this further, although information regarding the performance of AspectJ can be found on the AspectJ web-site (FAQ), see [19].

5.1 Cleanness of separation

As mentioned above, we believe that the cleanness of separation is a key factor for a system's effective maintenance and evolution. Object-oriented techniques already offer a widely accepted and largely effective form of separation through encapsulation and inheritance. However, feature-oriented systems implemented in this manner still tangle the resolution code, which allows a feature to work with other features, with a feature's core behaviour (e.g. as presented in figure 3).

The examples given in this paper have illustrated the cleanness of separation achieved with our two-level architecture. All other features in [11] also display this elegant separation when implemented. As the number of features increase, the importance of this separation also increases. For example, [11] has identified 26 interactions between the 10 features considered. Should an architecture based on separation of concerns *not* be employed, this clearly results in an unenviable number of extensions required to the basic feature code. A further valuable point is that not every interaction requires a separate resolution module (as illustrated by figure 7 above). Although the subject of future work, this also gives us incentive to look for more general interaction resolution *patterns*. Importantly, our approach will provide support in cases where patterns occur in, or cross-cut, multiple features.

As a further case for the cleanness of separation achieved in our approach, we have also re-factored some of ICEMail's GUI modules. The reason for this was our observation that, in the original modules, there was a great deal of code-tangling. For example, *FolderTableModel* is a GUI component for the display of messages in an email folder. It implements the *TableModel* interface in javax.swing.table and users can read their incoming message via this component (supported by most of today's email clients) [15]. However, the *FolderTableModel* should only be responsible for the displaying of messages; the actual management of messages should be carried by another management component, namely the content provider for *FolderTableModel*. Similarly, the *Fetchlet* class, a class used to get messages from a user's mailbox and place them in a local folder, need not know that the messages it fetched will be displayed by *FolderTableModel* class.

In our re-factoring, a simple interaction resolution called *Compose_FolderTableModel _Fetchlet* has been implemented as an aspect. This aspect knows about the interactions between the two classes, significantly reducing the size of each feature module. For example, the initial *FolderTableModel* contained 800 lines of source code, whilst our re-factored version contains only 70 lines. This reduction is achieved because, in our opinion, the original classes contained too many tangled concerns or interactions. The re-factored code demonstrates a much cleaner separation of concerns and hence, we believe, will permit more effective maintenance and future evolution.

5.2 Re-use

Interaction resolution modules such as those identified in figures 4 and 5 above (i.e. aspects representing very specific resolutions), perhaps only offer limited opportunities for re-use. This is mainly because the chosen resolution, of which there may be a number of options, depends on factors such as an ever-improving technological base that permits improvements to existing features, the rapid development of new features, a subjective judgement as to the best resolution, etc. Hence, we believe it is more likely that, for specific resolution modules, new modules will be developed rather than re-using existing ones.

However, our two-level architecture has the best opportunities for re-use at the base level, rather than the meta-level. We have experimented with the two GUI modules mentioned above, and have found our re-factored modules easier to re-use than the original modules. The reason for this is that the interaction concerns (soft logic) have been extracted from the core behaviour (hard logic), leaving a more generic feature component. For example, the *FolderTableModel* class can be easily re-used in another application with very minimal changes.

Importantly, if we can achieve one of our goals of identifying interaction resolution *patterns*, these obviously have much greater potential for re-use. In investigating this further, we have found that many of the interaction resolutions (like those we have mentioned above) involve boundary condition checking. Furthermore, all of the resolutions for the feature interaction cases of [11] can be fitted into some kinds of boundary checking patterns. Amongst the resolutions for the 26 feature interactions identified in [11], as well as an additional feature interaction identified by ourselves (see section 5.5 below), more than half can be generalised as generic resolutions that can be applied to other feature interaction problems. Hence, the identification of resolution patterns appears a realisable aim, and has obvious benefits for re-use.

5.3 Faithfulness of implementation to specification

Our two-level architecture has been designed to facilitate keeping the feature's implementation faithful to its specification. Not only does this improve the readability and simplicity of the resulting code, but also allows the feature's specification to map to the implementation more directly, thus facilitating better reasoning about the mapping. This, in turn, opens the door to generative programming techniques, as widely used in component-based development to generate code (or code templates) automatically from the specification.

5.4 Adaptability to requirement change

There are two key points of our architecture that help a developer to react to changing requirements. The first of these is that the separation we provide allows the developer to integrate new features into the system, without needing to consider, or worse rewrite, the existing features.

Secondly, and the point that to our knowledge makes our approach unique, is that by adopting aspect-oriented programming for the separate resolution modules, the developer can implement a feature without considering the interactions with other features, then focus on the interaction issues separately. The power of aspect-oriented techniques makes this last step viable and effective, as has been illustrated by the examples above.

Similarly, the removal of features from a system is just as clean and effective because of the separation we provide, meaning that any interactions have been made explicit. This helps to avoid redundant code being left embedded in feature boxes, a situation that leads to unnecessary complexity and lowers efficiency.

5.5 Support for interaction avoidance, detection and resolution

The central objective in this paper has obviously been in the provision of an architecture to cleanly handle the interactions between features. Our implementation of the email system has shown how our two-level architecture and the use of AspectJ can be used very effectively to handle feature interactions.

However, a crucial part of the paper that remains to be addressed is that, whilst our approach explicitly handles features interactions by providing separate resolution modules, what happens if our resolution modules themselves interact? Far from being a theoretical question, problems with resolution modules themselves interacting in undesirable ways have been identified in [28]. This is not an unexpected discovery, since resolutions themselves can be viewed as features, which, of course, are prone to interactions. A number of highly significant issues crop up with respect to this topic that we address in turn below.

Two-level versus multi-level architectures

By proposing a two-level architecture, we have effectively ruled out a multi-level architecture that is typical of many reflective architectures. One option would be to relax this condition, effectively allowing a meta-meta layer to handle the resolution interactions, and theoretically an infinite hierarchy to handle subsequent meta-meta level interactions and beyond. However, it should be noted that, in practice, reflective architectures rarely require more than three layers to be reified.

The problem with adopting this more relaxed approach is our choice of language. As AspectJ stands at the moment[1], the language does not support "aspects of aspects", i.e. aspects cannot be defined over other aspects, thus effectively preventing adequate support for beyond a two-level architecture. It should be noted however that other AOSD approaches do provide support for "aspects of aspects". In an email message posted on the Demeter web-site [6], a list of approaches supporting this is given, namely Incremental Programming [25], Aspectual Collaborations [22], Hyper/J [26] and DJ [24]. These require further investigation to determine if they would provide a more appropriate language choice for our work than AspectJ.

---

[1] To our knowledge, and after discussions in the AOSD community regarding this topic, AspectJ will remain without support for "aspects of aspects" in the foreseeable future, the developers believing the alternatives are unnecessarily complex.

Implicit versus explicit composition

A further issue relating to our language choice is the fact that, in AspectJ, composition of the resolution aspects with the features is done implicitly by the aspect weaver. As a result, there is no control over the composition operator used. Other languages provide explicit support for composition, a facility that we believe would be advantageous and give more flexibility in our solution. It is conceivable, in fact, that a custom-designed composition operator would only allow valid compositions of resolution modules, thus *avoiding* the problem of such modules interacting in undesirable ways. AOSD techniques that currently offer support for explicit composition include Hyper/J [26] and Composition Filters [1].

Detecting feature interactions

In our case study, we have relied heavily on the list of known interactions provided in [11], but what if these interactions are not known in advance?

In performing this case study, we have found that most of the feature subversions happen across boundary conditions. By explicitly separating out feature behaviour into our two-level architecture we are actually encouraging a systematic exploration of boundary conditions, although this is, of course, a manual process at present rather than benefiting from automated support. However, this systematic exploration led us to discover a further (undocumented) interaction as discussed below.

The *FilterMessage* feature ensures that "if a message comes from a *blacklisted address*, it must not reach the subscriber's *user domain*". There are two themes in this statement: "the source of the incoming address" and "the targeted user domain". For the former, there are many possible sources of the incoming address, either human originators or mechanical originators (e.g. auto-Responder, ForwardMessage, RemailMessage, etc.).

Now consider a message from a (mechanical) forwarding feature (e.g. from party A to party B to party C). A boundary condition for this is "What if the forwarding party (B) is not filtered, but the message it is forwarding is from a blacklisted party (A)". From this point, we identify a potential subversion of FilterMessage feature, as follows (undocumented in [11]):

- **Scenario:** Alice's domain is blacklisted and filtered by Carol's host (e.g. username carol@lancaster). Alice asks Bob, whose domain is not filtered by Carol's host, to forward her email to carol@lancaster, thus subverting the filtering mechanism.

From Alice and Bob's point of view, there is unlikely to be a problem here, since Bob knows Alice is using his account. However, from Carol's point of view, her domain has been blacklisted, presumably for good reason (e.g. virus control), yet this has been subverted by the message forwarding.

By continuing to explore the boundary condition of "the source of the incoming address", we can find other potential feature interaction problems involving *FilterMessage*.

In terms of automated support for the detection of feature interactions, there are, of course, many techniques that have been presented throughout the series of Feature Interaction Workshops, e.g. see [3]. The majority of these techniques have had a formal basis, and hence have worked from formal specifications/ models of the features. Such techniques are highly complementary to our approach.

Detecting resolution interactions

Effectively, our resolution interactions are actually aspect interactions, for which the AOSD community currently has very little support. We believe that there are four possible solutions here, all of which require further investigation:

- Avoiding interactions by applying a design by contract approach (e.g. [21]) or by utilising explicit composition operators (not available in AspectJ, see discussion above).
- Relaxing our two-level architecture to allow a meta-meta layer to handle resolution interactions (this also implies a language change since AspectJ does not support "aspects of aspects", also discussed above). Note that our two-level architecture is not a source of resolution interactions, it just restricts the way we can handle them.
- Model-checking certain properties to determine aspect interactions (in the same way that feature interactions have been identified by model-checking). This, of course, relies on the difficult problem of identifying appropriate properties for which to check.
- Investigate the recent work of [9] where execution monitors are deployed over aspects and "aspect laws" can be verified to determine the independence of aspects. Also relevant is the work of [31] where three different categorisations of aspects are identified. For certain types of aspect, they show how the proof of a particular property is relatively straightforward, yet for other types of aspect, this is more problematic. We have yet to investigate this further with respect to our work, particularly identifying which category our aspect resolutions fall into.

6. Conclusions

This paper has proposed a two-level architecture to enhance the development of feature-oriented software. For this architecture, we have proposed the separation of a feature's core behaviour (its hard logic) from extra behaviour required to allow it to adapt to other features (its soft logic). We have illustrated how the latter can be implemented at the meta-level using aspect-oriented programming techniques. Through a case study of an email system (initially very basic with just three features) we have shown the value of this clean separation.

Further, by incrementally evolving our email system from a system with just three features to a new system with ten features and a graphical user interface, we believe we have been able to critically evaluate our proposed architecture. This evaluation has raised a number of important benefits, but also areas that require further research. The most significant of these is the issue of detecting and resolving interactions between our resolution modules. These are effectively *aspect interactions*, an area that is currently under-researched in the AOSD community. Also highlighted in this evaluation has been our choice of language, AspectJ. Further investigations are required to determine whether other AOSD techniques would be more suited to our work.

References

[1] L. Bergmans and M. Aksit, "Composing Crosscutting Concerns using Composition Filters", Communications of the ACM, Vol. 44, No. 10, pp. 51-57, 2001.

[2] R.J.A. Buhr, D. Amyot, M. Elammari, D. Quesnel, T. Gray and S. Mankovski, "Feature-Interaction Visualization and Resolution in an Agent Environment", In Proceedings of the 5th International Workshop on Feature Interactions in Telecommunications and Software Systems (FIW'98), Lund, Sweden, IOS Press, Amsterdam, pp 135-149, October 1998.

[3]   M.Calder, E.Magill (eds), "Feature Interactions in Telecommunications and Software Systems VI", Glasgow, Scotland, IOS Press, Amsterdam, 2000.

[4]   R. Charette, "The Decision is in: Agile versus Heavy Methodologies", (results of a survey of 200 IS/ IT managers), Agile Project Management Executive Update, Vol. 2, No. 19, Cutter Consortium, see http://www.cutter.com/freestuff/epmu0119.html

[5]   M.A.Cusumano and R.W.Selby, "Microsoft Secrets: How the World's Most Powerful Software Company Creates Technology, Shapes Markets, and Manages People", Simon & Schuster, 1998. ISBN: 0684855313.

[6]   "Aspects of aspects", email correspondence on Demeter web-site, 2002. http://www.ccs.neu.edu/research/demeter/related-work/aspects-of-aspects/reading-list

[7]   J. de Luca, "Feature Driven Development: The Community Portal for all things FDD", Nebulon Pty Ltd, http://www.featuredrivendevelopment.com/. See also linked document "Agile Software Development using Feature Driven Development (FDD)", http://www.nebulon.com/fdd/index.html

[8]   E. W. Dijkstra, "A Discipline of Programming", Prentice Hall, 1976. (Now re-published, Prentice Hall PTR/Sun Microsystems Press, ISBN: 013215871X, 1997).

[9]   R. Douence, P. Fradet, and M. Südholt, "A Framework for the Detection and Resolution of Aspect Interactions", SIGPLAN/SIGSOFT Conference on Generative Programming and Component Engineering (GPCE'02), 2002, ACM.

[10] FDD (Feature Driven Development) Process #5: Build By Feature, A Practical Guide to Feature-Driven Development, Step-10 Pte Ltd., 2002. See http://www.step-10.com/FDD/FDD5BuildFeatures.html (web site associated with [27])

[11] R.Hall, "Feature Interactions in Electronic Mail", In [3], pp 67-82, 2000.

[12] W. Harrison and H. Ossher, "Subject-Oriented Programming (a critique of pure objects)", in Proceedings of OOPSLA'93, pp 411-428, 1993. See also http://www.research.ibm.com/sop/

[13] J. Highsmith and A. Cockburn, "Agile Software Development: The Business of Innovation," IEEE Computer, September 2001.

[14] J. Highsmith, "Agile Software Development Ecosystems", Addison Wesley, 2002. IBSN 0-201-76043-6.

[15] ICEMail email client, http://www.icemail.org. See also D. Nourie, "The Java™ Technologies Behind ICEMail: An Open-Source Project", June 2001, see http://developer.java.sun.com/developer/technicalArticles/javaopensource/icemail/.

[16] M. Jackson and P. Zave, "Distributed Feature Composition: A Virtual Architecture for Telecommunications Services", IEEE Transactions on Software Engineering, 24(10):831-847, October 1998.

[17] G. Kiczales, "Towards a New Model of Abstraction in Software Engineering", International Workshop on New Models in Software Architecture, Reflection and Meta-Level Architecture, 1992.

[18] G. Kiczales, "Aspect-Oriented Programming", ACM Computing Surveys, 28(4es), Article number 154, December 1996. ISSN:0360-0300.

[19] "Getting Started with AspectJ", Communications of the ACM, Vol. 44, No. 10, pp. 59-65, 2001. See also http://aspectj.org/

[20] G. Kiczales (ed), 1st International Conference on Aspect-Oriented Software Development, Enschede, The Netherlands, ACM Press, April 2002. ISBN: 1-58113-469X.

[21] H. Klaeren, E. Pulvermueller, A. Rashid, and A. Speck, "Aspect Composition Applying the Design by Contract Principle", 2nd International Symposium on Generative and Component-based Software Engineering (GCSE), Lecture Notes in Computer Science 2177, pp. 57-69, Springer-Verlag, 2000.

[22] K. Lieberherr, D.H. Lorenz and J. Ovlinger, "Aspectual Collaborations: Combining Modules and Aspects", Technical Report NU-CCS-02-?, 2002. Available from http://www.ccs.neu.edu/research/demeter/papers/publications.html

[23] J.Myers and M.Rose, Internet Request for Comments 1725: Post Office Protocol -- version 3, 1994. See http://www.ietf.org/rfc.html

[24] D. Orleans and K. Lieberherr, "DJ: Dynamic Adaptive Programming in Java", Proceedings of Reflection 2001: Meta-level Architectures and Separation of Crosscutting Concerns, Kyoto, Japan, Springer Verlag, 2001. See also http://www.ccs.neu.edu/research/demeter/biblio/DJreflection.html

[25] D.Orleans, "Incremental Programming with Extensible Decisions", First International Conference on Aspect-Oriented Software Development, Enschede, The Netherlands, G. Kiczales (ed), ACM Press, 2002. See also http://www.ccs.neu.edu/home/dougo/papers/aosd02/

[26] H. Ossher and P. L. Tarr, "Hyper/J: multi-dimensional separation of concerns for Java", International Conference on Software Engineering, pp. 734-737, ACM, 2001. See also http://www.research.ibm.com/hyperspace/

[27] S. Palmer and M. Felsing, "A Practical Guide to Feature-Driven Development", Prentice-Hall, 2002. ISBN: 0130676152. See also http://www.step-10.com/FDD/index.html

[28] J. Pang and L. Blair, "Resolving Feature Interactions with AOP and Feature Driven Software Development", draft paper available from Computing Department, Lancaster University, Lancaster, LA1 4YR, December 2002.

[29] D.L. Parnas "On the Criteria to be used in Decomposing Systems into Modules", Communications of the ACM, Vol. 15, No. 12, 1972.

[30] R. Pawlak, L. Seinturier, L. Duchien, and G. Florin, "JAC: A Flexible Solution for Aspect-Oriented Programming in Java", 3rd International Conference on Meta-Level Architectures and Separation of Concerns (Reflection'01), Lecture Notes in Computer Science 2192, pp 1-25, Springer-Verlag, 2001.

[31] M. Sihman, S. Katz, "Superimpositions and Aspect-Oriented Programming", to appear in The Computer Journal (British Computer Society), special issue on Aspect Oriented Software Development, 2003.

[32] P. Zave, "FAQ Sheet on Feature Interactions", AT&T, 2001. See
http://www.research.att.com-/~pamela/faq.html

Feature Interactions in Telecommunications
and Software Systems VII
D. Amyot and L. Logrippo (Eds.)
IOS Press, 2003

Feature Interaction Detection in Building Control Systems by Means of a Formal Product Model

Andreas METZGER, Christian WEBEL
Department of Computer Science, University of Kaiserslautern
P.O. Box 3049, 67653 Kaiserslautern, Germany
{metzger, webel}@informatik.uni-kl.de

Abstract. The complexity of present-day software systems has reached dimensions that require systematic approaches for coping with it. In addition to traditional domains, such as telecommunications or business applications, complex software systems can be found in the domain of reactive systems, of which building control systems are an interesting example. The extension and reuse of these systems have become important activities in the respective development processes. To be able to correctly execute these activities, the developers need to be aware of interactions that might exist between different features of a system.

In this paper, an approach for the systematic detection of feature interactions in building control systems is presented, which allows the automatic identification of such interactions based on existing requirements specification documents. This is realized by the application of a formal model of the development products, which includes traceability relations between these products.

Introduction

The complexity of modern reactive systems is on a steady rise. An interesting example of this is integrated building control systems, which are essential for optimizing the total building performance [1]. To provide optimal results, such systems have to consider the interaction of different physical effects, which most often requires a strong coupling of different parts of such control systems. Further, creating building control systems exposes a huge complexity due to the sheer number of objects that needs to be dealt with; e.g., the integrated control system that was developed for a floor of our university building consists of 920 objects [2].

Because of this complexity, a number of problems arise. One of these problems is the extension of such systems with additional functionality. This is typically required if new user needs should be considered. Besides the desired behavior, this new functionality might introduce *undesirable* interrelationships with old parts of the system, and thus an unwanted system behavior might be the result. To illustrate this, we assume an exemplary building control system that allows the automatic control of the illumination inside a room. In this system, the cost of artificial lighting is reduced by employing daylight whenever possible. If this system is extended to satisfy the user need of avoiding glare at the workplace, an undesirable interrelation might be introduced, such that daylight might be used in situations that produce glare.

Another problem can be discovered when complex reactive systems should be reused. Simply eliminating the parts that are not needed in the new context can result in faulty systems, because *required* interrelations to the parts that have been eliminated might not have been considered. To give an example for such a condition, we assume that the building control system above additionally realizes an alarm system. One functionality of this alarm system is to turn on all lights when an alarm has been triggered. If this system should be reused as an alarm system only, all parts that are relevant for lighting (lights, blinds, etc.) will probably be removed. However, if this removal is exercised without considering the required interrelations, a malfunctioning system is the result as the lights are also needed for the alarm system.

Although the above examples appear to be trivial, in real control systems, which typically realize between 200 [2] and 8000 [3] requirements, this detection cannot reasonably be performed manually. Therefore, we propose an approach for the *automatic* detection of such interactions. Our detection concept is based on a formal *product model*, which reflects each type of development product as well as its relations with other types of products. These relations are an important means for establishing traceability between products.

For the purpose of interaction detection, we employ a product model that is available as part of a precise definition of the requirements specification method *PROBAnD* [2]. After this product model has been instantiated using the set of available requirements documents, interaction information can be automatically derived from model data.

Related Work

In our opinion, the above problems are similar to the feature interaction problem, which has received considerable attention in the telecommunications domain [4]. Consequently, the product model-based approach presented in this paper might not only provide solutions for the depicted problems in the building control context, but could also prove to be valuable in the domain of telecommunication systems.

Several feature interaction detection approaches have been presented in the literature. In [5] an approach that detects undesirable interactions by employing Use Case Maps [6] and LOTOS [7] is introduced. After features have been semi-formally described with Use Case Maps, these features are formally defined with LOTOS, which then can be used for verification. Other authors also propose using formal description techniques for the purpose of interaction detection. In [8] and [9] temporal logic and model checking are suggested, in [10] a special heuristic is chosen, and in [11] as well as in [12] description logic and logic meta-programming are applied.

All of the above approaches require the modeler to explicitly and formally specify features. In contrast to this, our concept does not require the requirements engineer to provide any *additional* information during the specification process. The existing documents can be used exactly as they have been described in our requirements specification method *without* considering feature interaction. Therefore, this approach can provide an increase in product quality with only little additional effort.

However, this simplification does prevent us from distinguishing between interactions that are undesirable or required. Therefore, this has to be determined by the developers on the basis of the automatically analyzed interaction information and the existing documents. If an automation of this activity should be achieved, an extension of our requirements specification method would become necessary such that required interactions can be specified.

Where many approaches like [8] or [13] use actual implementation code or extended implementation code (e.g., meta-programming [11]) for detecting feature interaction, we are able to detect such interactions very early in the development process, i.e., during requirements specification. This has the major benefit of eliminating undesirable interactions before they find their way into the final product, where removing these interactions can become very costly.

In the remainder of this paper, the requirements specification method and its product model are described. Then, the concepts for applying this product model for the purpose of interaction detection are presented, which are illustrated by an example of a small building control system. Finally, the results of two complex case studies are discussed.

1. The Product Model

When defining a detailed product model that reflects all necessary development products, the model itself might become very complex. In our case, the complete model currently contains 80 different entities. Therefore, we structure our model by classifying the different types of entities according to Fig. 1.

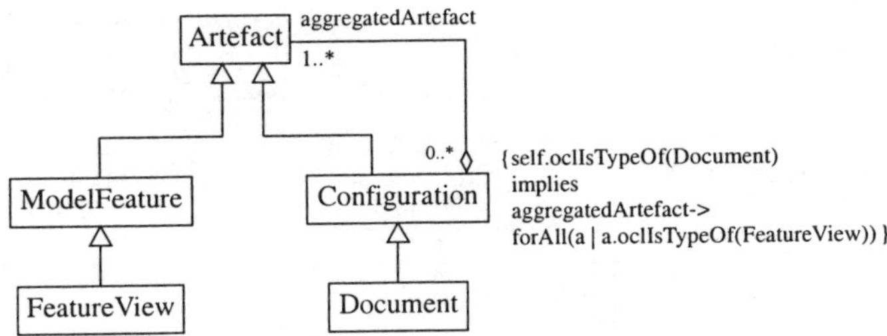

Figure 1. Classification of Development Products

The most general type of entity is called *artefact*, which can be identified with any development product. Each artefact has a unique name and can optionally be described by informal text.

More specialized types of entities are named *model-features*, which represent *atomic* development products. Note, that the notion of a model-feature is different from the general meaning of a feature, which is a self-contained functional part or aspect of a specification or system [14]. We will consistently be using the term model-feature in the remainder of this paper to make this distinction clear.

Model-features contain the actual development information independent of its representation. Thus, for each model-feature different *feature-views* exist, which contain model-feature information in a concrete representation, e.g. in HTML or SDL [15]. This is why feature-views can be conceived as specializations of model-features.

Additionally, *configurations* are introduced, which aggregate less complex artefacts. This is reflected by the aggregation relation in Fig. 1.

Usually, developers will be working on *documents*, which are composed of feature-views. Therefore, documents are classified as special configurations that are only allowed to aggregate feature-views. As documents inherit the aggregation relation of configurations, the constraint in Fig. 1 states that only feature-view instances can be at the *aggregatedArtefact* end of the relation.

This classification of product model entities further provides a means for reducing the number of artefacts that need to be handled for interaction detection. As model-features contain the same information as feature-views do, the detection concept can work on the level of model-features without neglecting any information. For that purpose, the set of available development documents is used for instantiating the model-features as well as the relations between these model-features. This step is carried out by decomposing the documents into feature-views and extracting model-feature information from these feature-views. Details on how this is performed can be found in [16].

With these prerequisites, our interaction detection approach is independent of the notation of the specification documents. As a consequence, the model-features of the product model, which will be described in Sect. 1.2, suffice as input for feature interaction detection.

1.1 The PROBAnD Method

Before the concrete model-features are depicted, an introduction to the PROBAnD requirements specification method [2][17] seems appropriate.

In Fig. 2, an overview of the documents and activities of the PROBAnD method is provided. The method takes the *problem description* as an input, which is divided into the *building description* and a collection of *needs*.

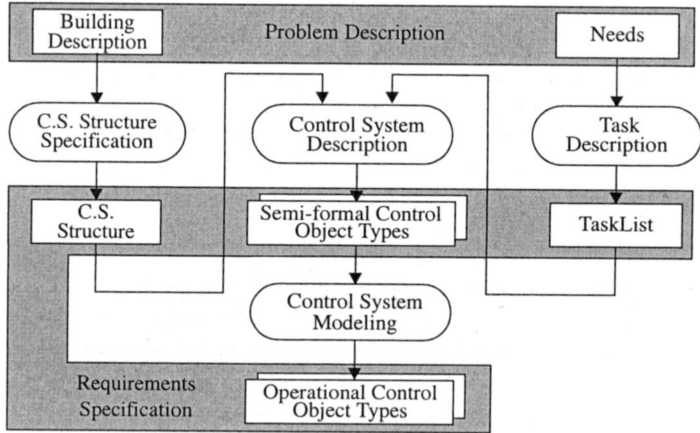

Figure 2. Overview of the PROBAnD Method

The building description contains a description of the building's structure; e.g., a floorplan that shows that the building is made up of one floor with three rooms. Further, the building's installation is depicted; e.g., in an informal text that states: "There are one light, one illumination sensor, and one motion-detector in each room".

From this building description, the *control system structure* is derived by considering the structure of the building. In the building automation domain this has proved to be a suitable approach as *control objects* (the objects of the control system) are most often identical to the building's objects or are a subset of these; e.g., lighting needs to be controlled only within the boundaries of one room, and therefore the control object for lighting can be derived from the room object. Additionally, sensors and actuators represent the interface of the control system to its environment and as such should be reflected in the specification.

To handle the huge number of control objects that need to be considered for large building control systems, *control object types* are formed, which are aggregated according to the hierarchy of the building's objects.

The other part of the problem description is made up by a collection of needs (i.e., user requirements). These needs informally describe the control system from the point of view of the users in natural language.

These needs have to be split into less complex *tasks* (i.e., developer requirements) such that they can be assigned to a single control object type.

After the control object types have been identified and tasks have been assigned, *strategies* for realizing the control tasks are informally given in natural language, leading to a collection of *semi-formal control object types*. From these control object types, *operational control object types* are specified in SDL, from which control system prototypes can be generated.

It should be noted that control object types are introduced during the requirements specification phase solely for the purpose of structuring the specification in a suitable way. In the design phase, this structure can be re-arranged. Furthermore, the specification of strategies should be understood as an exemplary solution of the developer requirements. This is needed for generating prototypes, which are an important means for the early validation of user requirements [18]. During the design phase, different solutions for the strategies are possible and legitimate.

1.2 The Product Model of PROBAnD

As it has been introduced in the previous section, an input to the *PROBAnD* requirements engineering method are the user requirements, called *needs*, which are shown in Fig. 3 as a specialization of the model-feature *requirement*. Only needs that represent functional requirements, i.e., *functional needs*, are regarded during our requirements specification process. Because of their granularity, we interpret these functional needs as features according to [14].

Needs are realized by *tasks*, which themselves might be realized by other tasks. This dependency is modelled by the *realizedBy* relation between *requirement* and *task* in Fig. 3. This relation—like all other relations of the product model—are explicitly reflected in the development documents, which are used for instantiating the model-features.

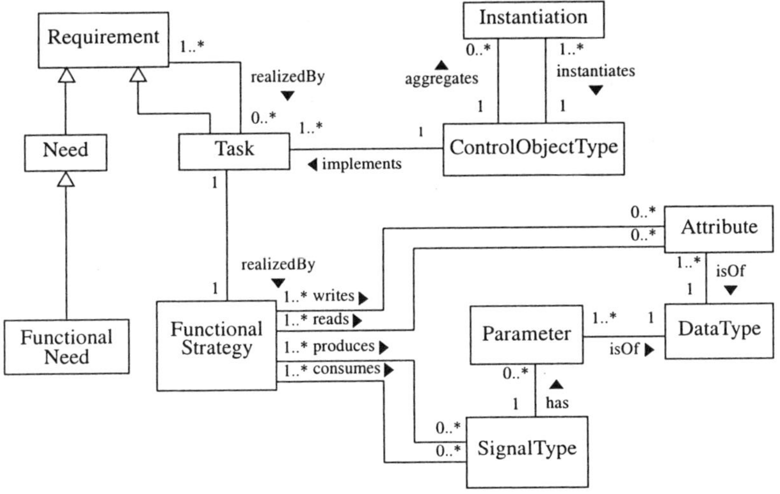

Figure 3. Types of Model-Features

Each task describes a responsibility that has to be fulfilled by the one *control object type* that implements this very task. This is reflected by the one-to-many *implements* relation from the model-feature *control object type* to the model-feature *task*. As it has been depicted in Sect. 1.1, these control object types are always instantiated in a strict aggregation hierarchy, which leads to a tree of instances. The strict aggregation is modelled by the one-to-many *aggregates* relation between the model-feature *control object type* and the model-feature *instantiation*.

For each task, a *functional strategy* is specified. As it has been pointed out, such a strategy describes a possible solution for realizing the responsibility of the task. To establish communication between strategies of different tasks, strategies can read and write *attributes* as well as produce and consume signals that are of globally defined *signal types*, which can possess *parameters*. Where attributes are used for the communication between tasks of the *same* control object type, signals are used for exchanging data between tasks of *different* object types. It is important to note that signals are only allowed to travel along the aggregation hierarchy. This for example implies, that if control objects at the same level of the hierarchy need to communicate, the signals have to be routed through the parent instance, which aggregates the communicating instances.

To illustrate the model-features of the product model and to provide an example for the remainder of this paper, a small building control system is presented in the following subsection.

1.3 Small Building Control System Example

Three needs of our small control system example have already been described in the introduction. The first need ($N1$) can be expressed as "Provide required illumination in a room if it is occupied". The second need ($N2$) is "Use daylight to reduce energy consumption", and the third need ($N3$) is "Avoid glare at the workplace". To make this system a little more complex, a further need is added. This need ($N4$) states "Provide required temperature in a room". Table 1 lists these needs.

Table 1. Needs for a Small Building Control System

Need	Description
N1	Provide required illumination in a room if it is occupied.
N2	Use daylight to reduce energy consumption.
N3	Avoid glare at the workplace.
N4	Provide required temperature in a room.

According to the *PROBAnD* method, these needs have to be refined, leading to a collection of tasks, which are assigned to specific control object types. A possible structure for the small building control system derived from the respective building description is presented in Fig. 4.

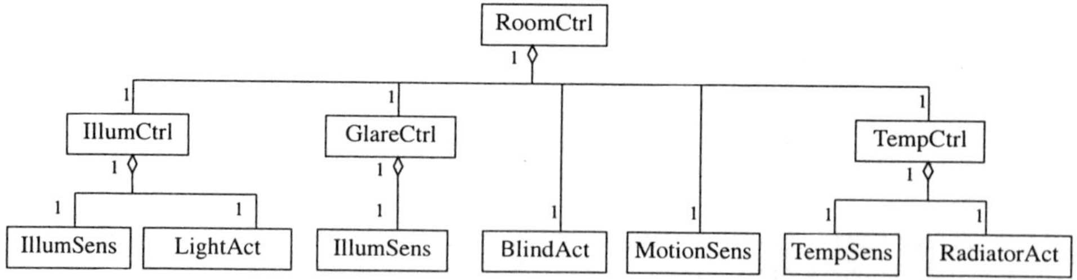

Figure 4. Structure of a Small Building Control System

The sensors and actuators, which represent the interface to the environment, can typically be found as leaf nodes in such a structure; e.g., *IllumSens* (IlluminationSensor) or *BlindAct* (BlindActuator) in Fig. 4.

Table 2 lists suitable tasks for refining the above needs. One such refinement for need $N3$ is provided by tasks $T6$ and $T3$, which describe the developer requirement of avoiding glare by employing the blinds. For this example, it is assumed that the building description provides blinds as the only means of shading.

The complete set of development documents of this small example can be found online at [19].

Table 2. Task List of a Building Control System

Task	Description	realizes	implementedBy
T1	If room is occupied, control indoor illumination with available light sources (blind and light), taking energy consumption into account.	N1, N2	IllumCtrl
T2	Turn light on or off on request.	T1	LightAct
T3	Open or close blind on request.	T1, T6	BlindAct
T4	Determine and report motion.	T1, T6	MotionSens
T5	Determine and report current illumination.	T1, T6	IllumSens
T6	Avoid glare at the workplace by using the blind if room is occupied.	N3	GlareCtrl
T7	Control room temperature by using the radiator.	N4	TempCtrl
T8	Open or close radiator valve on request.	T7	RadiatorAct
T9	Determine and report current temperature.	T7	TempSens

2. Feature Interaction Detection

In Sect. 1 we have outlined that functional needs can be interpreted as features. So, the goal of detecting feature interactions in building control systems can be reached by identifying dependencies between functional needs. These dependencies can be extracted by following the relations between model-features in an instantiation of the product model. Depending on the available level of information, results with different precision can be attained. In the following subsections, we will present how four different levels of information can be used for feature interaction detection.

2.1 Detection at Requirements Level

Early in the requirements specification process, needs and tasks are the only model-features that will have been specified. Therefore, interactions at this level can only be identified by employing these model-features together with the *realizedBy* relation between them. From this type of relation, a dependency graph between requirements (i.e., needs and tasks) can be attained. Fig. 5 shows such a graph, which depicts the dependencies between the requirements of the building control example. In such a graph, *points of interaction* can be identified, which are nodes that have more than one direct parent.

According to Table 2, the needs $N1$ and $N2$ are realized by task $T1$, where $T1$ is realized by $T2$ to $T5$. Additionally, tasks $T3$ to $T5$ are needed for fulfilling need $N3$. This leads to four points of interaction that can be identified: $T1$, $T3$, $T4$, and $T5$.

To formalize how such points of interaction can be found, a few basic definitions will be provided in the following paragraphs.

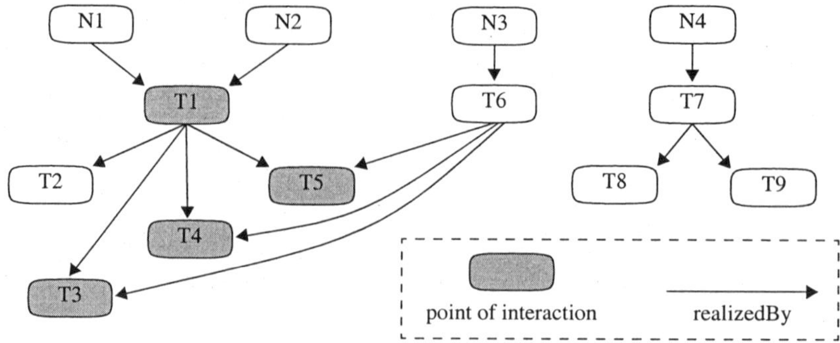

Figure 5. Dependency Graph for Requirements

Let $N = \{N_1, ..., N_n\}$ be the set of all needs and $T = \{T_1, ..., T_m\}$ be the set of all tasks. As both needs and tasks are requirements, the set of all requirements consequently is defined by $R = \{N_1, ..., N_n, T_1, ..., T_m\} = \{R_1, ..., R_{n+m}\}$. Further, we define a *realizedBy* predicate such that

$$R_i \rightarrow R_j \quad \textbf{iff} \quad \text{requirement } R_i \text{ is realized by requirement } R_j.$$

Obviously, this predicate is transitive; e.g., $N1 \rightarrow T4$ holds in Fig. 5. With these definitions, we can define a predicate φ for the point of interaction with

$$\varphi(T_i) \quad \textbf{iff} \quad \text{there exists a set } N(T_i) = \{N_{i,1}, ..., N_{i,k}\} \subseteq N \text{ with } |N(T_i)| > 1 \text{ such}$$
$$\text{that } N_j \rightarrow T_i \text{ for each } N_{i,j} \in N(T_i).$$

Accordingly, in our building control system $\varphi(T1)$, $\varphi(T3)$, $\varphi(T4)$, $\varphi(T5)$ hold.

To algorithmically determine these points of interaction, we initialize $N(T_i)$ with the empty set for $i = 1, ..., m$. Then, for each need $N_t \in N$ ($t = 1, ..., n$), we follow all *realizedBy* relations and add N_t to the set $N(T_i)$ for each task T_i that is traversed. As a result, each task T_i will have been marked by a set $N(T_i)$, from which $\varphi(T_i)$ can be computed according to the above formula.

Because our approach only employs needs and tasks at this level, it can be conceived as being domain and method independent. Especially, this implies that whenever a traceability relation from features (i.e., needs) to responsibilities (i.e., tasks) exists in a given specification method or language, the above algorithm can be applied. An example for such a language is the User Requirements Notation (URN [20]), which provides the necessary means for requirements traceability.

With the knowledge of the points of interaction, we are only able to systematically determine *possible* feature interactions, because dependencies at the level of needs and tasks do not necessarily imply that there will be interactions in the final system. For example, the task at a given point of interaction might never be used for realizing more than one need simultaneously. Therefore, from the points of interaction of Fig. 5, only a possible feature interaction between the needs $N1$ and $N2$ (at $T1$) as well as a possible interaction between $N1$, $N2$, and $N3$ (at $T3$, $T4$, and $T5$) can be derived.

2.2 Detection at Strategy Level

In order to refine this set of possible feature interactions, dependencies between tasks by means of their strategies can be examined. This dependency occurs because strategies are coupled by signal types or attributes (see Fig. 3). Two observations can be made when using this information.

The first observation is that, in spite of a possible interaction detected on the level of the requirements, an interaction between two tasks cannot occur if these tasks only consume (resp. read) signals (resp. attributes) that are produced (resp. written) by the task at the point of interaction. An example for this is task $T4$ of our small control system. As this motion sensor only produces signals, which are consumed (by $T1$ and $T6$), there will be no interaction.

The detection of such situations can be carried out easily by following the *produces/consumes* (resp. *writes/reads*) relations from the model-feature *functional strategy* to the model-feature *signal type* (resp. *attribute*).

It should be mentioned that we are working on requirements specification documents, only. Therefore, problems that could arise in the actual implementation of the system are not considered. One of these problems could be that the communication between objects is not reliable, which would lead to an unwanted interaction if the information from the motion sensor could not be received.

The second observation that can be made when employing strategy information is that, with the introduction of strategies, additional interactions can originate. It is possible for the developer to have the strategies of two independent requirements operate on the same set of data (e. g., by using the same attribute), thus introducing a coupling between the tasks. In our small example, this situation does not occur.

To elaborate this point, depending on the chosen realization of a strategy, the number and types of interactions can vary. This especially implies that when the development process enters the design phase, some of the interactions that have been identified during the requirements specification phase need to be re-evaluated.

2.3 Detection at Object Structure Level

After the above levels of information have been considered, the list of possible interactions is already refined. In order to further improve the set of possible feature interactions, interactions that cannot occur should be eliminated.

A step towards this goal can be taken by considering the aggregation hierarchy of the control object types. This information provides clues as to whether an interaction is possible, because control object types, which implement the respective tasks, might be instantiated in such a way that dependent tasks will never be used at the same point of instantiation. An example for this can be given for task $T5$, which is implemented by the control object type *IllumSens*. The depending tasks $T1$ and $T6$ are implemented by different control object types (*IllumCtrl* and *GlareCtrl* respectively). Because both individually aggregate an instance of *IllumCtrl* there will be no interaction between tasks $T1$ and $T6$ at the point of $T5$.

To formalize this fact, let $C = \{C_1, ..., C_l\}$ be the set of all control object types and $C(T_i)$ the control object type that implements task T_i. Further, $P(C_k, C_l)$ should be the set of all control object types that are instantiated on the *shortest paths* from the control object type C_k to C_l. For instance, in the small building control example

$$P(IllumSens, GlareCtrl) = \{IllumSens, GlareCtrl\}.$$

For evaluating $P(C_k, C_l)$, the *aggregates* and *instantiates* relations of the product model are considered. The calculation of the shortest paths starts at C_k, from which all control object types that can be found along the aggregation hierarchy to the point(s) where C_l is instantiated are determined. Typically, this requires passing through the hierarchy towards the root object type using the reverse direction of the *instantiates* and the *aggregates* relation. Once the root is reached, the tree is searched in a downwards fashion using the forward direction of the *aggregates* and *instantiates* relation. Of all possible paths that have been determined with this algorithm, the shortest ones are calculated and the all object types along these shortest paths form the required set $P(C_k, C_l)$.

To illustrate, Fig. 6 shows an excerpt of an instantiation of the product model for the above example.

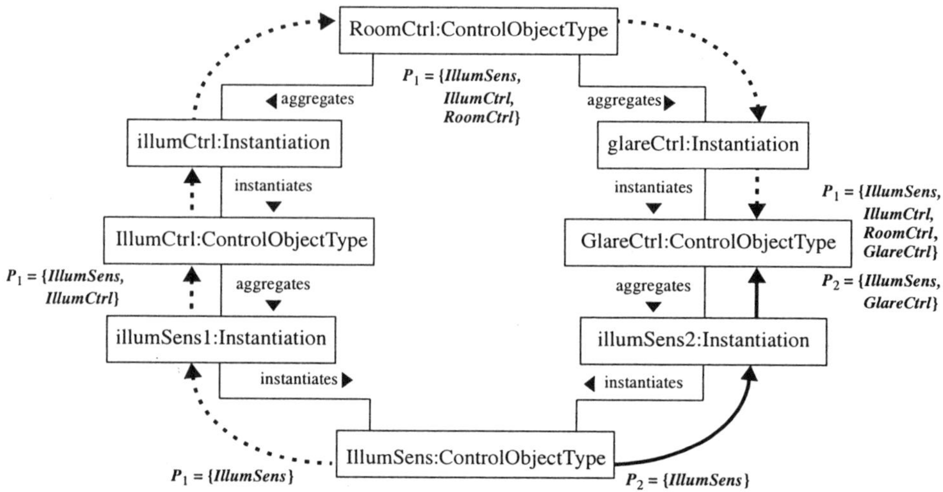

Figure 6. Calculation of Paths in an Instantiation of the Product Model

The calculation of the paths starts at the control object type *IllumSens*. Because two instantiations of this control object type exist, two possible paths (P_1 and P_2) can be identified. As P_2 is the shorter one, it can be deduced that

$$P(\textit{IllumSens}, \textit{GlareCtrl}) = P_2 = \{\textit{IllumSens}, \textit{GlareCtrl}\} \, .$$

With the technique for determining such paths, we can now eliminate the interactions that cannot occur because of the aggregation hierarchy.

A modeling guideline of the PROBAnD method requires that control object types that implement tasks that are connected by *realizedBy* relations should be instantiated as closely as possible, because this reduces the number of signals that need to be routed through other instances in the aggregation hierarchy. Therefore, only tasks that are directly and *not* transitively realized by the task at the point of interaction need to be considered.

If we assume that $T_d(T_i) = \{T_{d,1}, ..., T_{d,r}\}$ is the set of all tasks that are directly realized by T_i. The set of all control object types that are on the path between the object type that implement tasks T_i and the one that implements $T_{d,k}$ can be attainted by

$$P_{d,k} = P(C(T_i), C(T_{d,k})) \, .$$

If $P_{d,s} \cap P_{d,l} = \{C(T_i)\}$, then there can be no interaction between tasks $T_{d,s}$ and $T_{d,l}$, because there are no common object types along the paths and as such different instantiations of the object type that implements the conflicting task are used.

If the intersections of all possible combinations of $P_{d,k}$ are computed, all interactions that cannot occur because of the instantiation hierarchy can be eliminated.

For the above example of task $T5$, $T_d(T5)$ is $\{T1, T6\}$, and thus

$$P_{d,1} = \{\textit{IllumSens}, \textit{IllumCtrl}\} \text{ and } P_{d,2} = \{\textit{IllumSens}, \textit{GlareCtrl}\} \, ,$$
which leads to

$$P_{d,1} \cap P_{d,2} = \{\textit{IlltumSens}\} = \{C(T5)\} \, .$$

Hence, there is no interaction at point $T5$. Whereas in the case of $T3$ the following sets are computed:

$$P_{d,1} = \{\textit{BlindAct}, \textit{RoomCtrl}, \textit{IllumCtrl}\} \text{ and } P_{d,2} = \{\textit{BlindAct}, \textit{RoomCtrl}, \textit{GlareCtrl}\} \, ,$$
which leads to

$$P_{d,1} \cap P_{d,2} = \{\textit{BlindAct}, \textit{RoomCtrl}\} \neq \{C(T3)\} \, ,$$

and therefore this interaction cannot be eliminated from the list of possible interactions. The same is valid for the point of interaction at $T4$.

Our approach at this level is strongly domain and method specific, because the static nature and the strict aggregation hierarchy of the building structure are important prerequisites for the detection algorithm. This implies that the application of this reduction concept is possible only for closely related domains, which expose similar properties; e.g., automotive control systems.

If all considerations that have been presented in Sect. 2.1 to Sect. 2.2 are exercised for our small building control system, the refined set of interactions that can be deduced contains the interaction of $\{N1, N2\}$ at $T1$ and the interaction of $\{N1, N2, N3\}$ at $T3$, which—at a closer look—can be identified as real feature interactions.

2.4 Detection at Environment Level

As our approach is targeted towards reactive systems, the environment of such systems has to be considered. This is especially important for the detection of feature interactions, because the physical environment can be the source of an implicit coupling between different parts of the control system.

This fact can already be discovered in our small building control example. The dependency graph in Fig. 5 shows no dependencies between need $N4$ (temperature control) and the other needs (lighting control). However, there exists a physical link between the room temperature and the amount of daylight that comes into the room. The reason for that is that sunlight can considerably heat up a space. Therefore, an interaction between $N4$ and the other needs should be noted.

To systematically determine such implicit interactions, the links in the physical environment must be made explicit for the detection process.

During the development of each reactive system, a simulator of the system's environment is needed for testing the dynamic behavior of the system. As such a simulator has to consider the physical links, the simulator's models can be used for making the links explicit, thus making this information available for the detection process.

In our case, building performance simulators for testing control systems have been modelled using the PROBAnD method [21][22]. By merging the product model instantiation for the control system and the product model instantiation for the simulation, we achieve a combined model instance. This combined model instance can be employed for uncovering feature interactions by applying the above algorithms.

To merge the two product model instances, *connection points* must be identified. As it was already pointed out, control object types that represent sensors and actuators are the interface to the environment. These control object types can now be used for identifying their counterparts in the simulator, thus establishing the required connection points.

This solution is sketched in Fig. 7 for our small control system. We already know that $T3$ is realized by the control object type *BlindAct* and that $T9$ is realized by *TempSens*. If we assume that the task for simulating the blind actuator is named Ta and that the task for simulating the temperature sensor is named Tb, and if we further assume that both Ta and Tb are realized by task Tc, this very task can be identified as a point of interaction. This leads us to the conclusion that $N1$, $N2$ and $N3$ expose an interaction with $N4$ through the physical environment.

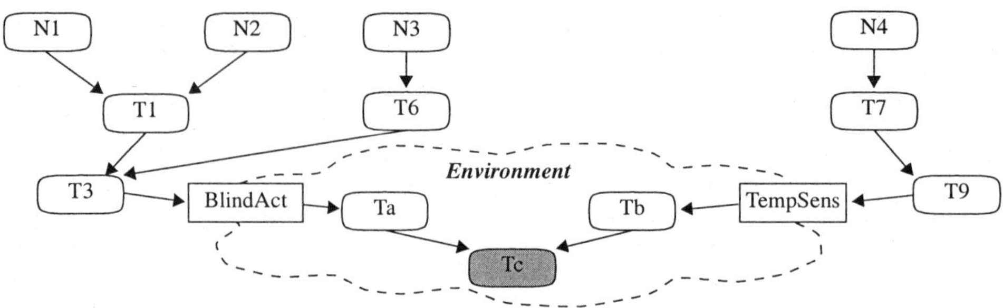

Figure 7. Dependency Introduced by Physical Environment

3. Results

To evaluate the above concepts and examine their feasibility in a real development context, tool prototypes have been implemented and applied in two case studies for different versions of a large building control system. The results of these case studies and a closer examination of the code as well as the run-time complexities of our tool will be provided in the following sections.

3.1 Case Studies

In the first case study, a heating and lighting control system, which we call *Floor32* in the remainder of this paper, was used as an example. A detailed description of this system and an analysis of qualitative and quantitative development data can be found in [23]. To give an impression of this system, a few of its 67 needs for heating and illumination are provided in Table 3.

Table 3. Needs for Large Heating and Illumination Control System (Excerpt)

Domain	Need	Description
Illumination	U2	As long as the room is occupied the chosen light scene has to be maintained.
	U3	If the room is reoccupied within t_1 minutes, the last chosen light scene has to be re-established
	U4	If the room is reoccupied after t_1 minutes, the default light scene has to be set.
	FM1	Use daylight to achieve the desired illumination whenever possible.
	FM6	The facility manager can turn off any light in a room or hallway section.
Heating	UH2	The comfort temperature shall be reached as fast as possible during heating up and maintained as best as possible afterwards.
	UH6	The user can manually move each sun blind up or down. This manual override holds until he/she leaves the room for a longer time period.
	FMH2	The use of solar radiation for heating should be preferred against using the central heating unit.

An extension of this system was used as a second example, which we call *Floor32X*. The extension of this system has been obtained by adding the functionality of an alarm system [24]. This is reflected by 12 additional requirements, some of which are listed in Table 4.

Table 4. Needs for Large Alarm System (Excerpt)

Domain	Need	Description
Alarm	UA2	If a person occupies a room with an activated alarm system, he/she can deactivate the alarm system by identifying himself/herself within t_2 seconds. Otherwise, the alarm must be triggered.
	UA3	If a room is unoccupied out of working hours for more than t_3 minutes, the alarm system must be activated automatically.
	UA10	If an alarm is triggered, all lamps in the corresponding sections are turned on. When the alarm is interrupted, the lamps are reset to their previous state.
	FA1	The facility manager can switch off an alarm, deactivate an alarm system, and activate the alarm system for an individual room or for all rooms of the building.

One might observe that these exemplary needs are very fine-grained and almost seem to describe the solutions rather than the actual requirements. This can be attributed to the fact that the persons that were in the role of the "customers" when specifying this system, were domain-experts, which are better at expressing solutions rather than actual problems.

To illustrate the structural complexity of *Floor32*, the simplified hierarchy of its 37 control object types is shown in Fig. 8. These control object types are instantiated to 920 control objects. when the system is executing.

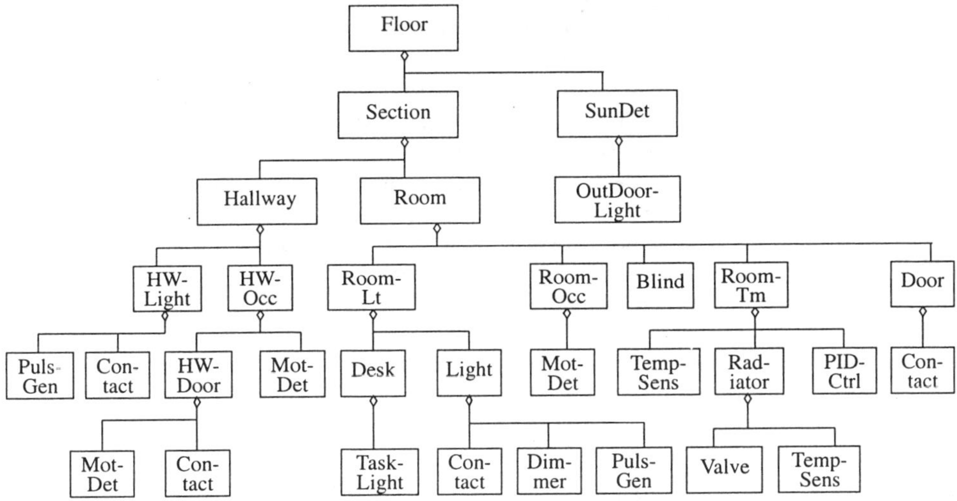

Figure 8. Object Structure of Complex Building Control System (*Floor32*)

To summarize, in *Floor32X*, a total number of 316 functional requirements (64 needs and 252 tasks) are realized, where in *Floor32* only 285 functional requirements (52 needs and 233 tasks) are implemented.

3.2 Detected Interactions

The application of our tool prototypes to the above specifications has lead to the results that are shown in Table 5 for the four different levels of available information (cf. Sect. 2).

Table 5. Results of Feature Interaction Detection for Complex Building Control Systems

Available Information	Number of Feature Interactions			Number of Points of Interaction		
	Floor32	Floor32X	Δ	Floor32	Floor32X	Δ
1. Requirements	34	41	7	52	58	6
2. Strategies (Signal Types only)	34	41	7	52	58	6
3. Object Structure	32	38	6	47	53	6
4. Environment	38	44	6	63	69	6

Typical interactions that have been identified in *Floor32* at the requirements level were between $\{U2, U3, U4, FM1, FM6\}$ as well as between $\{UH2, FMH2\}$. These interactions are fairly obvious as the interacting needs describe different aspects of a common feature, which is a consequence of the fact that needs were specified very fine-grained and solution-oriented. This also explains the relatively huge number of interactions that can be identified.

When the system was extended, seven new feature interactions have been introduced on the requirements level. One of these interactions occurs between $\{UA2, UA3, FA1\}$, which is inside the alarm system domain. Additionally, interactions between needs of different domains have been identified; e.g., $\{U2, UA10\}$. This interaction occurs, because both need $U2$ and need $UA10$ employ means of lighting for their realization.

The number of interactions that have been identified at level 1 and level 2 is the same. This can be attributed to the fact that our tool prototypes currently do not consider additional interactions caused by a coupling on the strategy level. Further, no interactions from level 1 can be eliminated because no point of interaction exists that just *produces* signals. One reason for that is that all sensors in our system can receive special signals that change a sensor's mode from polling to event-based communication.

From level 2 to level 3, the number of possible interactions is reduced by three and the number of the particular points of interactions is lessened by five. One example for an interaction that is eliminated is the one between $\{U2, U3, U4, UA2\}$. This interaction is detected at levels 1 and 2 because the lighting control object type *RoomLt* as well as the control object type *RoomOcc*, which is used for the alarm system, each aggregate an instance of *Contact* (cf. Fig. 8).

Finally, by using information about the system's environment at level 4, six additional interactions have been uncovered. As an example, one of these interactions occurs between $\{FM1, U2, UH6\}$, because an interaction between the control object type *TempSens*, which is needed for realizing the temperature control need ($UH6$), and the control object type *BlindAct*, which realizes the lighting control needs ($FM1$ and $U2$), is present.

3.3 Tool Complexities

The tool prototype that has been employed for the above calculations is part of our *PROTAG-OnIST* tool set [25]. It has efficiently been developed using the programming language Java. Each type of entity is represented by a Java class and each type of relation is implemented by attributes as well as the required accessor and mutator methods. Thus, operations on product model data, like following the *realizedBy* relations for identifying the point of interaction, can easily be implemented by directly mapping abstract concepts from the model level to Java code.

For performing the above case studies, tools with a total number of 10 700 lines of code have been developed. Approximately 1 900 lines of this code are needed for the actual interaction detection. An additional 4 100 lines of code are required for parsing the development documents [26]. The remaining code, which has automatically been generated from UML class diagrams, realizes the class frames as well as the attributes and relations of the product model.

An evaluation of the run-time complexity of this tool, supported by statistical measurements, shows very moderate processing power requirements.

For example, to determine feature interactions at the requirements level (see Sect. 2.1), for each of the n needs, all tasks that realize this need have to be examined. Based on exemplary evidence, this number does not seem to correlate with the number of needs. This implies that a linear run-time complexity of $O(n)$ can be attained for this activity.

For the concrete example of *Floor32X*, the actual detection of the interactions that are depicted in Table 5 has consumed 2.5 s of processing time on a 440 MHz HP-PA RISC workstation.

4. Conclusion and Perspectives

The extension and reuse of existing systems are important activities in software development. To correctly carry out these activities, the developers need to be aware of the interactions that exist between features. This paper has shown an approach for the detection of feature interactions that is based on a formal model of the development products. With this approach, detailed information about feature interactions can automatically be derived from existing requirements specification documents at each stage of the proposed requirements engineering method. The more complete this development information is, the more refined the information about the interactions will be.

Our approach can also be used for guiding the developer in such a way that undesirable interactions between features can be avoided. In detail, this means that after each important step in the requirements specification process (e.g. after important tasks have been elicited), possible interactions can be computed, and the developers can decide on how to handle undesirable interactions.

At the time of writing, our approach has been realized for four important activities in our requirements specification method: the specification of needs and tasks, the creation of the object structure, the specification of the communication between strategies, which realize tasks, and the modelling of the environment.

A further activity in this method is the precise specification of the behavior of the strategies through extended finite state machines, which basically are finite state machines that have been extended by variables and control constructs. With an extension of the product model to reflect these types of state machines, a far more refined examination of feature interactions might become feasible. Currently, we are in the process of refining our product model for that purpose, which will be followed by an evaluation of a possible refinement of our feature interaction detection approach.

Acknowledgements

Parts of this work have been funded by the Deutsche Forschungsgemeinschaft DFG in the Special Collaborative Research Center SFB 501 "Development of Large Systems with Generic Methods".

Finally, we want to thank the anonymous referees, who—by providing detailed and constructive comments—have contributed to the improvement of this paper.

References

[1]   V. Hartkopf, V. Loftness. "Global Relevance of Total Building Performance" in *Automation in Construction*. 8. New York; Amsterdam: Elsevier Science. 1999

[2]   A. Metzger, S. Queins. "Specifying Building Automation Systems with PROBAnD, a Method Based on Prototyping, Reuse, and Object-orientation" in P. Hofmann, A. Schürr (Eds.) *OMER – Object-Oriented Modeling of Embedded Real-Time Systems*. GI-Edition, Lecture Notes in Informatics (LNI), P-5. Bonn: Köllen Verlag. 2002. pp. 135–140

[3]   B. Ramesh, M. Jarke. "Towards Reference Models for Requirements Traceability" in *IEEE Transactions on Software Engineering*, 27(1). 2001. pp. 58–93

[4]   E. Magill, M. Calder (Eds.) *Feature Interactions in Telecommunications and Software Systems VI*. Amsterdam: IOS Press. 2000

[5]   D. Amyot, L. Charfi, N. Gorse et al. "Feature Description and Feature Interaction Analysis with Use Case Maps and LOTOS" in *Sixth International Workshop on Feature Interactions in Telecommunications and Software Systems FIW '00*. Glasgow, Scotland. 2000

[6]   R.J.A. Buhr. "Use Case Maps as Architectural Entities for Complex Systems" in *IEEE Transactions on Software Engineering, Special Issue on Scenario Management*, 24(12). 1998. pp. 1131–1155

[7]   T. Bolognesi, E. Brinksma. "Introduction to the ISO Specification Language LOTOS" in *Computer Networks and ISDN Systems*, 14. 1986. pp. 25–59

[8]   L. du Bousquet. "Feature Interaction Detection using Testing and Model-checking – Experience Report" in *World Congress in Formal Methods*. Toulouse, France. 1999

[9]   M. Calder, A. Miller. "Using SPIN for Feature Interaction Analysis – A Case Study" in M.B. Dwyer (Ed.) *Model Checking Software 8th International SPIN Workshop*. Toronto, Canada. 2001. pp. 143–162

[10]   M. Heisel, J. Souquieres. "Detecting Feature Interactions – A Heuristic Approach" in G. Saake, C. Turker (Eds.) *Proceedings 1st FIREworks Workshop*, Preprint 10/98. Fakultät für Informatik, University of Magdeburg. 1998. pp. 30–48

[11]   R. van der Straeten, J. Brichau. "Features and Feature Interactions in Software Engineering using Logic" in *Feature Interactions in Composed Systems*. 2001. pp. 79–88

[12]   C. Araces, W. Bouma, M. deRijke. "Description Logics and Feature Interaction" in *Proceedings of the International Workshop on Description Logics DL-99*. 1999

[13]   E. Ernst. "What's in a Name?" in *Feature Interactions in Composed Systems*. European Conference on Object-Oriented Programming ECOOP 2001. Workshop #8. 2001. pp. 27–33

[14]   E. Pulvermüller, A. Speck, J.O. Coplien et al. "Feature Interaction in Composed Systems" in (Eds.) E. Pulvermüller, A. Speck, J.O. Coplien et al. (Eds.) *Feature Interactions in Composed Systems*. European Conference on Object-Oriented Programming ECOOP 2001. Workshop #8. 200. pp. 1–6

[15]   A. Olsen, O. Færgemand, B. Møller-Pedersen et al. *System Engineering Using SDL-92*. 4th Edition, Amsterdam: North Holland. 1997

[16]   A. Metzger, S. Queins. "Early Prototyping of Reactive Systems Through the Generation of SDL Specifications from Semi-formal Development Documents" in *Proceedings of the 3rd SAM (SDL And MSC) Workshop*. Aberystwyth, Wales: SDL Forum Society; University of Wales. June, 2002

[17]   S. Queins. *PROBAnD – Eine Requirements-Engineering-Methode zur systematischen, domänenspezifischen Entwicklung reaktiver Systeme*. Ph.D. Thesis. Department of Computer Science. University of Kaiserslautern. 2002

[18]   R. Budde, K. Kautz, K. Kuhlenkamp, et al. *Prototyping: An Approach to Evolutionary System Development*. Berlin; Heidelberg: Springer-Verlag. 1992

[19]   A. Metzger. *Small Building Control Example*. On-line Development Documents. Department of Computer Science. University of Kaiserslautern. 2003
       http://wwwagz.informatik.uni-kl.de/d1-projects/ResearchProjects/InteractionExample/

[20]   ITU-T. *User Requirements Notation (URN) — Language Requirements and Framework*. Recommendation Z.150. Geneva: Iternational Telecommunication Union, Study Group 17. 2003
       http://www.usecasemaps.org/urn/

[21]   A. Mahdavi, A. Metzger, G. Zimmermann. "Towards a Virtual Laboratory for Building Performance and Control" in R. Trappl (Ed.) *Cybernetics and Systems 2002*. Vol. 1. Vienna: Austrian Society for Cybernetic Studies. 2002. pp. 281–286

[22]   G. Zimmermann. "Efficient Creation of Building Performance Simulators Using Automatic Code Generation" in *Energy and Buildings*, 34. Elsevier Science B.V. 2002. pp. 973–983

[23]   S. Queins, G. Zimmermann: *A First Iteration of a Reuse-Driven, Domain-Specific System Requirements Analysis Process*. SFB 501 Report 13/99. University of Kaiserslautern. 1999

[24]   S. Queins, M. Trapp, T. Brack et al. *Floor 32*. On-line Development Documents. Department of Computer Science. University of Kaiserslautern. 2002
http://wwwagz.informatik.uni-kl.de/d1-projects/ResearchProjects/Floor32/

[25]   A. Metzger et al. *PROTAGOnIST – Tools for Automated Software Development*. Web Site. Department of Computer Science. University of Kaiserslautern. 2003
http://wwwagz.informatik.uni-kl.de/staff/metzger/protagonist/

[26]   A. Metzger. "Requirements Engineering by Generator-Based Prototyping" in H. Alt, M. Becker (Eds.) *Software Reuse – Requirements, Technologies and Applications*. Proceedings of the International Colloquium of the SFB 501. Department of Computer Science. University of Kaiserslautern. 2003. pp. 25–35

*Feature Interactions in Telecommunications
and Software Systems VII*
D. Amyot and L. Logrippo (Eds.)
IOS Press, 2003

Representing New Voice Services
and Their Features

Kenneth J. TURNER

Computing Science and Mathematics, University of Stirling, Scotland
`kjt@cs.stir.ac.uk`

Abstract. New voice services are investigated in the fields of Internet telephony (SIP – Session Initiation Protocol) and interactive voice systems (VoiceXML – Voice Extended Markup Language). It is explained how CRESS (Chisel Representation Employing Systematic Specification) can graphically represent services and features in these domains. CRESS is a front-end for detecting feature interactions and for implementing features. The nature of service architecture and feature composition are presented. CRESS descriptions are automatically compiled into LOTOS (Language Of Temporal Ordering Specification) and SDL (Specification and Description Language), allowing automated analysis of service behaviour and feature interaction. For implementation, CRESS diagrams can be compiled into Perl (for SIP) and VoiceXML. The approach combines the benefits of an accessible graphical notation, underlying formalisms, and practical realisation.

1 Introduction

1.1 Motivation

The representation of services has been well investigated for traditional telephony and the IN (Intelligent Network). Feature interaction in these domains is also well researched. However the world of communications services has moved rapidly beyond these into new applications such as mobile communication, web services, Internet telephony and interactive voice services.

This paper concentrates on developments in new voice services. Specifically, it addresses Internet telephony with SIP (Session Initiation Protocol [15]) and interactive voice services with VoiceXML (Voice Extended Markup Language [5]). Such developments have been mainly driven by commercial and pragmatic considerations. Research and theory for their services have lagged behind practice. For these new application areas, the work reported here addresses questions like:

- What is a service, and how might it be represented?
- What service architecture is needed, and what does feature composition mean?
- How can services and features be analysed and implemented?
- What properties should services have, and how does feature interaction manifest itself?

1.2 Relationship to Other Work

The author's approach to defining and analysing services is a graphical notation called CRESS (Chisel Representation Employing Systematic Specification). CRESS was originally based on the Chisel notation developed by BellCore [1]. The author was attracted by the simplicity, graphical form, and industrial orientation of Chisel. However, CRESS has considerably advanced from its beginnings. Although it lives naturally in the communications world, CRESS is not tied to this. Indeed, CRESS supports the notion of plug-in domains. That is, the vocabulary used to talk about services is defined in a separate and modular fashion. Applying CRESS to a new application mainly needs a new vocabulary for events, types and system variables.

CRESS is also neutral with respect to the target language. It can, for example, be compiled into LOTOS (Language Of Temporal Ordering Specification [7]) and SDL. This gives formal meaning to services defined in CRESS, and allows rigorous analysis of services. For direct implementation, CRESS can also be compiled into SIP CPL (Call Processing Language), SIP CGI (Common Gateway Interface, realised in Perl) and VoiceXML.

CRESS is thus a front-end for defining, analysing and implementing services. It is not in itself an approach for detecting feature interactions. Rather it supports other detection techniques. LOTOS and SDL, for example, have been used in several approaches to detecting interactions. Among these the author's own approach is applicable [16], but so are a number of others like [4, 6]. CRESS is also a front-end for implementing features. It is possible to translate LOTOS and SDL to implementation languages such as C, thus realising features. However the CRESS tools also support more direct implementation through SIP CPL, SIP CGI and VoiceXML.

Industry seems to prefer graphical notations. Several graphical representations have been used to describe communications services. SDL (Specification and Description Language [9]) is the main formal language used in communications. Although it has a graphical form, SDL is a general-purpose language that was not designed particularly to represent services.[1] As a result, SDL service descriptions are not especially convenient or accessible to non-specialists. MSCs (Message Sequence Charts [8]) are higher-level and more straightforward in their representation of services. However neither SDL nor MSCs can readily describe the notion of features and feature composition.

Several notations have been specially devised for communications services. Among these, UCMs (Use Case Maps [2, 3]) and DFC (Distributed Feature Composition [19, 20]) are perhaps most similar in style to CRESS. Both lend themselves well to describing features and their composition, though the mechanisms are quite different from CRESS (plug-in maps for UCMs, pipe-and-filter composition for DFC). Both have been used successfully to model features and analyse their interactions. Unlike UCMs, CRESS allows both plug-in and triggered features. UCMs have been translated into LOTOS, but CRESS is explicitly designed to support translation into a number of target languages. DFC is primarily a software architecture, similar to work on ADLs (Architecture Description Languages [13]). However the means of feature composition has a natural graphical form.

The need for additional services was recognised early in SIP's development. SIP supports several mechanisms for user control of calls. CPL (Call Processing Language [11]) allows the user to manage call preferences, such as rejecting calls from certain addresses or forwarding calls based on caller and time of day. SIP also supports a web-like CGI (Common Gateway

---

[1]SDL does have something called a service, but this is not the usual kind of communications service.

Interface [12]) that is normally deployed in a server to intercept and act on SIP messages. A further solution is the SIP servlet, patterned after the idea of a Java servlet.

However, all of these are rather pragmatic. In the author's opinion, CPL is too high-level and (intentionally) too restricted to allow a full range of services to be created. On the other hand, SIP CGI is too low-level to allow services to be defined at an appropriate level of abstraction. Feature interaction in SIP has received limited attention, [10] being a notable exception. As will be seen, CRESS has been used to investigate services and features in SIP – their architecture, representation and analysis.

VoiceXML is aimed at IVR (Interactive Voice Response) systems. Being an application of XML, it is textual in form. However several commercial packages (e.g. Covigo Studio, Nuance V-Builder, Voxeo Designer) provide a graphical representation of VoiceXML. Some of these reflect the hierarchical structure of VoiceXML, while others emphasise the flow among VoiceXML elements. In the author's experience, these packages are (not surprisingly) very close to VoiceXML and do not give a sufficiently high-level view of VoiceXML services. More seriously, VoiceXML takes a pragmatic and programmatic approach. There is no way of formally analysing the correctness and consistency of a VoiceXML description. Interestingly, VoiceXML does not have the usual view of a feature (though it has roughly equivalent mechanisms). As will be seen, CRESS has been applied to VoiceXML services – graphical description, formalisation, feature composition, and analysis of services.

1.3   Overview of The Paper

The goal of this paper is to demonstrate that CRESS is a flexible notation of value in a number of domains. As background, Section 2 summarises the CRESS diagram format. Of necessity, the description is brief and condensed. Refer back to it when studying the diagrams that appear later. More on CRESS appears in [17, 18]. To complement previous CRESS work on IN services [17], Sections 3 and 4 show how CRESS can be applied to SIP and VoiceXML. It will be seen that SIP has affinities with IN telephony, but that VoiceXML supports very different kind of services. Nonetheless, the same notation can be used in all three domains for various purposes: representing services and their architecture, composing features, analysing features, and implementing features.

2   The CRESS Notation

At first sight, it might seem that CRESS is just another way of drawing state diagrams. However it differs in a number of important respects. State is intentionally implicit in CRESS because this allows more abstract descriptions to be given. It follows that arcs between nodes should not be thought of as transitions between states. Arcs may be guarded by event conditions as well as value conditions. Perhaps most importantly, CRESS has explicit support for defining and composing features. CRESS has plug-in vocabularies that adapt it for different application domains. This allows CRESS diagrams to be thoroughly checked for syntactic and static semantic correctness. CRESS is also neutral with regard to target language (whether formal or programmatic), and can be translated into a number of languages.

2.1 Diagram Elements

A CRESS diagram is a directed, possibly cyclic graph. The oval nodes contain events and their parameters (e.g. *StartRing A B*). Events may also occur in parallel ($|||$). Events may be signals (input or output messages) or actions (like programming language statements). A **NoEvent** (or empty) node can be useful to connect other nodes. An event may be followed by assignments optionally separated by '/' (e.g. *Event / Busy A* $<-$ *True* sets *Busy(A)* to true). A node is identified by a number that may be followed by a symbol to indicate its kind, e.g.:

- '$<$' denotes an input node, while '$>$' denotes an output node (required only if the same signal can be received as well as sent)
- '+' starts a template that is appended to a matching node, while '$-$' indicates a template that is prefixed (used when describing features triggered by other behaviour).

The arcs between nodes may be labelled by guards. These may be either value expressions (imposing a condition on the behaviour) or event handlers (that are activated by dynamic occurrence of a condition). Event handlers are distinguished by their names (e.g. **NoInput**, triggered when the user does not respond to a VoiceXML prompt). If a graph is cyclic, it may not be possible to uniquely determine the initial node. In such a case, an explicit **Start** node is given; this is otherwise implicit. Comments may take several forms: text between parallel lines, hyperlinks to files, and audio commentary.

A CRESS diagram may contain a rule box (a rounded rectangle) that defines things like:

- the diagram variables and their types (e.g. **Uses Address** *A* **Value** *V*); temporary variables are predeclared for each type (e.g. addresses *A0..A9*, messages *M0..M9*, values *V0..V9*)
- the other services or features on which the diagram depends (e.g. **Uses** ... */ PROXY*)
- assignments triggered by signals (e.g. *Off-hook P / Busy P* $<-$ *True*, meaning that when phone *P* goes off-hook then it is noted as busy)
- macros (e.g. *Free P* $<-$ $\sim$ *Busy P*, defining free as not busy for phone *P*)
- configuration information like the chosen features, translator options and user profiles.

Ultimately, CRESS deals with a single diagram. However it is convenient to construct diagrams from smaller pieces. A multi-page diagram, for example, is linked through connectors. More usefully, features are defined in separate diagrams that are automatically included by either cut-and-paste or by triggering.

2.2 Service Architecture

CRESS diagrams usually rely on some infrastructure. For example, IN billing is handled by a separate subsystem with which call control cooperates. Similarly, call processing in the IN collaborates with the SCP (Service Control Point). It is therefore normal for CRESS to define a framework for each application domain. Such a framework is specified using the same target language as the one to which diagrams are compiled (e.g. LOTOS, SDL, VoiceXML). Although the framework is specific to a domain and a target language, it is independent of the particular services or features deployed. The framework includes macro calls that activate the CRESS preprocessor. This automatically generates configuration-specific information such as the translated diagrams, user profiles, and feature-dependent data types.

A main CRESS diagram defines root behaviour. Although this may be the only diagram, CRESS supports feature diagrams that modify the root diagram (or other features).

A spliced (plug-in) feature is inserted into a root diagram by cut-and-paste. The feature indicates how it is linked into the original diagram by giving the insertion point and how it flows back into the root diagram. This may lead to nodes and guards being inserted, existing nodes and guard being replaced, and portions of the root diagram being deleted. This style of feature is appropriate for a one-off change to the original diagram. Suppose that a feature requires a PIN or password to be given before a call can proceed. This is a once-only action at the start of a call, and is appropriate for a spliced feature. The main disadvantage of this kind of feature is that it may have to duplicate large portions of the original diagram.

A template (macro) feature is triggered by some event in the root diagram. The triggering event is given in the first node of the feature. Feature execution stops on reaching a **Finish** (or empty) node. At this point, behaviour resumes from the triggering node in the original diagram. A template feature is realised statically by instantiating it using the parameters of the triggering event. The instantiated feature may be appended or prefixed to the triggering node. Since it is common for several features to be triggered by the same event, a number of feature instances may be chained. To help resolve certain categories of feature interaction at design time, CRESS supports priorities among features to control their order of application. For example after dialling, Abbreviated Dialling must be applied before Originating Call Screening. Some features are also cyclic, e.g. call forwarding may yield a new address that is itself subject to call forwarding. A loop back to the beginning of a feature is treated as a return to the start of the feature chain. This correctly handles billing, for example, if there are multiple call forwarding legs.

Although CRESS is mainly concerned with user services, it also supports ancillary aspects such as user profiles and billing. The services chosen by each user may be defined (e.g. call forwarding to a particular number, or call screening for particular callers). CRESS also supports billing. This might appear to be little more than logging the start and finish of calls. However, CRESS has explicit support for features like Charge Card, Freephone, Split Charging, and independent billing for each call leg.

2.3 Tool Support

The CRESS toolset has the form of a conventional compiler but is unusual in some respects. For portability it is written in Perl, comprising about 9000 lines of code. Java would also have been a reasonable choice, but Perl was chosen because of its excellent support for systems programming. Although it might have been desirable to use a parser generator (e.g. Antlr), parsing is only a small part of what CRESS has to do. A traditional compiler deals with textual languages. CRESS, however, deals with a graphical language. This creates interesting challenges, e.g. compiling cyclic rather than hierarchical constructs.

The CRESS toolset consists of five main tools, supported by seven underlying modules plus ancillary scripts. Internally the CRESS toolset comprises a preprocessor (that instantiates the specification framework), a lexical analyser (that deals with various diagram formats), a parser (that performs syntactic analysis), and several code generators.

Figure 1 summarises CRESS application domains and target languages. For interactive voice services, CRESS uses VoiceXML as the implementation language. For SIP CGI, CRESS makes specialised use of Perl as the primary implementation language. (In general, a SIP CGI script may be almost any executable code.) Preliminary work has been undertaken on compiling into SIP CPL, but this is possible for only very limited forms of feature diagram.

Domain	Target Language				
	LOTOS	SDL	VXML	CGI	CPL
IN	✓	✓	✗	✗	✗
IVR	✓	✓	✓	✗	✗
SIP	✓	✓	✗	✓	✓✗

Figure 1: CRESS Language Support

3 SIP Services

3.1 Introduction to SIP

SIP [15] is an Internet standard for controlling sessions. In the context of Internet telephony, SIP is used to control voice calls. However SIP is a more general-purpose protocol that can be used to establish multimedia sessions such as video-conferences. SIP has also been adopted for use in call control for 3G (third generation mobile communication).

SIP is patterned after HTTP. The main requests are *Invite* (propose a session), *Cancel* (abort a request), *Bye* (close a session) and *Ack* (acknowledge the response to an *Invite*). Responses are identified by numeric code. CRESS identifies specific responses such as *Busy-Here*, *Ringing* and *Success*, as well as classes of response like *Failed* (error) and *Terminal* (unrecoverable error).

To establish a session, the caller sends an *Invite* to the callee. An *Invite* response such as *Success* elicits an *Ack* from the caller. To close a session, either party sends *Bye* and waits for the response. *Cancel* may be used to abort a previous request, mainly to cancel an *Invite* because a call attempt has been abandoned. Once a session is established, data flows directly between the users. That is, SIP is concerned only with session control and not session content.

Although superficially a straightforward protocol, SIP contains hidden complexity in its use of header fields. The SIP standard is also vague about unusual cases like cancelling an *Invite*, or crossover situations like receiving *Cancel/Bye* after sending *Cancel/Bye*. The standard has little formal definition of SIP, so its formalisation via CRESS is a useful clarification.[2]

3.2 CRESS Root Diagrams for SIP

Ideally, SIP services would be described purely from a user standpoint (i.e. without internal details of the protocol). This, for example, is how IN services are often formalised. A CRESS description of this external SIP behaviour *is* available as a single root diagram.

However it is in the nature of SIP services that they build on key events in the protocol (e.g. receiving an *Invite* or sending a *Bye*). It was also a goal of using CRESS to translate SIP service descriptions into CGI (Perl) and CPL. CRESS is therefore obliged to have some knowledge of protocol activities. A reasonable compromise has been reached by defining an abstract protocol interface. For example, this interface hides timeouts and the processing of header fields. Instead, the essential aspects of the protocol are made visible: requests and their main parameters (URIs, i.e. user addresses), responses and their codes. This abstract protocol interface is easily mapped onto the actual protocol. In OSI terms, CRESS maps user service

---

[2]Or, at least, it formalises what the author thinks SIP is meant to do!

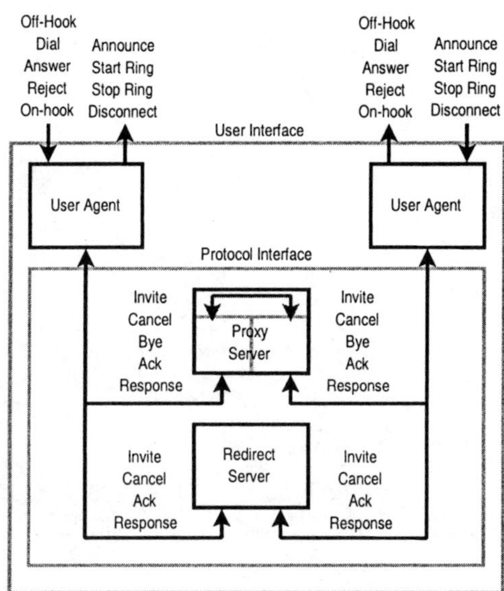

Figure 2: SIP Elements

primitives to the underlying SIP service primitives as shown in figure 2. For example, a user *Dial* request causes a SIP *Invite*.

SIP services can be deployed in three places: User Agents (which support the user interface to SIP), Proxy Servers (which relay and may manipulate requests), and Redirect Servers (which indicate how an *Invite* should be redirected to reach a user). As a consequence, the CRESS model of services exposes all three elements as shown in Figure 2. For familiarity, service primitives follow telephony terminology. Thus a user is said to go *Off-hook* or *On-hook*, though an actual phone may not be in use. Similarly a call results in *Start Ring* or *Stop Ring*, though this might be just a visual indication. A user may *Answer* or *Reject* an incoming call. *Announce* sends a call progress signal to the user. *Disconnect* means the other user has hung up.

The model of Figure 2 requires separate root diagrams for a User Agent, Proxy Server and Redirect Server; features modify each separately. For brevity, only a part of one root diagram is given in this paper. The terminating call side of a User Agent is shown in Figure 3; a separate diagram omitted here describes an originating call. The normal sequence of behaviour is as follows (with reference to node numbers in the figure). User *A* is the local callee, and user *B* is the remote caller. If an *Invite* is received by *A* from *B* (50), *A* starts ringing and *B* is notified of this (51). If *A* now answers (57), ringing stops and *B* is notified of successful setup (58). *B* acknowledges this with an *Ack* (59). Now both users can communicate. If *B* hangs up, a *Bye* is received (63) and *A* is notified of disconnection (64). *A* now hangs up (65) and *B* is informed of successful disconnection (66).

The rule box on the left defines how session status is maintained. For example, a user is noted as busy when going off-hook or as not busy when going on-hook. Observe that the notion of busy is *defined* and is not intrinsic to CRESS. For a conventional telephone or SIP phone, the definition of busy in Figure 3 is appropriate. However CRESS allows other definitions of busy. For example, a softphone is essentially never busy – a new call window

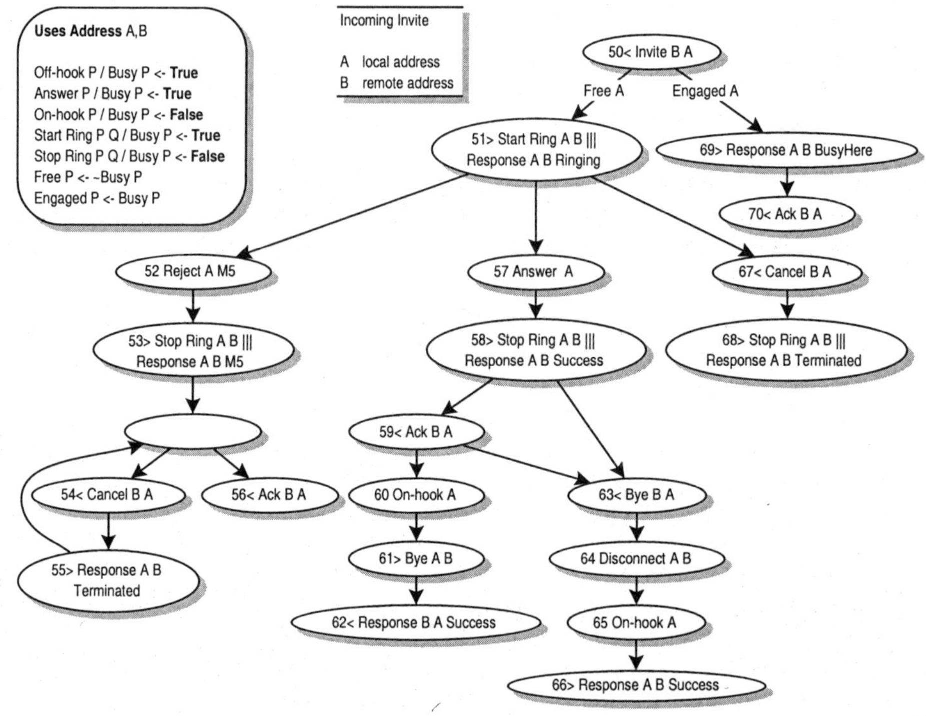

Figure 3: User Agent Root Diagram (Incoming *Invite*)

can be popped up at any time. A user may also be busy to certain callers (e.g. friends while at work) but free to others (e.g. managers). The time of the call might also influence whether the user is considered to be busy or not. Such factors can built into the definition of busy, or could be implemented as separate features. If appropriate, features may also define their own individual notion of busy.

3.3   *CRESS Feature Diagrams for SIP*

A SIP feature modifies the corresponding root diagram. Unfortunately, SIP features may vary in their definition according to where they are deployed. For example User Agents and Proxy Servers differ in their environment, and what they may initiate is also different.

Figure 4 shows a call forwarding template for a User Agent (left) and a Proxy Server (right). The '<' in template node 1 means it is triggered by input of an *Invite*, while the '+' means the template is appended to the matching node (e.g. figure 3 node 50). After substitution of *B* for parameter *P* and *A* for parameter *Q*, the template is copied and inserted in the root diagram. If the callee is busy but has a forwarding number (*ForwardBusy*), a User Agent reports that the callee has temporarily moved to this address (node 2). A Proxy Server, however, issues a new *Invite* (node 3) and handles the response (node 4).

Figure 5 shows that Terminating Call Screening is the same for a User Agent or a Proxy Server. If the caller is in the callee's screening list (*ScreenIn*), the call is declined (node 2).

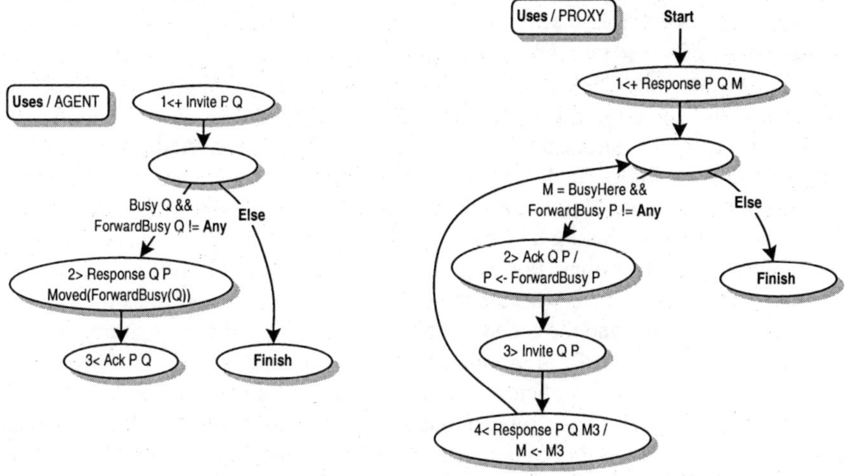

Figure 4: User Agent and Proxy Server Feature Diagrams (Call Forward Busy Line)

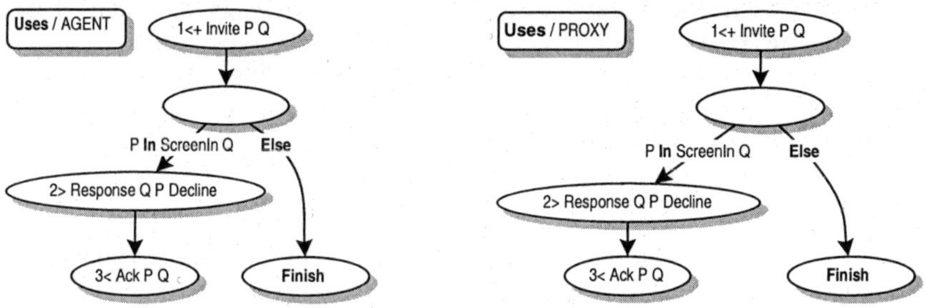

Figure 5: User Agent and Proxy Server Feature Diagrams (Terminating Call Screening)

3.4 CRESS as A Front-End for detecting SIP Interactions

When used for Internet telephony, SIP immediately lends itself to IN-like features. CRESS can readily be used to model SIP features such as Automatic Call-Back, Call Forwarding (several varieties), Call Screening (several varieties), Camp On Busy and Return Call.

As noted earlier, CRESS is a front-end for analysing and implementing features. Feature interactions are detected using separate techniques. For example, the author's approach in [16] is applicable to SIP. This considers an interaction to have occurred if a feature's behaviour changes in the presence of other features. Each feature is characterised by use-case scenarios derived from automatic simulation of the feature description in CRESS. The scenarios are represented as processes when using LOTOS or as MSCs when using SDL. The scenarios are not simply traces, but can include non-determinism, parallelism, and dependency on the presence of other features. A feature may be validated in isolation by running the scenarios on the feature combined with the corresponding root diagram. More usefully, a feature may be validated in combination with all other features. An interaction manifests itself as deadlock (because the feature cannot proceed as expected) or as non-determinism (because a triggering condition can lead to behaviour that is unexpected for the feature).

IN-like features such as the above can be readily represented using CRESS. Using the author's approach (or several others), it is easy to demonstrate feature interactions in SIP that are well known from the IN. For example, Call Forward Busy Line (Figure 4) interacts with Terminating Call Screening (Figure 5): trying to forward a call to a screened number will fail.

Certain kinds of IN interaction have different (or no) manifestations in SIP. As noted in [10], Internet telephony (including SIP) also introduces the possibility of new kinds of features and interactions. CRESS can discover the technical interactions in [10] (e.g. between Outgoing Call Screening and Call Forward). However, a number of the interactions in [10] concern user intentions. For example forking by a Proxy Server may lead to a call being picked up by voicemail, whereas the caller may prefer to wait for a person to answer. Such interactions are beyond the scope of CRESS (and indeed of most feature interaction approaches). The author is involved in separate work to tackle this [14]. The idea is to capture user intentions in the form of policies, and to perform resolution based on these.

Many features centre on busy, for example Automatic Call-Back, Call Forward Busy Line, Call Waiting and Camp On Busy. As noted already, busy may have a very different interpretation in SIP. As a result, busy-related features may not interact or may become irrelevant. Features related to call charging may also not apply. In a local or research environment, SIP calls are likely to be free. Features such as Charge Card, FreePhone and Split Charging are therefore irrelevant. However as SIP moves into a commercial phase, such features will become important. Calling Number Delivery is often a separate IN feature. However the address of the caller is in principle always available in SIP.

3.5   *CRESS Translation of SIP Services*

SIP CGI allows arbitrary features to be written. For example, a CGI script could query a database or invoke a complex algorithm. CRESS for SIP, however, describes features that perform only input, output, conditional tests and assignment. This is sufficient for many features but does not, of course, allow everything a CGI script might do. If it were necessary, SIP-specific actions could be included in CRESS much as VoiceXML-specific actions have been included. Whereas VoiceXML has a well-defined repertoire of actions that can be supported in CRESS, this is not feasible for SIP.

To give a feeling for how CRESS translates SIP services, the following LOTOS is an extract of what is generated for a User Agent incoming call (figure 3) as modified by Call Forward Busy Line (figure 4). After an *Invite*, the *Busy* and *ForwardBusy* values for *A* are determined. If *A* is busy but has a forwarding address, a *Moved* response is sent with this forwarding address. Following the *Ack* from *B*, the session instance ends. As seen below, the CRESS translator automatically adds comments so the LOTOS can be related directly back to diagrams. The diagram label *AGENT_CFBL_1* refers to the User Agent's Call Forward template, instance 1.

```
            Recv !Invite ?B:Address ?A:Address;                        (* AGENT input 50 *)
            Stat !Read !Busy !A;                                        (* read status *)
            Stat !Read ?Busy_A:Bool;
            Stat !Read !ForwardBusy !A;                                 (* read status *)
            Stat !Read ?ForwardBusy_A:Address;
            (
                [Busy_A And (ForwardBusy_A Ne AnyAddress)] ->           (* condition valid? *)
                Stat !Read !ForwardBusy !A;                             (* read status *)
                Stat !Read ?ForwardBusy_A:Address;
```

```
        Send !Response !A !B !Moved(ForwardBusy_A);        (* AGENT_CFBL_1 output 2 *)
        Recv !Ack !B !A;                                   (* AGENT_CFBL_1 input 3 *)
        Stop                                               (* end of behaviour *)
    []
      [Not (Busy_A And (ForwardBusy_A Ne AnyAddress))] ->  (* condition invalid *)
        ...
    )
```

4 VoiceXML

4.1 Introduction to VoiceXML

VoiceXML [5] derives from earlier work on scripting languages for interactive voice services. VoiceXML is designed to support what users wish to do in a call – talk, as opposed to choosing selections by using a keypad. VoiceXML is receiving impetus from widespread use of mobile telephony (where a user on the move may not have web access). The need for access by the partially sighted or disabled is also a strong motivation for voice services.

VoiceXML is a mixture of the declarative and the procedural, the event-driven and the sequential. The underlying model is that the user completes fields in forms (or menus) by speaking in response to prompts. Each field is associated with a variable that is set to the user's input, using speech recognition to extract digital data. Some actions may be governed by a condition or a count that specifies when the action is permitted. For example a different prompt may be given on the third attempt at input, or a field may be selected only when some condition is true. VoiceXML also supports a hierarchical event model. A script may throw an event, aborting the current behaviour and activating a matching event handler.

The goal of using CRESS with VoiceXML is to define the key aspects of an interactive voice service. The advantages of CRESS over using VoiceXML directly are:

- Services are represented at a more abstract level. VoiceXML is very close to the realisation of a service. As a result, it is easier to grasp the essence of a service described in CRESS.

- There is no formal definition of VoiceXML. Indeed, some concepts in VoiceXML are only vaguely described (e.g. the meaning of events) and some are defined loosely (e.g. the semantics of expressions and variables). CRESS thus contributes to a more precise understanding of VoiceXML.

- A large VoiceXML script typically consists of many documents with many parts. It can be difficult to check whether the script is self-consistent, e.g. will not loop or end prematurely. As far as the author can tell, VoiceXML in practice is developed by manual debugging. CRESS gives the immediate benefit of translation to a formal language (LOTOS, SDL). The resulting specification can be rigorously analysed, e.g. automated techniques can be used to detect unspecified receptions, unreachable states, deadlocks and livelocks.

Speech synthesis markup can be included in a prompt, e.g. for emphasis or to spell out a word. The markup is preserved on translation to VoiceXML, but ignored on translation to LOTOS or SDL. Variable values may also be interpolated, using *$variable* to say the value of this variable. As a special case, *$enumerate* is used to speak the options of the current field. VoiceXML can also interpolate variable values, but the syntax is more awkward than in CRESS.

In practice, VoiceXML applications are often written as a number of documents containing a number of forms. This is the most natural form of modularity in VoiceXML. However

CRESS	VoiceXML	Interpretation
Audio *message*	<**audio**> *message*	Speak message
Clear *variables*	<**clear**> with namelist *variables*	Reset prompt counter, undefine variables
Option *variable prompt options*	<**field**> name *variable*, <**prompt**>, <**option**>s	Start new field, prompt for input, analyse input using options, set variable from input
Prompt *message*	<**prompt**> *message*	Speak prompt
Reprompt	<**reprompt**>	Re-process current form, usually causing most recent prompt to be re-issued
Request *variable prompt grammar*	<**field**> name *variable*, type *grammar*, <**prompt**>	Start new field, prompt for input, analyse input using grammar, set variable from input
Retry	Undefine current <**field**> variable, <**reprompt**>	Restart current form, re-inputting most recent field
Submit *URI variables*	<**submit**> to *URI* the namelist *variables*	Send values to web server URI (usually an executable script)

Figure 6: CRESS-VoiceXML Correspondence

a VoiceXML application can be considered as a single document with a single form, and this is how it is represented in CRESS. The fields of a form can be mimicked as separate sections or pages of a CRESS diagram, using connectors to join them. For a large application this is convenient and more modular. However for a small application it is sufficient to represent the form as a single integrated whole. For this reason, fields are deliberately not prominent in CRESS and are instead introduced implicitly.

For those unfamiliar with VoiceXML, Figure 6 outlines its main capabilities. For those familiar with VoiceXML, this figure indicates the correspondence with CRESS. In a number of cases, an optional condition or count may be given after the action. For example, a different prompt can be issued for the fourth attempt[3] at entering a field: **Prompt** *"State your name"* 4.

Actions appear in CRESS diagram nodes. Plain arcs between nodes are used for sequences of actions. Arcs may also be qualified by guards that are value expressions or event handlers. Events include **Error** (run-time script error), **Exit** (script exited), **NoInput** (no user response to prompt) and **NoMatch** (user response did not match expected form of input). These are all shorthands for the more general form of **Catch** plus an event list. **Filled** acts like an event handler, though it is not treated as such in VoiceXML. If the user responds appropriately to a prompt, the input is stored in the field variable and the **Filled** 'event' occurs.

4.2   *CRESS Root Diagrams for VoiceXML*

CRESS is not a direct graphical representation of VoiceXML. This would be pointless since most commercial tools for VoiceXML do this anyway. In fact the structuring mechanisms of CRESS are completely different from those of VoiceXML. Both support actions, sequences and alternatives. The flow of control in CRESS is quite visible; in VoiceXML it can be hard to determine, because it is sometimes implicit (e.g. transitioning to the next field on completion of the current one) and sometimes buried (e.g. an embedded **GoTo**). CRESS supports cyclic structures such as loops in a diagram. These have to be coded indirectly in VoiceXML, so

---

[3]Actually the prompt is issued if the count is $>= 4$, but less than the next highest prompt count for that field.

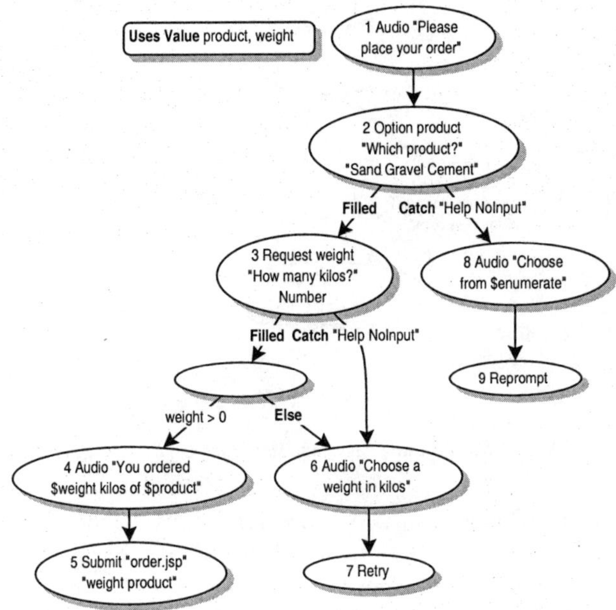

Figure 7: VoiceXML Root Diagram for Quarry Ordering Application

the CRESS structure is clearer. Although VoiceXML supports a **GoTo** which appears to be equivalent, this can be used only for transitions to another field or document. As a result, it is not obvious how to translate a directed cyclic graph like a CRESS diagram into VoiceXML.

CRESS expects to have a definition of root behaviour. There appears to be something similar in VoiceXML – an application root document. However this is very restrictive, and may contain only variables, event handlers and elementary definitions that are common to the documents of a VoiceXML application. As a result, a CRESS root diagram is taken to be the core behaviour of a VoiceXML application. This is not unique, in the way that POTS is the obvious choice for the IN or a User Agent for SIP. Every VoiceXML application therefore has its own root diagram.

For concreteness, Figure 7 shows a root diagram for a VoiceXML application. It is supposed that the hypothetical Quarry Inc. requires interactive telephone ordering of its products. The description invites the caller to order a *product* (sand, gravel, cement) and the required *weight*. These items are then submitted to the *order.jsp* Java servlet. If the user asks for help (simply by saying 'Help') or says nothing, an explanation is given and the user is re-prompted. In the case of *weight*, the user is reprompted if the value is not positive. **Retry** is used to clear the value entered for *weight*, otherwise the field will be ignored on the reprompt because it has already been filled. (This is how VoiceXML behaves. The programmer must force a field to be re-entered if it has already been filled.)

4.3 *CRESS Feature Diagrams for VoiceXML*

VoiceXML lacks the telephony concept of feature as a behaviour that may be triggered by some condition. The nearest equivalent in VoiceXML is a subdialogue (resembling a subrou-

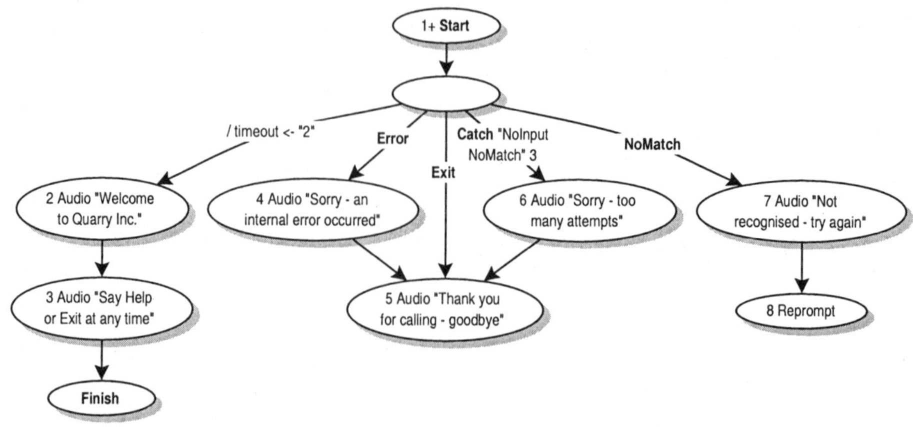

Figure 8: VoiceXML Feature Diagram to introduce Quarry Inc. Applications

tine). A subdialogue may have parameters and return results. Subdialogues are executed in an independent interpreter context, making it difficult to share certain information (such as prompt count). This limits the value of subdialogues as features. In VoiceXML, the best that can be done is to *explicitly* invoke some code as a 'feature'. This means that any such 'feature' needs to be written into the original code. It is not possible to invoke features automatically in the way that is common with the IN (or SIP). A VoiceXML programmer would probably regard this a good thing since there are no hidden surprises. Triggered features have, however, proven their worth in telephony. They are therefore supported by CRESS even though the concept of feature is unknown to VoiceXML.

As examples of desirable features, suppose that Quarry Inc. has a range of applications besides the ordering application in Figure 7. There might, for example, be separate applications to modify an order, pay an account, or change the delivery address. It would be desirable to ensure a consistent VoiceXML treatment of these applications. For example, there should be the same default handling of events and a common introduction to the applications. It would also be worthwhile to request confirmation before anything is submitted to a web server.

Figure 8 defines an introductory environment to modify any root diagram. Common handlers are defined for various events. The feature is placed just after the **Start** node in the root diagram (implicit before Figure 7 node 1). In the absence of event handlers like those in Figure 8, platform-defined handlers are used that may not be suitable in general. Although an application is likely to deal with **NoInput** and **NoMatch** on a per-field basis, figure 8 ensures that after three such failures the user is disconnected. Figure 8 shows that general VoiceXML properties can be defined; here the timeout for no input is set to two seconds (*timeout* $<- 2$). Welcome messages are also spoken before executing application-specific code.

Figure 9 defines a confirmation feature that will ask the user to proceed before submitting information to a web server. This feature is triggered by a **Submit** action, but executed before it (as indicated by the '−' in the first template node number). If the user says 'yes', execution continues with submission. Otherwise, the current fields are cleared and the user is reprompted.

Even small VoiceXML features can be useful. Figure 10 shows one that inhibits input timeout (value 0), and one that prevents prompts from being interrupted (so-called barge-in).

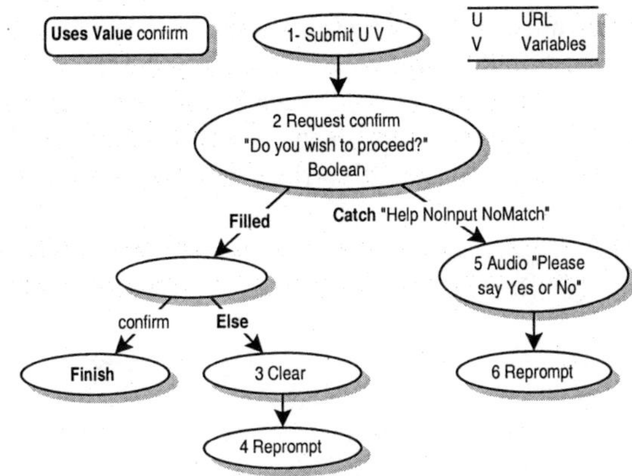

Figure 9: VoiceXML Feature Diagram for Confirmation

Figure 10: VoiceXML Feature Diagrams for No Timeout and No Interruption

4.4 CRESS as A Front-End for detecting VoiceXML Interactions

As for SIP, CRESS is merely a way of representing services and features. Separate detection techniques must be used on the chosen formalism for the diagrams. In point of fact, CRESS is perhaps most useful for checking the integrity of a VoiceXML description (freedom from deadlock, etc.). Just as features are foreign to VoiceXML, so is feature interaction. However several general categories of 'feature interactions' can be identified:

- Platform properties may be defined hierarchically. For example, the generic timeout in Figure 8 may be overridden within a field by some feature. From the user's point of view, this could lead to a small but observable change in behaviour.

- Two features may change an application variable inconsistently, leading to an interaction.

- Event handlers are defined in a hierarchy. When an event occurs, the VoiceXML interpreter looks upwards in the hierarchy for the appropriate handler. For example, consider Figures 7 and 8. If there is no input in response to the *product* prompt, execution follows the field handler (figure 7 node 8). However after three failures to input, the generic handler will be invoked (figure 8 node 6). A consequence of this is that features may unexpectedly override the usual handling of an event. From the user's point of view, a certain combination of features could result in different behaviour.

- A more subtle interaction can arise if several input grammars are active at once. User input may therefore be parsed in a different way if certain features are combined.

- Another indirect interaction arises with use of DTMF (Dual-Tone Multi-Frequency) responses. VoiceXML allows these in place of voice input, e.g. 1 might select the first choice from a menu. By default, DTMF digits are allocated in sequence to choices. If a feature introduces another choice earlier in the menu, the numbering will be completely altered.

All these cases lead to unexpected changes in behaviour when certain features are present, so feature interaction is detected as normal in CRESS. VoiceXML may also be involved indirectly in conventional feature interactions since scripts are allowed to initiate phone calls. These may suffer from the usual telephony interactions, for example call screening might interfere with call forwarding. Since such interactions are external to VoiceXML and therefore to CRESS, the VoiceXML diagrams themselves will not help to find the problems.

4.5   CRESS Translation of VoiceXML Services

VoiceXML is a large language embedded in an even larger framework. For example, VoiceXML includes support for ECMASCRIPT (JavaScript). It also supports complex grammars for speech recognition and markup for speech synthesis. VoiceXML is integrated with other technologies such as database access and web servers. It is not feasible to represent the entirety of such voice-based services. Instead, CRESS concentrates on the essential aspects of VoiceXML control. Limited support is provided for ECMASCRIPT (specifically for numerical, string and logical expressions). External aspects such as databases and the web are outside CRESS.

CRESS focuses on VoiceXML control. Special parameters relevant only to a VoiceXML interpreter can be given in a diagram at the end of an action. They are copied literally when CRESS is converted to VoiceXML, but are ignored for translation to other target languages. For example, a diagram usually just contains audio prompt such as **Audio** "State your name". However optional VoiceXML parameters may also be given, such as the URI of a source sound file and a timeout (2 sec.) for fetching this:

Audio "State your name" src="http://www.server.org/name.wav" fetchtimeout="2"

VoiceXML allows application-specific speech grammars to be defined. It is not practicable to translate these into, say, LOTOS or SDL. Instead, CRESS supports only standard VoiceXML grammars such as *boolean*, *number* and *time*. A CRESS specification framework includes the ability to parse such inputs.

To give a feeling for how CRESS translates IVR services into VoiceXML, the following corresponds part of the quarry ordering application (figure 7) where it is modified by the confirmation feature (figure 9). The CRESS **Request** becomes a VoiceXML field that fills in the *confirm* value. If this is assigned *true*, the order values are submitted to the server. Otherwise the form that invoked confirmation is cleared and the user is prompted for new values. If the user asks for help, does not say anything or says something invalid, an event handler catches this and reprompts the user. The diagram label *CONFIRM.1* means the *Confirm* template, instance 1.

```
<field name='confirm' type='boolean'>       <!-- CONFIRM.1 node 2 field 'confirm' -->
  <prompt>                                   <!-- CONFIRM.1 node 2 prompt -->
    'Do you wish to proceed?'
  </prompt>
  <catch event='help noinput nomatch'>                        <!-- catch event -->
    <audio>                                  <!-- CONFIRM.1 node 5 audio -->
      'Please say Yes or No'
    </audio>
    <reprompt/>                              <!-- CONFIRM.1 node 6 to form top -->
  </catch>                                                     <!-- end catch -->
  <filled>                                                     <!-- filled event -->
    <if cond='confirm'>                                        <!-- check confirm -->
```

```
        <submit expr='order.jsp' namelist='weight product'/>     <!-- ORDER node 5 to server -->
        <exit/>                                                   <!-- exit script -->
      <else/>                                                     <!-- else after confirm -->
        <clear/>                                                  <!-- CONFIRM.1 node 3 clear -->
        <reprompt/>                                               <!-- CONFIRM.1 node 4 to form top -->
      </if>                                                       <!-- else after confirm -->
    </filled>                                                     <!-- end filled -->
  </field>                                                        <!-- end field -->
```

5 Conclusions

It has been shown that CRESS is a flexible notation that can describe a variety of voice services
and features – the IN in previous work, and now SIP and VoiceXML. SIP lends itself to a
telephony treatment, so many IN-like features can be described and many IN-like interactions
can be detected. As has been seen, VoiceXML is rather different in character. Nonetheless
VoiceXML services can usefully be described in CRESS, and a meaningful interpretation can
be given of features in this context.

In all cases, CRESS is the front-end that describes services/features, composes them, and
translates them to a target languages for analysis or implementation. CRESS thus separates
representation from analysis, and supports a variety of specification languages. CRESS complements existing interaction detection techniques, enabling them to be applied in new areas.

The plug-in architecture of CRESS has now been demonstrated in three different domains.
Although these are all examples of voice services, the approach is generic and should be
relevant to non-voice applications such as web services. For example, it is hoped in future to
apply CRESS to services for WSDL (Web Services Description Language).

Acknowledgements

The author is grateful to Stephan Reiff-Marganiec (University of Stirling) for discussions on
VoiceXML, and for reviewing a draft of this paper.

References

[1] A. V. Aho, S. Gallagher, N. D. Griffeth, C. R. Schell, and D. F. Swayne. SCF3/Sculptor with Chisel:
Requirements engineering for communications services. In K. Kimbler and W. Bouma, editors, *Proc. 5th.
Feature Interactions in Telecommunications and Software Systems*, pages 45–63. IOS Press, Amsterdam,
Netherlands, Sept. 1998.

[2] D. Amyot, R. J. A. Buhr, T. Gray, and L. M. S. Logrippo. Use case maps for the capture and validation
of distributed systems requirements. In *Proc. 4th. IEEE International Symposium on Requirements Engineering*, pages 44–53. Institution of Electrical and Electronic Engineers Press, New York, USA, June
1999.

[3] D. Amyot, L. Charfi, N. Gorse, T. Gray, L. M. S. Logrippo, J. Sincennes, B. Stepien, and T. Ware. Feature
description and feature interaction analysis with use case maps and LOTOS. In M. H. Calder and E. H.
Magill, editors, *Proc. 6th. Feature Interactions in Telecommunications and Software Systems*, pages 274–
289, Amsterdam, Netherlands, May 2000. IOS Press.

[4] M. H. Calder, E. H. Magill, and D. J. Marples. A hybrid approach to software interworking problems: Managing interactions between legacy and evolving telecommunications software. *IEE Software*, 146(3):167–
180, June 1999.

[5] V. Forum. *Voice eXtensible Markup Language*. VoiceXML Version 1.0. VoiceXML Forum, Mar. 2000.

[6] Q. Fu, P. Harnois, L. M. S. Logrippo, and J. Sincennes. Feature interaction detection: A LOTOS-based approach. *Computer Networks*, 32(4):433–448, Apr. 2000.

[7] ISO/IEC. *Information Processing Systems – Open Systems Interconnection – LOTOS – A Formal Description Technique based on the Temporal Ordering of Observational Behaviour*. ISO/IEC 8807. International Organization for Standardization, Geneva, Switzerland, 1989.

[8] ITU. *Message Sequence Chart (MSC)*. ITU-T Z.120. International Telecommunications Union, Geneva, Switzerland, 2000.

[9] ITU. *Specification and Description Language*. ITU-T Z.100. International Telecommunications Union, Geneva, Switzerland, 2000.

[10] J. Lennox and H. Schulzrinne. Feature interaction in internet telephony. In M. H. Calder and E. H. Magill, editors, *Proc. 6th. Feature Interactions in Telecommunications and Software Systems*, pages 38–50, Amsterdam, Netherlands, May 2000. IOS Press.

[11] J. Lennox and H. Schulzrinne, editors. *CPL: A Language for User Control of Internet Telephony Services*. Internet Draft CPL-05. The Internet Society, New York, USA, Nov. 2001.

[12] J. Lennox, H. Schulzrinne, and J. Rosenberg, editors. *Common Gateway Interface for SIP*. RFC 3050. The Internet Society, New York, USA, Jan. 2001.

[13] N. Medvidovic and R. N. Taylor. A framework for classifying and comparing architecture description languages. In *Proc. 6th. European Software Engineering Conference/Proc. 5th. Symposium on the Foundations of Software Engineering*, pages 60–76, Zurich, Switzerland, Sept. 1997.

[14] S. Reiff-Marganiec and K. J. Turner. Use of logic to describe enhanced communications services. In D. A. Peled and M. Y. Vardi, editors, *Proc. Formal Techniques for Networked and Distributed Systems (FORTE XV)*, number 2529 in Lecture Notes in Computer Science, pages 130–145. Springer-Verlag, Berlin, Germany, Nov. 2002.

[15] J. Rosenberg, H. Schulzrinne, G. Camarillo, A. Johnson, J. Peterson, R. Sparks, M. Handley, and E. Schooler, editors. *SIP: Session Initiation Protocol*. RFC 3261. The Internet Society, New York, USA, June 2002.

[16] K. J. Turner. Validating architectural feature descriptions using LOTOS. In K. Kimbler and W. Bouma, editors, *Proc. 5th. Feature Interactions in Telecommunications and Software Systems*, pages 247–261, Amsterdam, Netherlands, Sept. 1998. IOS Press.

[17] K. J. Turner. Formalising the CHISEL feature notation. In M. H. Calder and E. H. Magill, editors, *Proc. 6th. Feature Interactions in Telecommunications and Software Systems*, pages 241–256, Amsterdam, Netherlands, May 2000. IOS Press.

[18] K. J. Turner. Modelling SIP services using CRESS. In D. A. Peled and M. Y. Vardi, editors, *Proc. Formal Techniques for Networked and Distributed Systems (FORTE XV)*, number 2529 in Lecture Notes in Computer Science, pages 162–177. Springer-Verlag, Berlin, Germany, Nov. 2002.

[19] P. Zave. Architectural solutions to feature-interaction problems in telecommunications. In K. Kimbler and W. Bouma, editors, *Proc. 5th. Feature Interactions in Telecommunications and Software Systems*, pages 10–22. IOS Press, Amsterdam, Netherlands, Sept. 1998.

[20] P. Zave and M. Jackson. New feature interactions in mobile and multimedia telecommunications services. In M. H. Calder and E. H. Magill, editors, *Proc. 6th. Feature Interactions in Telecommunications and Software Systems*, pages 51–66, Amsterdam, Netherlands, May 2000. IOS Press.

*Feature Interactions in Telecommunications
and Software Systems VII*
D. Amyot and L. Logrippo (Eds.)
IOS Press, 2003

A Framework on Feature Interactions in Optical Network Protocols

Caixia CHI, Dong WANG, Ruibing HAO
Lucent Technologies, Bell Labs

Abstract. This paper studies the feature interaction problems in optical network protocols. A framework is defined, in which the basic services needed by optical network protocols are discussed, and the feature interaction problems in optical network protocols are analyzed. Especially, the interactions occurred in or between the control and management protocols of optical networks are firstly studied in this paper and some guidelines to detect and resolve these problems are proposed.

1   Introduction

Nowadays fiber optics is being widely used for establishing telecommunication networks as well as data communication networks throughout the world because of its high bandwidth and very low received bit error rate [1]. ASON (Automatic Switched Optical Network) is the core optical networks with high-speed, high-bandwidth, flexibility and reliability to meet the demands for capacity, speed and survivability from applications such as video, audio and the other emerging ones [2].

ASON consists of a number of Optical Cross-connects (OXCs), each of which consists of a switching fabric and a separate control plane processor. The control plane processors in an ASON, communicate with each other over data communication network whose control and management tend to be IP centric. Switching fabrics of OXCs forms the data plane of ASON, and the topology of control plane and data plane need not be identical. The architecture of ASON is different from the traditional telephone networks in that its control plan can run over IP networks which provides transparent connectivity between two nodes in the network, while equipments in traditional networks can only communicate with their physical neighbors. The differences between IP network and traditional telephone network described in [3] also exist between ASON and the traditional telephone network, so does the impact of these differences on the feature interaction problems.

But different from the traditional Internet, the services of optical networks are provided by the cooperation of both control plane and data plane which can be physically separated with each other and have fairly different characteristics and features. The IP-centric control plane and connection-oriented data plane make the operation of ASON have the characteristics of both PSTN and Internet software and become more complicated. The standardization of optical control plane protocols is still undergoing and most of the services provided by ASON are still proprietary, the feature interaction problems in optical network have not been studied yet. In this paper, we define a framework for feature interaction problems in optical network management protocols.

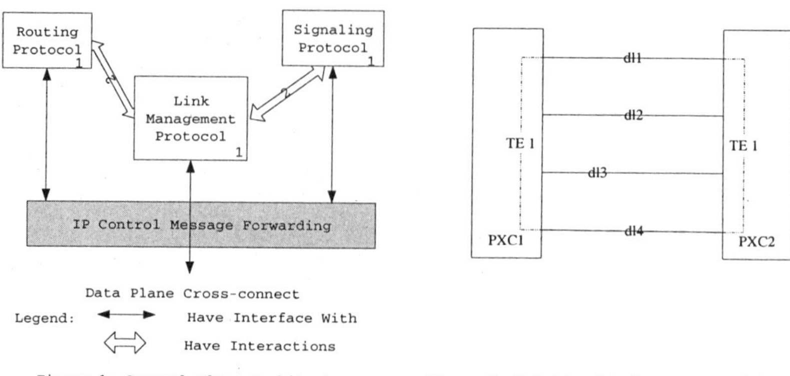

Figure 1: Control Plane Architecture　　　Figure 2: Relationship between TE Link and Data Link

This paper is organized as follows: section 2 introduces the framework of optical network control and management, in section 3, the feature interactions in optical network are analyzed. The techniques to detect and resolve the feature interaction problems in optical network are introduced in section 4 and this paper concludes in section 5.

2　Model of System and Services

In this section, we define the kernel functions provided by optical networks at first, then briefly introduce the intelligent features of optical network protocols and their relationship.

2.1　Kernel Functions of ASON

In ASON, the data plane and control plane can be separated and each has its own topology such that ASON is taken as two layers: optical transport layer and protocol control layer. Optical transport layer provides the management of the hardware equipment, handles the signals from the hardware and generates indication to upper layer protocols. Protocol control layer increases the intelligence of ASON, provides the functions such as resource discovery, connection management, topology/state dissemination.

Different transport equipment has different capabilities such that services provided by data plane to control layer protocols are different. The generic data link layer and physical layer features have been defined in ITU-T X.211 and X.212, which can be defined as the basic services provided by data plane. The capabilities of SONET/SDH equipment, specified by ITU-T G.780 series, are also the basic services provided by data plane when analyzing the control and management protocols of optical networks. Of course, only part of these capabilities are needed by a specific protocol.

Figure 1 illustrates the simple architecture of a control plane node. It shows that routing and signaling protocols only need the kernel functions provided by IP control module, but protocols that fulfill the resource management functions need to interact with the cross-connect hardware. So the kernel functions needed by optical network management protocols and provided by optical transport layer and protocol control layer are different with respect to the protocols for different purpose.

2.2 Features of Optical Network Protocols

Given the kernel functions of ASON, which are provided by the IP control message forwarding and the data plane, lots of features can be developed to increase the intelligence of ASON. As indicated in figure 1, control and management features of ASON are provided by three kinds of protocols, that is, routing protocol, link management protocol and signaling protocol.

The routing protocol provides the functions to collect, disseminate network resource and topology information, which is maintained by link management protocol. The link management protocol manages local resource and summarizes it for traffic engineering (TE) purpose. The TE link information can be used by routing protocols to generate their Link State Advertisements (LSAs); it also maps TE links and control channels which can be used by signaling protocols to set up a Label Switched Path (LSP). Based on the topology or network resource information learned by routing protocol, a path can be calculated for a connection request and signaling protocol is used to create, maintain, restore, and delete the connection. As a result of processing of the signaling protocol messages, the state of local resource can be changed, that is, connections can be established or removed, corresponding to allocation or deletion of resources. In the whole architecture, link management protocol provides the fundamental functions to support routing and signaling protocols.

All the features of optical network protocols are being standardized by IETF under the umbrella of GMPLS framework [5]. Link Management Protocol (LMP)[6], enhancements to OSPF/IS-IS, and enhancements to RSVP/CR-LDP are the most promising protocols to provide these features to optical network.

3 Feature Interactions in Optical Network Protocols

The ever-increasing features of optical network protocol and its openness in control plane invariably result in many new feature interaction problems. As indicated in figure 1, feature interactions of optical network protocols can occur in a single protocol as shown at location 1, or between different protocols as shown at location 2. The latter case is usually referred to as protocol interactions and not strictly be *feature* interactions [3]. Intelligence of ASON is provided by the cooperation of several protocols, such that all the features provided by protocols are also features of ASON intelligence. From the level of ASON intelligence, we define that when the presence of a feature of a protocol results the behavior of another protocol different from what is desired when they are designed independently, feature interactions between different protocols occur.

Feature interactions in ASON is classified into two categories: feature interactions between the features of a single protocol and feature interactions between the features of different protocols. For each category, we illustrates the scenarios that will result in the feature interactions in the following. Table 1 summaries the reasons of these feature interactions and possible resolution techniques.

3.1 Interactions Between Features of a Single Protocol

To reduce the number of protocols running in networks, there is a tendency to extend existing protocols to provide new services rather than to develop new protocols. So more and more features can be developed via the same suite of messages or some new messages are added to

Table 1: Feature Interactions in Optical Network Protocols and Their Resolution Techniques

Resolution Reasons of FIs	Policy Definition 4.2	Behavior Restriction 4.3	Advanced Scheduling 4.4	On Line Control 4.5
Coexistence of incompatible multiple capabilities 3.1.1	✓			✓
Requirement violation 3.1.2		✓		
Race condition 3.2.1			✓	
Single trigger event and multiple inconsistent responses 3.2.2	✓			
Performance degradation 3.2.3	✓			

the existing protocol to provide new functions. These features of a single protocol can interact with each other.

3.1.1 Feature Interactions in Label Distribution Protocol

Label distribution protocol defines a set of procedures by which one Label Switched Router (LSR) informs another of the meaning of labels used to forward traffic between and through them [4].

LDP provides two label advertisement modes, Downstream on Demand mode and Downstream Unsolicited mode, for each interface on an LSR to initiate mapping requests and mapping advertisements. It is possible for neighboring LSRs with different advertisement modes to get into a live lock situation where everything is functioning properly, but no labels are distributed.

3.1.2 Feature Interactions in Link Management Protocol

The features of LMP include: control channel management, link property correlation, link connectivity verification, and fault management. It is required that when a TE link becomes up, any data link in up/free state in the TE link can be allocated to a request. The requirement can be satisfied when only link property correlation and control channel management features are invoked in LMP. But when link connectivity verification process is added, the property fails to hold sometimes.

For example, in figure 2, four data links are configured to a TE link. The initial state of TE link is Init, which is changed to up when at least one data link in the TE link becomes up. When link connectivity verification is not invoked, both TE link and data link state are changed to up when link property correlation completes its operation. But when link connectivity verification process is invoked in the nodes, link connectivity verification process can change the state of data link to up once it finishes verification process. For example, in figure 2, the state of dl1 is changed to up when link connectivity verification finishes its operation, so the state of TE link 1 is also changed to up. In this case, dl1 can not be allocated to a request though dl1 and the TE link including dl1 are both in up state, because the properties of dl1 has not been exchanged yet, and the incompatible properties in both sides of the dl1 can result in the failure of the path set up. The inconsistent operation of link connectivity

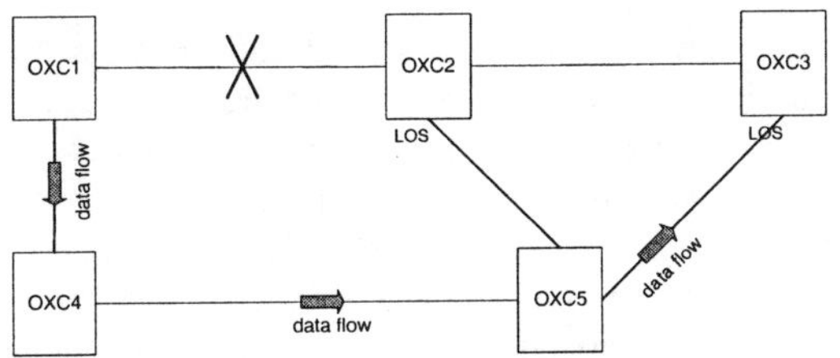

Figure 3: Interactions between Fault Detection and Path Restoration.

verification and link property correlation makes them interact and result in the violation of the requirement.

3.2   Interactions between Features of Different Protocols

3.2.1   Feature Interactions between Path Setup and Link Connectivity Verification

Both LMP and LDP can change the state of a data link between two OXCs. For an up and free data link, LDP can request LMP to change its state to be allocated once a lightpath is set up along the data link; link management protocol can change its state to test if it needs to monitor the status of the link. Interaction occurs when LDP gets an up/free data link whose state is changed to test before LDP succeeds its operation and changes the data link to be allocated.

The communication of two protocols in optical control plane is usually asynchronous and interaction can be resulted by the race conditions among the operations of different protocols. To be more clarified, consider that LDP and LMP are two processes. LDP sends a request to LMP to get an up/free data link, and LMP response LDP with an up/free data link with the attributes of the data link. LDP checks that the attributes of the data link is compatible with the lightpath requirement, it sends LMP an indication to set the state of the data link to be allocated. But before LDP sends the indication to LMP, the monitor timer of LMP expires and the state of data link returned to LDP is changed to test by the link connectivity verification function of LMP. If LMP discards the indication of LDP, the lightpath set up by LDP is not usable for it's illegal to transmit traffic while a data link is under test. If LMP accepts the indication of LDP and stops the test process, the incomplete monitor process can result a data link to be down. Such a dilemma indicates that interactions occur between path setup function of LDP and link connectivity verification function of LMP.

3.2.2   Feature Interactions between Fault Detection and Path Restoration

Several layers in optical networks, such as physical layer, data layer, network layer and service layer, have their own fault management mechanisms, each focusing on a different aspect. These fault mechanisms can interact with each other, especially when fault occurs in lower layer and the fault mechanisms in all upper layers are triggered. Some inconsistent behaviors in different layer can result in the failure of some features.

Figure 3 shows a network composed of several optical cross-connects. Before errors occur between oxc1 and oxc2, user traffic is transmitted along the primary path oxc1, oxc2 and oxc3. When fault occurs between oxc1 and oxc2, the fault can be detected by oxc2 and oxc3 as a result of loss of light. In this case, several fault management mechanisms can be triggered. If LMP is running on each node, fault localization process of LMP will be triggered to identify the location of the fault; if 1:n or 1+1 protection mechanism is deployed to the primary path, the protection mechanism can be triggered to recover the user traffic from the fault. If the protection mechanism is triggered and completes its operation before fault localization finishes its function, the fault localization feature of LMP fails to hold. Because when restoration mechanism is triggered, the user traffic is switched to path oxc1, oxc4, oxc5 and oxc3, there is no traffic carried along the primary path at this case. Fault localization process of LMP has supposed that the user traffic is always carried along the primary path such that it can find oxc1 is ok and oxc2 has detected the loss of signal, then decide that fault occurs between oxc1 and oxc2. With the user traffic switched over to the protection path, both oxc1 and oxc2 detect the loss of signal, so it cannot decide the location of the fault.

3.2.3   Feature Interactions between Routing and Link Management Protocols

In optical network, when a lightpath is set up and torn down, the status of all the involved channels is changed from being available to occupied and vice versa, such a network resource information is flooded throughout the control plane by routing protocols. Link management protocol maintains the status of channel status and triggers the flooding of the resource information once it is changed.

With a large number of links between two devices, flooding information of routing protocols will become tremendous, which can lead to control network congestion and instability. Link management protocol bundles these links into TE links, it triggers routing protocol to flood resource information until the resource change reaches certain points. But too infrequent flooding makes the network resource information inaccurate in each node, which can result in the inefficient utilization of the network resource and degrade network performance.

4   Feature Interactions Detection and Resolution Techniques in Optical Network

The causes of feature interactions in intelligent networks that are identified as limitations on network support, intrinsic problems in distributed systems and violation of feature assumptions [7], can also result in feature interactions in optical networks, such that the feature interaction resolution techniques developed in traditional telecommunication network can be applied to detect and resolve the problems in optical networks. But the interactions occurred in optical control and management protocols can result in much serious effect on the network operation, most interaction should be predicated before the deployment and more efficient resolution should be provided in run-time to prevent the service failure. Because of this, formal methods plays a much more important role in the design period to verify protocol properties, and some new methods are developed to resolve the interaction problems in run-time. We have used the automatic protocol validation tool Spin [8] to detect the feature interaction problem in the automatic neighbor discovery protocol [2] of optical networks and the on-going work is applying testing techniques to detect feature interaction problems.

4.1   Interaction Resolution by Policy Definition

Policy definition is much more important in protocol design and deployment of optical network protocols for its simplicity in solving possible interaction problems.

To solve the live lock between two LSRs operating in different label advertisement modes as indicated in section 3.1.1, a rule should be defined in an LSR operating in Downstream Unsolicited mode that an LSR operating in Downstream on Demand mode should not be expected to send unsolicited mapping advertisements. Therefore, if the downstream LSR is operating in Downstream on Demand mode, the upstream LSR is responsible for requesting label mappings as needed.

A single failure in optical networks can trigger the operation of multiple protocols with different behaviors, which results in the interactions between fault detection and path restoration in section 3.2.2. To solve this problem, a policy can be defined for the control path restoration process such that it can wait a period of time before beginning its operation. Such a policy is usually implemented by a timer in a real system.

4.2 Interaction Resolution by Behavior Restriction

For the interactions that are resulted by violation of the properties of a feature when adding or invoking a new one, behavior restrictions on the new features can solve them.

The invoking of link connectivity verification of LMP having violated the requirement of link property correlation as indicated in section 3.1.2 can be resolved by restricting the behavior of link connectivity verification. That is, when a data link passes the verification, its state is not changed to up until a link summary message on this data link is received and the property of the data link in both ends are checked to be compatible with each other.

4.3 Interaction Resolution by Advanced Scheduling Techniques

Many interactions in optical network protocols are resulted by the inconsistent operation on the shared resource. Data links in optical network are shared resource by all the protocols, and inconsistent operation on its state can result in the failure of some features, example given in section 3.2.1 is for this case.

A shared resource is usually associated with a resource management mechanism. Based on operation mechanism of the management system associated with a resource, the resource can be classified as: time-shared resource and space-shared resource. A space-shared scheduler executes a request for the resource by running it on a dedicated part of the resource when allocated; a time-shared scheduler starts its requests immediately upon its arrival and share resources among all jobs. Data link database can not be accessed simultaneously, so a space-shared systems can associate with it to handle its access request. The space-shared systems use resource allocation policies such as first-come-first-served (FCFS), shortest-job-first served (SJFS). This system must guarantee that once a request from a protocol has not been served completely, no other request can be adopted.

For the example given in section 3.2.1, the resource management system of the data link should be designed to control that once a read request has been received from LDP, and before receiving the response from LDP, no other access can be allowed to the same data link. If the data link database is implemented with shared memory instead of an independent database management system, a semaphore can be associated with each data link, and each protocol should get the semaphore no matter it needs to read or write the state of the data link. Once LDP gets the semaphore, it should not release it until it completes the operation on this data link.

4.4   On-line Detection and Resolution Techniques

Many on-line detection and resolution techniques can be applied to ASON. But to detect and resolve the interactions in control and management protocols in optical network, the speed and efficiency is of the first priority. In many protocols, interactions are detected by predication in design and resolved in run-time, that is, during design period, the possible interaction problems are predicated and some fields in the messages are reserved for negotiation to reduce the possible interactions in run-time. Such an on-line negotiation process can be used to solve many interactions resulted by the coexistence of multiple capabilities.

5   Conclusion

The feature interaction problems in ASON control and management protocols are very important for their severe impact on the network operation. We have proposed a framework and discussed the feature interaction problems in optical network management protocols. Our analysis shows that feature interactions can not only occur between features of a single protocol, but also occur between features of different protocols which can have much more serious effect on the network performance and operation. Some new resolution techniques which are particularly suitable for optical control and management protocols are proposed.

Adopting the characteristics of optical network control and management protocols to develop more techniques to detect and solve the feature interaction problems in optical networks is an interesting topic. What kind of techniques developed in traditional telecommunication network can be applied to optical network protocols and what's the barrier of using such techniques need further study.

References

[1] Dabashis Saha, Debabrata Sengupta, "An Optical Layer Lightpath Management Protocol for WDM AONs", Photonic Network Communications, 2:2, 185-198, 2000.

[2] Caixia Chi, et al., "Automatic neighbor discovery protocol for optical networks", Proceeding of Asia Pacific Optical Conference, Nov.2001.

[3] L. Blair, J. Pang, "Feature Interactions - Life Beyond Traditional Telephony", Feature Interaction in Telecommunications and Software Systems VI, M.Calder and E.Magill, IOS Press, 2000, pp.83-93.

[4] LDP Specification, RFC 3036.

[5] Ayan Banerjee et al., "Generalized Multiprotocol Label Switching: An Overview of Routing and Management Enhancements", IEEE Communication Magazine, vol.39, no.1, pp. 144-150, January 2001.

[6] Jonathan P. Lang, "Link Management Protocol (LMP)", Internet draft, draft-ietf-ccamp-lmp-05.txt, August 2002, work in progress.

[7] Cameron E J, Griffeth N D, et al., "A feature interaction benchmark for IN and beyond", in Proceedings of Second International Workshop on Feature Interactions in Telecommunications Systems, Bouma L G and Velthuijsen H (Eds), IOS Press, 1994.

[8] Gerard J. Holzmann, " Design and Validation of Computer Protocols", PRENTICE-HALL, Englewood Cliffs, New Jersey 07632.

*Feature Interactions in Telecommunications
and Software Systems VII*
D. Amyot and L. Logrippo (Eds.)
IOS Press, 2003

Feature Interactions in Web Services

Michael WEISS
School of Computer Science, Carleton University
Ottawa, ON K1S 5B6, Canada

Abstract. Web services promise to allow businesses to adapt rapidly to changes in the business environment, and the needs of different customers. However, the rapid introduction of new services paired with the dynamicity of the business environment also leads to undesirable interactions that negatively impact service quality and user satisfaction. Whereas most current work on web services has focused on low-level standards for publishing, discovering, and invoking services, our research looks at the problems that can arise from service integration, and how to manage them. In this paper we argue that service integration issues can be understood as feature interactions, and identify an initial set of feature interactions between web services. We also highlight the importance of considering the non-functional requirements of services (such as privacy or security) for understanding feature interactions between web services.

1. Introduction

The recent interest in web services as a paradigm for constructing distributed applications is motivated by the lack of centralized control, the diversity of technology platforms, and the rapid evolution of the business environment into which these services are deployed [9]. As the number of partners a business needs to interact with grows, the assumption of a single center of control that imposes a common technology platform becomes untenable.

Without a central point of control, businesses need to connect their own platform to an increasing number of other platforms. This means that connections between applications must be established without being tailored to a specific platform. The new business environment for services is also highly dynamic. New services become available at a fast rate, while the number of service providers is constantly growing. Increased competition forces businesses to provide customized services as a way of differentiating themselves from competitors. This requires a capability to adapt rapidly to changes in the environment, and the needs of different customers.

Much of the work on web services to date has focused on low-level standards for publishing, discovering, and invoking services. The current set of standards includes WSDL, UDDI, and SOAP. A web service can be described using WSDL (Web Service Description Language), published using a UDDI (Universal Discovery and Directory Interface) registry, and invoked using the SOAP (Simple Object Access Protocol) protocol [8].

More recent work has looked at how higher-level services can be composed dynamically from lower-level services [3][5]. Service integration raises a number of difficult challenges, including the description, discovery, composition, and execution of services. At each stage we may experience undesirable interactions that prevent the proper performance of the service. However, there has been little research on managing such interactions between services. Most existing work is limited to managing the mechanics of the interaction (for example, how to enforce a legal sequence of messages exchanges between parties).

Service integration issues can be understood as feature interactions. The feature interaction problem has been first formally studied in the telecommunications domain [18], but the phenomenon of undesirable interactions between components of a system can occur in any software system that is subject to changes. In the rest of the paper, we first present a model of web services in order to establish a common ground. Then we list the types of feature interactions we identified, and describe them in terms of the non-functional requirements violated including privacy, security, usability, predictability, interoperability, correctness, integrity, and manageability. Understanding these interactions, and developing techniques for their detection and resolution will be critical for deploying web services at any significant scale.

2. Web Service Integration

A web service is a network accessible interface to application functionality, built using standard Internet technologies like HTTP, XML, or SMTP [8]. Web services act as an abstraction layer that separates the platform and language-specific details of how the application logic is actually invoked. A key feature is that web services communicate via XML documents, and are not tailored to a specific API (that is, using method calls of a specific interface).

It is common for web services to aggregate and coordinate the execution of other web services. For example, a composite web service could provide a high-level trip planning service by coordinating lower-level web services for airline, hotel, and car rental search. From this perspective, it is more appropriate to think of web services as a web *of* services [8].

As shown in Figure 1, a service is defined as a set of endpoints. An endpoint groups service operations. Each operation is defined by the messages exchanged to perform it. The Web Services Flow Language (WSFL) [10] defines the notions of activities and flows. Each organization owns processes (or flows) that themselves consist of links and activities. Each activity can be implemented by an operation in a service of this or another organization.

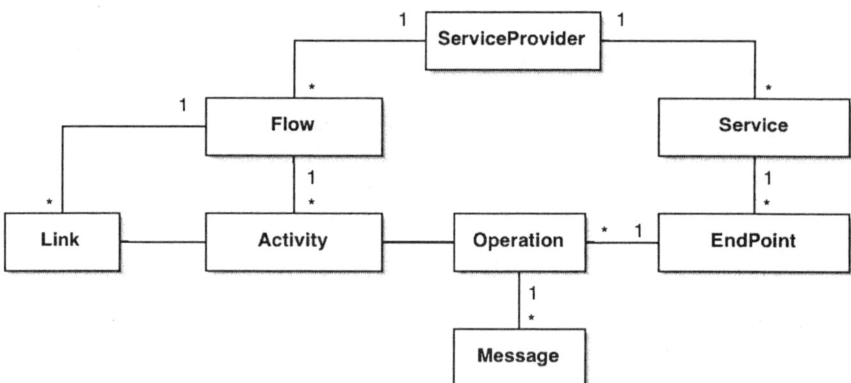

Figure 1: Object Model of Web Services (derived from [15])

The appropriate metaphor for thinking about the services an organization provides is therefore not the pipe-and-filter model of traditional telephony systems that chains together features in their order of precedence [17], but that of a flow system. Flow systems have three types of components: processing stages that can be connected in various ways (not just sequences), data representations that are exchanged between stages, and orchestrators (engines) that coordinate the flows. Their behavior is richer than that of pipe-and-filter systems.

When composing web services, the functionalities provided by the component services must be considered. We also need to ensure that data and message types, sequence logic etc. are compatible. However, as stated in [11], service composition amounts to much more than functional composition. Consideration must also be given to non-functional requirements such as privacy, security, predictability, and interoperability. For example when composing a personalized purchasing service, we must also consider utility services such as identity management, and user profiling. However, sharing and maintaining sensitive user information at the service provider's end raises privacy concerns. With some identity management services such as Microsoft's proposed Passport, the user has little control over how information is shared with a service provider [4]. In fact, once the Passport service authenticates the user, it gives the provider complete access to the user profile. In most cases, this is clearly not what the user had anticipated, and it violates their right to (selective) privacy.

In light of this, we like to refer to web service composition as web service *integration*. Functional composition is only one aspect of integration. Integration also includes the consideration of non-functional aspects of the resulting systems. The flow model of web services in Figure 1 does therefore not provide a sufficient basis for analyzing feature interactions. Likewise a composition calculus derived from the flow model of web services will only be able to deal with functional composition. Different techniques will be required to analyze integration issues caused by non-functional requirements. In the following we will create a taxonomy of such integration issues. Their formal analysis is subject of further research.

3. Feature Interaction Problem in Web Services

The feature interaction problem concerns the coordination of features or services (we won't distinguish between features and services) such that they cooperate towards a desired result at the application level. The root causes for feature interaction in telephony systems are [18]:

- Conflicting goals (services with the same preconditions but incompatible goals are in conflict, for example, services triggered by a busy extension)
- Competition for resources (services compete with each other for limited resources, which need to be partitioned among the services)
- Changing assumptions on services (services make implicit assumptions about their operation, which can become invalid when new services are added)
- Design evolution (services need to be added to meet new customer needs, and the system will need to interoperate with other vendors' systems)

A classical feature interaction is the interaction between Call Waiting and Call Forwarding on Busy. Both features trigger when the receiver of a call is busy, but only one of them should become active. This type of problem is usually resolved by introducing priorities. A slightly more complex example is the interaction between Outgoing Call Screening and Call Forwarding on No Answer. Assume Alice is on Bob's outgoing call screening list (for instance, Alice could be the teenaged girlfriend of Bob's son Mark, and Bob does not want him to call her). However, Mark quickly learns that he only needs to call his friend Joe who temporarily forwards incoming calls to Alice. The solution to this type of problem involves confirming with the originating party, Bob, if Joe's forwarding the call to Alice is acceptable.

However, the feature interaction problem is not limited to the telecommunications domain. As stated in the workshop call, the phenomenon of undesirable interactions between components of a system can occur in any software system that is subject to changes. This is certainly the case for service-oriented architectures. While web services promise to allow businesses to adapt rapidly to changes in the business environment, and the needs of different customers, the rapid introduction of new services paired with the dynamicity of the environment may also lead to undesirable interactions.

The issues of changing assumptions and design evolution are, indeed, exacerbated in the context of web services, since web services are by definition developed by third parties (a service knows only about itself, not its users or clients), and there is a strong incentive to personalize web services, in particular in the context of context-aware services. Goal conflicts and competition for resources take on a new dimension of complexity, as the execution model for web services differs from that of telephony systems, and we are dealing with a much more diverse set of shared resources (anything that can be identified by an URL).

In addition, in an environment where applications will be assembled dynamically from services, we also have to consider the issue of service maintenance [12]. The business processes and web services need to be monitored, so that system failures can be detected and localized. Expected service characteristics are usually specified in the form of service level agreements [15].

4. Types of Feature Interactions in Web Services

The following classification of feature interactions in web services is based on the non-functional requirements violated if an undesirable interaction occurs. Figure 2 shows the non-functional requirements (e.g. privacy, security, and usability) we considered, and the stakeholders (users, providers, and deployers of web services) concerned with each. The annotations in parentheses restrict the involvement of a stakeholder to a specific aspect of a non-functional requirement. For example, Users [Access] Security means that users are concerned about security only as far as access to services is concerned.

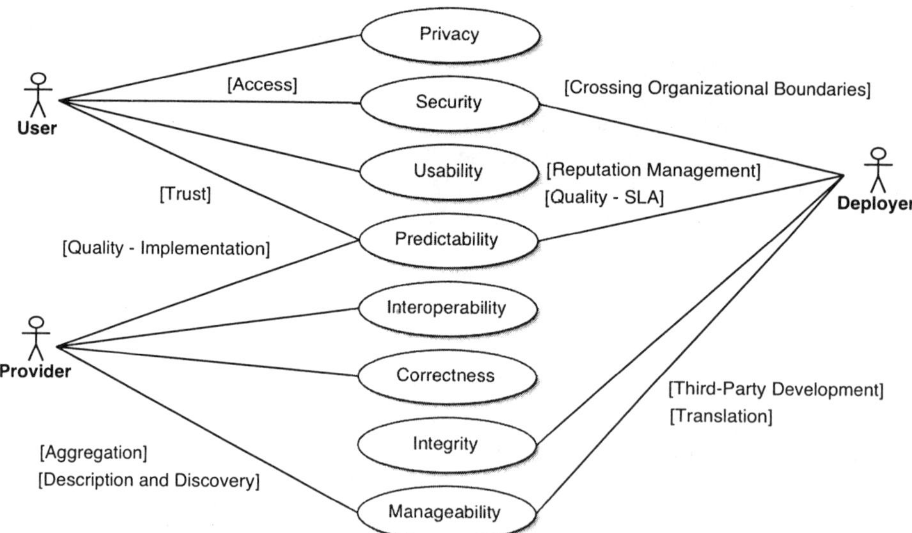

Figure 2: Non-functional requirements affected by web services and their stakeholders

Privacy. Privacy means controlling who has access to personal information sent as part of a service request, and under what conditions. It is also concerned with in what form and over which time frame this information is collected and stored. For example, in order to receive personalized web services (e.g. movie recommendations), a user needs to disclose private information (e.g. a list of movies they liked or disliked in the past). This information is usually stored in raw form at the service provider. A recently proposed solution to aspects of this problem is based on sending personal information in encrypted form, and key sharing [2].

Privacy is also a concern when a request is routed through intermediate nodes before it arrives at the service provider. A common solution is to restrict access to individual fields of the SOAP message by encrypting them, so that nodes can only access a field for which they have the private key [8]. Another potential source for undesirable interactions is that current identity services such as Microsoft's Passport blur the line between authentication of the user's identity, and managing the user's personal information. Once the user authenticates with a Passport-enabled web site, all her information is shared with the site. In addition, it is unclear how to enforce privacy policies restricting the use of the user information [4] by the identify service providers themselves (as they have access to considerable information).

Security. Security is concerned with protecting a web service against unauthorized access. Each of the stakeholders (providers, deployers, and users) has distinct security needs [6]. Service providers only want authorized users to access their services. Service deployers need to provide secure storage for the web services (code and data) they host. Service users expect that their personal information is protected from interception and corruption. Web service registries must also ensure authorized access to service catalogs. While critical to the operation of services, security management is often orthogonal to the primary goals of service providers and deployers. If security measures are not properly managed, they may actually make a system more vulnerable than a system that does not use them.

One proposed approach is to free up providers and deployers from the task of managing the security of their services by delegating security functions (such as key management, and authentication) to web services that are professionally configured and managed [6]. Two solution candidates are the XML Key Management Specification (XKMS) [7], which provides key and certificate management services, and the Security Assertion Markup Language (SAML) [16], which provides a single sign-on to an administrative domain.

Usability. Personalizing web services enhances their usability. For example, the user's shipping address can be stored in a profile, and can be filled in automatically whenever the user sends a purchase order request and does not provide a shipping address. Maintaining and disclosing personal information to service providers raises privacy concerns as outlined earlier. There is also the issue of how to best acquire a user profile. The two approaches are to ask the user a set of questions upon first use of the service, or to learn the profile by observing the user's behavioral patterns. While not a feature interaction in the strict sense, asking users for personal information before they have experienced the value of a service can be annoying.

Another issue caused by personalization is that users may want to associate different profiles with different contexts they are in (e.g. their locations, the activities they are engaged in, as well as other context attributes and user preferences). The main difficulty is detecting the user's current context from the user's behavior. Another problem is that there is not a uniform notion of context, while current standards such as CC/PP [14] are too limited.

Predictability. The behavior of a web service must be predictable. Predictability includes trust, and quality. At a minimum, trust implies authentication. However, simply asserting the identity of a service provider does not provide the user with enough information about the experience they will get from using the service. There are two issues. First, trust is something that can only be built over time by using a service. For example, a service for restaurant recommendations becomes trustworthy only when its recommendations include restaurants that the user has liked in the past (these are not redundant, but on the contrary, the basis for trust).

Second, if there exists no prior history of interacting with a service, how can we provide a substitute for immediate trust? Consider a business traveler who visits a new city and is looking for a restaurant to have dinner tonight. She would have to consult a tourism web service with which she had not interacted before. The likely solution is that the user will interact with the service via an intermediary (portal) instead, with which she has a trusted relationship. Even in this case, an obvious cause for user annoyance that remains is bias: the service may only return restaurants that paid a fee to the service provider for being included in the restaurant directory.

This brings us to the quality component of predictability. A basic limitation of dynamic discovery as currently conceived of is that there is no means of ensuring that the service provider correctly implements the service contract. All we can ensure is that it responds to all of the messages defined in the service interface, including the syntax of the message content. Still, e.g. the restaurant service may miscalculate the distance between the user and a restaurant. As we have to treat the implementation of a service as a black box, we need to consult other services (reputation services) that give us a rating of the service provider. While current implementations of reputation mechanisms are geared towards human users, nonetheless, recent research into monitoring web service level agreements [15] might provide a solution.

Interoperability. The WSDL description of a web service does not only include the messages understood by the service, but also the XML schemas for the content of those messages. One would think that this would ensure that service users and service providers agree on the meaning of the terms in the schema. However, we cannot assume this. There is often enough semantic ambiguity due to differences in the versions of the schema used by user and provider, and divergent interpretations given to the same terms.

Consider the example of a `<price>` element in a purchase order document. If a new version of the schema were to add a `currency` attribute (e.g. `<price currency="cad">`), the service user could misinterpret the amount to pay in the invoice sent in reply to the purchase order. Even if the service user were to check for the version of the schema, it could interpret the price as to include shipping, whereas, in fact, it does not. In this case, the ambiguity arises from assumptions not documented in the schema. Some of these problems are currently studied as part of the Semantic Web effort.

Correctness. Web service composition can sometimes generate uncertain results, that is, multiple outputs where one output was expected, or conflicting outputs (e.g. due to outdated information) [13]. Consider a `name-to-phone` service that combines a `person-name-to-address` and an `address-to-phone` service. Assume that multiple people can share the same address, and different people living at the same address may use different phones. If Bob and Alice live at the same address but have different phone numbers, the `name-to-phone` service would return both their phone numbers when asked for Bob's phone number. A filter mechanism [13] can handle some of these interactions by modifying the output of a service.

Integrity. Our notion of integrity is very specific. It is concerned with the integration of utility services such as metering, billing, and payment that are required by businesses whose core business is the provision of web services (that is, where web services are not part of an existing broader business relationship). There are broader definitions of integrity that include such concerns as privacy, security, and data coherence, but we consider these as separate concerns. The main issue from the feature interaction perspective is that due to the decentralized nature of web services there is no single authority for metering, billing, and collecting payments. Rather, each service deployer must provide its own implementation. On the other hand, these are cross-cutting concerns, affecting all services, and utility services could be provided by specialized service providers in a similar way as discussed for security.

Manageability. Manageability is concerned with the creation, evolution, and deployment of services. It comprises aggregation, service description and discovery, deployment of third-party services, and translation between service schemas. The main benefit of aggregation is that is simplifies the creation of complex services. A complex service can be defined recursively from lower-level services provided by other parties. A purchasing service could be created by assembling cataloging, order management, payment, and shipping services.

In the eFlow system [3], the notion of a generic service node is used to support the dynamic creation of composite services. Generic service nodes are not statically bound, but include a configuration parameter that can be set to the list of actual service nodes to be executed. eFlow will detect service nodes whose input and output data are not compatible with the composite service definition. However, it appears that these service nodes are strict subflows, and that eFlow does not provide ways for detecting interactions of concurrently executing service flows.

Undesirable interactions due to non-determinism can occur at two levels in an architecture for web service composition like eFlow: (1) within a generic service node, service nodes with common preconditions will be selected for execution, but could have conflicting goals – even if one node were selected at random, the result would be non-deterministic; and (2) between concurrently executing flows in the same engine, if triggered by the same preconditions. It would seem beneficial to model composite web services, and the integration of multiple concurrent web services within the same system using the methodology proposed in [1], which describes services using UCM (Use Case Map) scenarios.

While aggregation helps manage complex service interactions, it also introduces new ones. Some are due to non-determinism, others to predictability, interoperability, and correctness concerns. The latter are largely a result of using third-party services that have been designed independently (a consequence of design autonomy, and the lack execution autonomy). Finally, translation services will be instrumental in the integration of heterogeneous (in particular, regional) web services. However, they are also faced with the problem that there may not always a unique correspondence between terms in the respective vocabularies of service requesters and providers.

5. Conclusion

The integration of web services raises a number of difficult challenges, including the description, discovery, composition, and execution of such composite services. At each stage we may experience undesirable interactions that prevent the proper performance of a service. These interactions take place when one service modifies or subverts the operation of another (e.g. personalized services, which require a system to store sensitive information, vs. an identify management service that passes on that information to service providers in an unconstrained manner and violates the user's expectation of privacy). However, there has been little research on managing such interactions between services. The research that exists is limited to issues regarding the mechanics of the interaction.

We argued that service integration issues can be understood as feature interactions, and presented an initial set of feature interactions between web services. We described these feature interactions in terms of the non-functional requirements violated if the interaction occurs. The non-functional requirements we included in our analysis were privacy, security, usability, predictability, interoperability, correctness, integrity, and manageability. For each concern we identified potential feature interactions, and provided examples. We also gave pointers to current solutions.

As a community, we are just at the beginning of understanding the implications of web services on the design of distributed applications. Understanding the interactions between web services, not just at the mechanistic level of service description languages, registries, and messaging, but also at the level where these services can affect the performance of each other in undesirable ways, will be one of the most important tasks. Without this understanding, we believe, it will be unlikely that we will be able to deploy web services at any significant scale.

Once we understand the sources of interactions between web services, we can also start designing mechanisms for detecting and resolving interactions in service-oriented architectures, for both design time and runtime use. Based on the experience with feature interaction management in the telecommunications domain (e.g. [18][12]), we can already say with certainty that agent-based approaches will play a significant role in their construction.

References

[1]   Amyot, D., Logrippo, L., et al, Use Case Maps for the Capture and Validation of Distributed Systems Requirements, Symposium on Requirements Engineering, 44-53, 1999.
[2]   Canny, J., Collaborative Filtering with Privacy, IEEE Conference on Security and Privacy, 45-57, 2002.
[3]   Casati, F., and Shan, M.C., Dynamic and Adaptive Composition of e-Services, Information Systems, 26:3, 143-163, 2001.
[4]   Conry-Murray, A., Microsoft's Passport to Controversy, Network Magazine, 2002, www.networkmagazine.com/article/NMG20020304S0003.
[5]   Constantinescu, I., et al, Abstract Behavior Representations for Service Integration, Workshop on AgentCities: Challenges in Open Agent Environments, at Conference on Autonomous Agents and Multiagent Systems (AAMAS), 2002.
[6]   Fernandez, E.B., Web Services Security, in: Fletcher, P., and Waterhouse, M. (eds.), Web Services Business Strategies and Architectures, 290-302, Expert Press, 2002.
[7]   Ford, W., et al, XML Key Management Specification (XKMS), W3C, 2001, www.w3.org/TR/xkms.
[8]   Glass, G., Web Services: Building Blocks for Distributed Systems, Prentice Hall, 2001.
[9]   Hagel, J., Out of the Box: Strategies for Achieving Profits Today and Growth Tomorrow through Web Services, Harvard Business School, 2002.
[10] Leymann, F., Web Services Flow Language (WSFL) 1.0, 2001, available from www.ibm.com/software/solutions/webservices/pdf/WSFL.pdf.
[11] O'Sullivan, J., Edmond, D., and ter Hofstede, A., What's in a Service? Towards Accurate Description of Non-Functional Service Properties, Distributed and Parallel Databases, 12, 117-133, Kluwer, 2002.
[12] Pinard, D., Gray, T., Mankovskii, S., and Weiss, M., Issues in Using an Agent Framework for Converged Voice and Data Applications, Conf. on Practical Applications of Intelligent Agents and Multi-Agent Technology (PAAM), 1997.
[13] Ponnekanti, S., and Fox, A., SWORD: A Developer Toolkit for Web Service Composition, World Wide Web Conference (WWW), 2002.
[14] Reynolds, F., et al, Composite Capability/Preference Profiles (CC/PP): A User Side Framework for Content Negotiation, W3C, 1999, www.w3.org/TR/NOTE-CCPP.
[15] Sahai, A., et al, An Adaptive and Extensible Model for Managing Web Services and Business Processes, HP Openview University, 2002.
[16] Security Assertion Markup Language (SAML), OASIS, 2002, xml.coverpages.org/saml.html.
[17] Utas, G., A Pattern Language of Feature Interactions, in: Rising, L. (ed.), Design Patterns in Communication Software, Cambridge University Press, 2001.
[18] Velthuijsen, H., Distributed Artificial Intelligence for Runtime Feature Interaction Resolution, IEEE Computer, 45-55, August 1993.

Human Factors

*Feature Interactions in Telecommunications
and Software Systems VII*
D. Amyot and L. Logrippo (Eds.)
IOS Press, 2003

On Preventing Telephony Feature Interactions which are Shared-Control Mode Confusions

Jan BREDEREKE

Universität Bremen, FB 3 · P.O. box 330 440 · D-28334 Bremen · Germany
brederek@tzi.de · www.tzi.de/~brederek

Abstract. We demonstrate that many undesired telephony feature interactions are also shared-control mode confusions. A mode confusion occurs when the observed behaviour of a technical system is out of sync with the behaviour of the user's mental model of it. Several measures for preventing mode confusions are known in the literature on human-computer interaction. We show that these measures can be applied to this kind of feature interaction. We sketch several more measures for the telephony domain.

1 Introduction

Automation surprises are ubiquitous in today's highly engineered world. We are confronted with *mode confusions* in many everyday situations: When our cordless phone rings while it is located in its cradle, we establish the line by just lifting the handset — and inadvertently cut it when we press the hook toggle button as usual with the intention to start speaking. We get annoyed if we once again overwrite some text in the word processor because we had hit the "Ins"-key before (and thereby left the insert mode!) without noticing. The American Federal Aviation Administration (FAA) considers mode confusion to be a significant safety concern in modern aircraft. For instance, consider the crash of an Airbus A320 near Strasbourg, France, in 1992 [1]. Probably due to heavy workload because of a last-minute path correction demanded by the air traffic controller, the pilots confused the "vertical speed" and the "flight path angle" modes of descent. As a result, the Air Inter machine descended far too steeply, crashed, and 87 people were killed.

Many safety-critical systems today are so-called embedded shared-control systems. These are interdependently controlled by an automation component and a user. Examples are modern aircraft and automobiles.

In telephony, call control is shared between users and many modern telephony features. Some examples are call screening, call forwarding, voice mail, and credit card calling. Multi-party features such as three-way calling let all users involved share call control to some extent. In contrast to many shared-control systems, telephones usually are not immediately safety-critical. Nonetheless, users expect a comparably high reliability which must not be impaired by undesired feature interaction.

Feature interaction occurs when one feature modifies or subverts the operation of another one. The above mode confusion example of the auto off-hook feature and the hook toggle button feature also demonstrates an undesired interaction between these two features. The feature interaction benchmark of Cameron *et al.* [2] presents examples for many different kinds of feature interactions (see Sect. 3).

We found that many undesired telephony feature interactions are also shared-control mode confusions. Several measures for preventing mode confusions are known in the literature on human-computer interaction. We now show that these measures can be applied to this kind of feature interaction.

This paper is organized as follows: Section 2 introduces to mode confusions. In Section 3, we investigate the well-known feature interaction benchmark [2] and discuss the mode confusions we found there. Section 4 presents a definition of mode confusion adapted to telephony, and in Section 5 we show how the notion of mode confusion can help against undesired feature interactions. Section 6 summarizes our paper and discusses future work.

2   Mode Confusions

2.1   What is a Mode Confusion?

Humans use *mental models* when they interact with technical systems in general, and with automated systems in particular. This is generally agreed upon in cognitive science [3]. Unfortunately there is an additional, completely different meaning for the notion "mental model" in the pertinent literature. We refer to the one introduced by Norman [4]: a mental model represents the user's knowledge about a technical system, it consists of a naïve theory of the system's behaviour.

An explicit description of a mental model can be derived, according to Rushby [3]. It can be represented, e. g., in form of a *state machine* with modes and mode transitions. We can extract it from training material, from user interviews, or by user observation. Cañas *et al.* [5] survey work on this and show in three experiments with 140 participants how exposing users to different knowledge elicitation tasks allows to figure out their mental models. Rushby qualifies his statement by conceding that it is difficult and expensive to extract the mental models of the individual users. Fortunately, this is neither necessary for his nor for our work. We are interested in representative examples of mental models. Our goal is to design the system such that as many users as possible will construct an adequate mental model.

A mental model can sometimes change over time, but this does not conflict with our goal of mode confusion prevention. The user's knowledge changes when he/she learns or forgets. When the user's mental model changes after we have extracted and documented it, we cannot predict the behaviour of the user correctly anymore. But we are not interested in any particular user's behaviour. Again, we are interested in improving the design of the technical system in order to ease its use in general.

The mental model which is the user's long-term knowledge is different from the user's current working abstraction of this mental model. When performing a task, the user concentrates on the part of his/her knowledge which he/she assumes to be relevant [5].

Interestingly, no publication defines the notions of "mode" and "mode confusion" rigorously [6, 7, 8, 9, 10, 11, 12, 13, 14, 15, 16] in a recent survey of ours [17]. We therefore propose such a rigorous definition of "mode" and "mode confusion" for safety-critical systems in that work. We sketch it in the following.

Some researchers in the Human Factors (HF) community will disagree with our rigorous definitions [17] at certain points, but this does not matter for the investigation of mode confusions in telephony feature interactions. Different researchers in the HF community have a slightly different intuitive understanding of mode and mode confusion (and we are part of that). We proposed our rigorous definitions as a base for discussion. They enable to pinpoint the differences and to resolve them by discussion in the HF community.

We must use a *black-box view* of a running technical system for the definition. This is because the user of such a system has a strict black-box view of it and because we want to solve the user's problems. As a consequence, we can observe (only) the environment of the technical system. When something relevant happens in the environment, we call this an *event*.

The technical system has been constructed according to some *requirements* document REQ. We can describe REQ entirely in terms of observable events, by referring to the *history* of events until the current point of time. For this description, no reference to an internal state is necessary. Usually, several histories of events are equivalent with respect to what should happen in the future. Such equivalences can greatly simplify the description of the behaviour required, since we might need to state only a few things about the history in order to characterise the situation.

During any run of the technical system, it is in one specific state at any point of time. The (possibly infinite) state transition system specified by REQ defines the admissible system runs.

We call the user's mental model of the technical system REQ^M. Ideally, REQ^M should be the same as REQ. During any run of the technical system, REQ^M is also in one specific state at any point of time. You may think of the behaviour of REQ^M as a "parallel universe" in the user's mind. Ideally, it is tightly coupled to reality.

Our approach is based on the motto *"the user must not be surprised"*. This is an important design goal for shared-control systems. We must make sure that the reality does not exhibit any behaviour which cannot occur according to the mental model of it. Additionally, the user must not be surprised because something expected does *not* happen. When the mental model prescribes some behaviour as necessary, reality must not refuse to perform it. For example after dialling a number, a phone must either produce an alert tone or a busy tone, and it must never ring itself.

The rule of non-surprise means that the relationship between the reality and the user's mental model of it must be a *relationship of implementation to specification*. The reality should do exactly what the mental model prescribes, no less and no more. In case that the user does not know what to expect, but knows that he/she does not know, then the reality is free to take any of the choices. A common example is that the user does not know the exact point of time at which the technical system will react to an event, within some limits.

We can describe such an implementation/specification relationship formally by a refinement relation. Many formalisms have been proposed for this. We use CSP [18, 19]. It is one suitable formalism with suitable tool support. In CSP, *failure refinement* is precisely the relation described above.

In CSP, one describes the externally visible behaviour of a system by a so-called process. Processes are defined over events. CSP offers a set of operators. One can use them to specify processes. CSP also provides a number of refinement operators. There is a precise formal semantics for all of this [18].

The user does not always notice when the behaviour of his/her mental model REQ^M is not the same as that of the reality REQ. This is because the user's mind does not take part in *any* event in the environment. The user perceives the reality through his/her senses only.

The user's senses SENSE translate from the set of events in the environment to a set of events in the user's mind. SENSE is not perfect. Therefore we must distinguish these two sets. For example, the user might not hear a signal tone in the phone due to loud surrounding noise. Or the user might not listen to all of a lengthy announcement text, or he/she might not understand the language of the announcement. At the very least, there is always a larger-than-zero delay between any environment event and the respective mental event. In all these cases, what happens in reality, as described by REQ, is different from what happens according to the user's perception of it, as described by $\text{SENSE}(\text{REQ})$.[1]

The user is surprised only if the *perceived* reality does not behave the same as his/her expectations. This is why the user does not always notice a difference between the actual reality REQ and the "parallel universe" REQ^M in his/her mind.

We cannot compare the perceived reality $\text{SENSE}(\text{REQ})$ to the mental model of the reality REQ^M directly. They are defined over different sets of events (mental/environment). We need a translation.

The user has a mental model of his/her own senses SENSE^M. SENSE^M translates the behaviour of the mental model of the technical system REQ^M into events in the user's mind. It does this in the same fashion as SENSE does it for REQ.

The user's knowledge about the restrictions and imprecisions of his/her own senses is also part of SENSE^M. Ideally, the user should know about them precisely, such that SENSE^M and SENSE are the same. The user is not surprised if the process $\text{SENSE}(\text{REQ})$ is a failure refinement of the process $\text{SENSE}^\text{M}(\text{REQ}^\text{M})$.[2]

Our definition of mode confusion [17] concentrates on the safety-relevant aspects of the technical system. This is because traditionally the safety-critical systems community has perceived mode confusions as a problem.

When the user concentrates on safety, he/she performs an on-the-fly simplification of his/her mental model REQ^M towards the safety-relevant part $\text{REQ}^\text{M}_\text{SAFE}$. This helps him/her to analyse the current problem with the limited mental capacity. Psychological studies show that users always adapt their current mental model of the technical system according to the specific task they carry out [5]. The "initialisation" of this adaptation process is the static part of their mental model, the so-called "conceptual model" [20]. This model represents the user's knowledge about the system and is stored in the long term memory.

Analogously to the abstraction performed by the user, we perform a simplification of the requirements document REQ to the safety-relevant part of it REQ_SAFE. REQ_SAFE can be either an explicit, separate chapter of REQ, or we can express it implicitly by specifying an abstraction function, i.e., by describing which aspects of REQ are safety-relevant. We abstract REQ out of three reasons: $\text{REQ}^\text{M}_\text{SAFE}$ is defined over a set of abstracted events, and it can be compared to another description only if it is defined over the same abstracted set; we would like to establish the correctness of the safety-relevant part without having to investigate the correctness of the entire mental model REQ^M; and our model-checking tool support demands that the descriptions are restricted to certain complexity limits.

We express the abstraction functions mathematically in CSP by functions over processes. Mostly, such an abstraction function maps an entire set of events onto a single abstracted event. Other transformations are hiding (or concealment [19]) and renaming. But the formalism also allows for arbitrary transformations of behaviours; a simple example being a certain

---

[1]One must also distinguish between what happens in reality and what happens according to the software's perception of it, due to imperfect input/output devices. But these software events are not relevant here since we must view the technical system as a black-box.

[2]In [17], we used the name MMOD for $\text{SENSE}^\text{M}(\text{REQ}^\text{M})$. We did not define SENSE^M and REQ^M separately. We now make a distinction between these two different kinds of mental model for clarity.

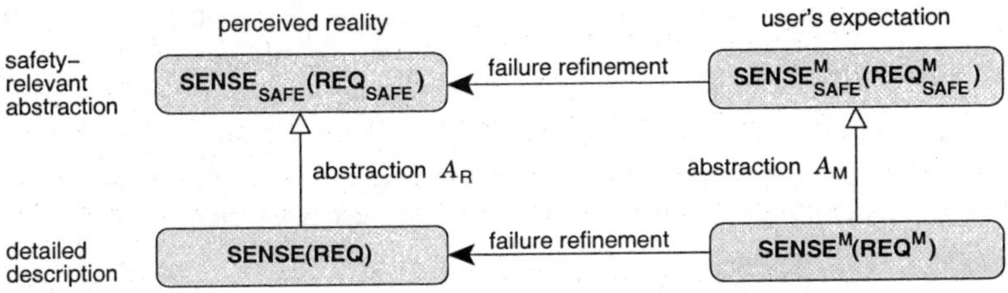

Figure 1: Relationships between the different refinement relations.

event sequence pattern mapped onto a new abstract event. We use the abstraction functions $\mathcal{A}_R$ for REQ and $\mathcal{A}_M$ for $\mathrm{REQ^M}$, respectively.

The relation SENSE must be abstracted in an analogous way to $\mathrm{SENSE_{SAFE}}$. They are relations from processes over environment events to processes over mental events. It should have become clear by now that $\mathrm{SENSE_{SAFE}}$ needs to be rather true, i. e., a bijection which does no more than some renaming of events. If $\mathrm{SENSE_{SAFE}}$ is "lossy", we are already bound to experience mode confusion problems. $\mathrm{SENSE^M_{SAFE}}$ accordingly is the user's mental model of $\mathrm{SENSE_{SAFE}}$.

Figure 1 shows the relationships among the different descriptions. In order not to surprise the user with respect to safety, there must be a failure refinement relation on the abstract level between $\mathrm{SENSE_{SAFE}(REQ_{SAFE})}$ and $\mathrm{SENSE^M_{SAFE}(REQ^M_{SAFE})}$.

We defined the notion of mode confusion rigorously [17]. The central definition is:

Definition 1 (Mode confusion) A *mode confusion* between $\mathrm{SENSE_{SAFE}(REQ_{SAFE})}$ and $\mathrm{SENSE^M_{SAFE}(REQ^M_{SAFE})}$ occurs if and only if $\mathrm{SENSE_{SAFE}(REQ_{SAFE})}$ is not a failure refinement of $\mathrm{SENSE^M_{SAFE}(REQ^M_{SAFE})}$, i.e., iff $\mathrm{SENSE^M_{SAFE}(REQ^M_{SAFE})}$ $\not\sqsubseteq_F$ $\mathrm{SENSE_{SAFE}(REQ_{SAFE})}$.

We also defined the notion of mode rigorously [17]. A mode is a just state, but we reserve the word for the "states" of abstracted descriptions, i. e., of $\mathrm{SENSE_{SAFE}(REQ_{SAFE})}$ and of $\mathrm{SENSE^M_{SAFE}(REQ^M_{SAFE})}$. A state (or mode) is a potential future behaviour. We can distinguish two states (or two modes) of a system only if the system may behave differently in the future. This is because of the black-box view. For two different modes, there must be a safety-relevant difference. We omit the further details here.

Every time $\mathrm{REQ^M_{SAFE}}$ changes, one must decide anew whether a mode confusion occurs. Our definition of mode confusion is based on the (rather strong) assumption that $\mathrm{REQ^M_{SAFE}}$ is stable over time. The user generates $\mathrm{REQ^M_{SAFE}}$ on-the-fly from $\mathrm{REQ^M}$ and must re-generate it later when he/she needs it again. This re-generation might lead to a different result. In particular, the re-generation requires the user's recollection of the current mode. A user's lapse [21] here can result in a mode confusion. This happens when the user selects a mode as initial mode which does not match the reality's current mode.

2.2 What Can We Do Against Mode Confusion Problems?

Rushby proposes a procedure to develop automated systems which pays attention to the mode confusion problem [3]. The main part of his method is the integration and iteration of a model-checking based consistency check and the mental model reduction process introduced by [22, 6].

Vakil and Hansman, Jr. [23] recommend three approaches to reduce mode confusion potential in modern aircraft: pilot training, enhanced feedback via an improved interface, and, most substantial, a new design process (ODP, for operator directed design process) for future aircraft developments. ODP aims at reducing the complexity of the pilot's task, which may involve a reduction of functionality.

Our definitions allow us to classify mode confusion problems, to derive recommendations from the classification for avoiding some of the problems, and we can also use the definitions as a foundation for detecting mode confusions [17].

We *classify* mode confusion problems into four classes:

1. Mode confusion problems which arise from an *incorrect observation* of the technical system or its environment. This may have *physical* or *psychological* reasons.

2. Mode confusion problems which arise from *incorrect knowledge* of the human. The incorrect knowledge may be either about the *technical system* or its environment, or about the human's own *senses*.

3. Mode confusion problems which arise from the *incorrect abstraction* of the user's knowledge to the safety-relevant aspects of it.

4. Mode confusion problems which arise from an *incorrect processing* of the abstracted mental model by the user. This can be a memory lapse or a "rule-based" mistake [21], i. e., a mode transition that is not part of the model.[3]

The above causes of mode confusion problems lead directly to some *recommendations for avoiding* them. In order to avoid an incorrect observation of the technical system and its environment, we must check that the user can physically observe all safety-relevant environment events, and we must check that the user's senses are sufficiently precise to ensure an accurate translation of these environment events to mental events. If this is not the case, then we must change the system requirements. We must add an environment event controlled by the machine and observed by the user which indicates the corresponding software input event. In order to avoid an incorrect observation because of a psychological reason, we must check that observed safety-relevant environment events become conscious reliably.

Establishing a correct knowledge of the user about the technical system and its environment can be achieved by documenting the requirements of them rigorously. This enables us to produce user training material, such as a manual, which is complete with respect to functionality. This training material must not only be complete but also learnable. Complexity is an important learning obstacle. Therefore, the requirements of the technical system should allow as little non-deterministic internal choices as possible, since tracking all alternative outcomes is complex. This generalises and justifies the recommendation by others to eliminate "implicit mode changes" [13, 11]. We can eliminate a non-deterministic internal choice by the same measure as used against an incorrect physical observation: we add an environment event controlled by the machine which indicates the software's choice.

---

[3]In [17], we viewed this cause as part of the "incorrect knowledge" cause. We now prefer to distinguish the different kinds of human error identified by Reason [21].

Improving the user's knowledge about his/her own senses has little potential for avoiding mode confusion problems. If the user knows that some things may happen, but he/she cannot perceive them, then they are non-deterministic choices to the user's mind. Again, the user will have difficulties with the complexity of tracking alternative outcomes. We can improve the technical system's design if we not only avoid incorrect observations but also such incomplete observations. The concrete measures are similar to those against incorrect observations.

Ensuring a correct mental abstraction process is mainly a psychological question and mostly beyond our scope. Our work leads to the basic recommendation to either write an explicit, rigorous safety-relevance requirements document or to indicate the safety-relevant aspects clearly in the general requirements document. The latter is equivalent to making explicit the safety-relevance abstraction function for the machine $\mathcal{A}_R$. Either measure facilitates to produce training material which helps the user to concentrate on safety-relevant aspects.

Incorrect processing of the abstracted model by the user is a *human error*. Reason [21] distinguishes three basic types of human error: skill-based slips and lapses, rule-based mistakes, and knowledge-based mistakes. Slips appear as incorrect observation for psychological reasons in our classification, and knowledge-based mistakes appear as incorrect knowledge. Lapses and rule-based mistakes cause incorrect processing. The design of a system that avoids them is again a psychological question and mostly beyond our scope. Reason [21] surveys advices for reducing the human error risk.

Our definitions form a *foundation for detecting* mode confusions by model-checking. We model-checked successfully an autonomous wheelchair for mode confusion problems [24].

3   Mode Confusion Problems in the Feature Interaction Benchmark

We found that a considerable number of undesired telephony feature interactions in the feature interaction benchmark of Cameron *et al.* [2] are also shared-control mode confusions. The benchmark was written by practitioners from the telecom industry. It provides representative examples of a broad range of undesired feature interactions.

The remainder of this section contains a description and a discussion of those feature interaction examples that are also mode confusion problems. All descriptions of feature interactions are quoted from Cameron *et al.* [2] (except one variant in Section 3.3 and one extra example of ours in Section 3.11). For completeness, we also briefly sketch the wide range of causes from the other feature interaction examples.

We use an informal notion of mode confusion in our discussion here that does not abstract the system to its safety-relevant aspects. This would not make sense for telephony. We discuss this aspect in Section 4 below.

Table 1 summarizes the mode confusions we found in the feature interaction benchmark. There are 12 mode confusion problems in 8 of the 22 examples of the feature interaction benchmark. This shows that mode confusions are definitely a relevant cause for feature interaction problems. (Obviously, there are other causes, too. We discuss them briefly in the end of this section.)

3.1   *Example 2 – Call Waiting and Three-Way Calling (CW & TWC)*

Description [2].   "The signalling capability of customer premises equipment (CPE) is limited. As a result, the same signal can mean different things depending on which feature is

Table 1: a summary of the mode confusions found in the feature interaction benchmark of Cameron *et al.*

benchmark example no.	benchmark example ID	number of mode confusions	benchmark example no.	benchmark example ID	number of mode confusions
1	CW&AC	–	12	OCS&CF/2	–
2	CW&TWC	2	13	CW&ACB	–
3	911&TWC	1	14	CW&CW	2
4	TCS&ARC	–	15	CW&TWC/2	1
5	OCS&ANC	–	16	CND&UN	–
6	Operator&OCS	–	17	CF&CF	–
7	CCC&VM	2	18	ACB&ARC	–
8	MBS-ED&CENTREX	–	19	LDC&MRC	1
9	CF&OCS	–	20	Hotel	2
10	CW&PCS	1	21	Billing	–
11	OCS&MDNL-DR	–	22	AIN&POTS	–

anticipated. For example, a flash-hook signal (generated by hanging up briefly or depressing a 'tap' button) issued by a busy party could mean to start adding a third party to an established call (Three-Way Calling) or to accept a connection attempt from a new caller while putting the current conversation on hold (Call Waiting). Suppose that during a phone conversation between **A** and **B**, an incoming call from **C** has arrived at the switching element for **A**'s line and triggered the Call Waiting feature that **A** subscribes to. However, before being alerted by the call-waiting tone, **A** has flashed the hook, intending to initiate a three-way call. Should the flash-hook be considered the response for Call-Waiting, or an initiation signal for Three-Way Calling?"

Discussion. This feature interaction can be resolved by a precedence rule. An activated Call-Waiting feature should have a higher precedence than the Three-Way Calling feature, with respect to the interpretation of the flash-hook. This allows both features to work most of the time without introducing a new user signal.

The mode confusion problem in the above particular situation remains, though. The mode of the switching element has changed, but the user issuing the flash-hook signal has not yet noticed it. He/she therefore will be surprised by an unexpected reaction of the system.

The cause of the mode confusion is a race condition between the notification tone for the mode change and the user signal. (A mode change not apparent to the user is called an *implicit mode change* in the literature on human-computer interaction [13, 11].) When **A** plans how to perform the three-way call, **A** will use his/her knowledge about the behaviour of the system, but he/she will concentrate on the "relevant" part for efficiency. This will probably make **A** exclude the Call Waiting feature, even when **A** usually is aware of it. The result is the mode confusion in which **A** expects a dial tone but is connected to the new party instead.

A careful user **A** can avoid the mode confusion, but still remains in an uncomfortable situation. **A** must expect a non-deterministic reaction of the system to the flash-hook signal. **A** can find out the actual current mode only by waiting a short amount of time whether a dial tone becomes audible. Only after this re-synchronization, **A** can either proceed with his/her plan, or adjust it to the new situation of an incoming call.

3.2 Example 2b – Plain Old Telephone Service and Plain Old Telephone Service (POTS & POTS)

There is another mode confusion problem in the same example of the benchmark. Cameron *et al.* do not consider it as a feature interaction for the technical reason that no *incremental* features to the Plain Old Telephone Service (POTS) are involved.

Description, continued [2]. "A similar situation occurs when lifting a handset is interpreted as accepting the incoming call, even though the user's intention is to initiate a call – remember the cases when one picks up the phone in the absence of ringing and somebody is already at the other end of the line. The call processing is behaving just as it was designed to, but some users may be momentarily puzzled."

Discussion. This mode confusion is very similar to the previous one. The cause for the mode confusion is the same.

3.3 Example 3 – 911 and Three-Way Calling (911 & TWC)

The following example has been presented by several other authors in a more dangerous and also more concise variant. We also include this variant, using our own words.

Description [2]. "A Three-Way Calling subscriber must put the second party on hold before bringing a third party into the conversation. However, the 911 feature[4] prevents anyone from putting a 911 operator on hold. Suppose that **A** wishes to aid a distressed friend **B** by connecting **B** to a 911 operator using the Three-Way Calling service. If **A** calls **B** first and then calls 911, **A** can establish the three-way call, since **A** still has control of putting **B** on hold before calling 911. However, if **A** calls 911 first, then **A** cannot put the 911 operator on hold to call **B**; therefore **A** cannot make the three-way call. [...]"

Description of a more dangerous variant (911 and Consultation Call). "**A** calls **B**; **B** suddenly gets distressed, and **A** does a successful Consultation Call to the 911 operator. Then **A** wants to switch back to **B** for further information, but **A** can't do this because **A** first must put the 911 operator on hold, which is prevented by the 911 feature."

Discussion. The second scenario in the upper description where **A** cannot make the desired Three-Way Call is a mode confusion. **A** would succeed in a normal call, but does not in a 911 call. It is likely that **A** is not aware of the mode change concerning call control in an emergency situation. Again, when **A** plans the three-way call, **A** will concentrate on the "relevant" part of the behaviour of the features for efficiency. **A** does not intend to play tricks on the 911 operator and therefore excludes call control aspects from his abstracted mental model of the system.

---

[4]In North America, dialling 911 connects to an emergency service.

Discussion of the variant. The system and the abstracted mental model of it of **A** are clearly in different modes here. The cause is the same as for the mode confusion above. A correct plan for **A** would have been to initiate a Three-Way Calling call instead of a Consultation Call.

3.4   Example 7 – Credit-Card Calling and Voice-Mail service (CCC & VM)

Description [2].   "Instead of hanging up and then dialling the long distance access code again, many credit-card calling services instruct callers to press [#] for placing another credit-card call. On the other hand, to access voice mail messages from phones other than his/her own, a subscriber of some Voice-Mail service such as *Aspen* can (1) dial the Aspen service number, (2) listen to the introductory prompt (instruction), (3) press [#] followed by the mail-box number and passcode to indicate that the caller is the subscriber, and then (4) proceed to check messages. However, when a customer places a credit-card call to Aspen, the customer does not know exactly when the Credit-Card Calling feature starts passing signals to Aspen instead of interpreting them itself. Suppose that **A** has frequently called Aspen and knows how to interact with Aspen. When **A** places a credit-card call to Aspen, **A** may hit [#] immediately without waiting for the Aspen's introductory prompt. However, the [#] signal could be intercepted by the credit-card call feature; hence it is interpreted as an attempt to make a second call."

Discussion.   The abstracted mental model of the system already has switched to the voice mail mode, while the system still is in credit-card mode. The user wanted to plan a shortcut in the voice mail service. In order to do this, he constructed an abstraction of the relevant parts of the telephone system. But in the abstraction step he made the mistake of dropping the entire Credit-Card Calling feature, even though it was still latently active.

Even with a correct abstraction, a potential for mode confusion remains. The user cannot observe when the system switches from credit-card mode to voice-mail mode. This happens somewhere between dialling the last digit of the Aspen access number and the end of Aspen's introductory prompt. We have an implicit mode change, again. The user can re-synchronize only by waiting, as above.

3.5   Example 10 – Call Waiting and Personal Communication Services (CW & PCS)

Description [2].   "Call Waiting is a feature assigned to a *directory number*. However, Call Waiting uses the status of the *line* with which the number is associated to determine whether the feature should be activated: at present in a public switched telephone network, if a non-ISDN line is in use, then it is busy; a second call to the same line will trigger the switching element to send out a call-waiting tone. PCS[5] customers may not all be subscribers of Call Waiting. Suppose that **X** and **Y** are both PCS customers currently registered with the same CPE[6]; **X** has Call Waiting but **Y** does not. We further assume that **Y** is on the phone when somebody calls **X**. Since **X** has Call Waiting and the line is busy, the new call triggers the Call Waiting feature of **X**. But is it legitimate to send the call-waiting alert through the line to interrupt **Y**'s call? If not, then **X**'s Call Waiting feature is ignored."

---

[5]personal communication services
[6]customer premises equipment

Discussion.　If **Y** is alerted, this will cause a mode confusion for **Y** due to incorrect knowledge about the system. **Y** has no Call Waiting and does not know about the additional alerting mode the system can get into. When **Y** is alerted, **Y** probably does not know how to leave the mode that causes this annoying signal.

Personal Communication Services (PCS) will show mode confusions when combined with several other features, too. When a user shares a line with other users through PCS, his/her mental model of the system must also comprise a significant part of the features of the other users. Since this often will not be the case, the system can easily get into modes he/she does not know of, similar as in the above example.

3.6　Example 14 – Call Waiting and Call Waiting (CW & CW)

We have two mode confusion problems here.

Description, first part [2].　"Call Waiting allows a subscriber to put the other party on hold. However, it does not protect the subscriber from being put on hold. Confusion can arise when two parties exercise this type of control concurrently. Suppose that both **A** and **B** have Call Waiting, and **A** has put **B** on hold to talk to **C**. While on hold, **B** decides to flash the hook to answer an incoming call from **D**, which puts **A** on hold as well. If **A** then flashes the hook expecting to get back to the conversation with **B**, **A** will be on hold instead, unless either **B** also flashes the hook to return to a conversation with **A** or **D** hangs up automatically returning **B** to a conversation with **A**."

Discussion.　**A**'s mental model of the system does not include the possibility of being put on hold while exercising the Call Waiting feature. This incorrect knowledge about the system causes the mode confusion.

Description, second part [2].　"An ambiguous situation arises, when **B** hangs up on the conversation with **D** while **A** is still talking to **C**; there are two separate contexts in which to interpret **B**'s action. Assume that CW1 refers to the Call Waiting call among **C-A-B** and CW2 refers to the one among **A-B-D**. According to the specification of Call Waiting, in the context of CW2 **B** will be rung back (because **A** is still on hold) and, upon answering, become the held party in the CW1 context and hear nothing. But in the context of CW1 the termination **B** will be interpreted as simply a disconnection, thus **A** and **C** are placed in a normal two-way conversation, and **B** is idled. The question is: Should **B** be rung back or should **B** be idled?"

Discussion.　**B**'s incorrect knowledge about the implemented system can cause a mode confusion. The incomplete requirements specification for this combinations of features must be disambiguated by the implementers, in one way or the other. The choice is not obvious. User **B** must also disambiguate the situation, but may well take the opposite choice, even without noticing that there are others. In case that **B** expects to be idled after hanging up, but actually is rung back, this results in a mode confusion for **B**. **B** interprets the ringing as a new incoming call but then hears nothing when answering. In case that **B** is idled but expects to be rung back, this results in a period of mode confusion while **B** waits to be called back in vain.

3.7 *Example 15 – Call Waiting and Three-Way Calling (revisited) (CW & TWC / 2)*

Description [2].   "Consider how Call Waiting and Three-Way Calling might interact in the situations where a user can exercise both features simultaneously on the same line. The call control relationship can now become quite complicated. Suppose that **A** has both Call Waiting and Three-Way Calling, and **A** is talking to **B**. Now **C** calls **A**, so **A** uses Call Waiting to put **B** on hold and talks to **C**. **A** may decide to have **B** join his conversation with **C**, so he puts **C** on hold, makes a second call to **B**, and after **B** answers the call with Call Waiting, **A** brings **C** back into the conversation to establish a three-way call. There are three contexts in this establishment: a Call Waiting call and a Three-Way Calling call, both established by **A** among **B-A-C**, and a Call Waiting call established by **B** as **A-B-A**. Now, if **B** hangs up, then according to the contexts established by **A**, the session becomes a two-way call between **A** and **C**; according to the contexts established by **B** though, **B** should get a ring-back because **B** still has **A** on hold."

Discussion.   **B**'s incorrect knowledge about the implemented system can cause a mode confusion exactly as in the previous example.

3.8 *Example 19 – Long distance calls and Message Rate Charge services (LDC & MRC)*

Description [2].   "Each long distance call consists of at least three segments – two local accesses at each end and one provided by an interexchange carrier in between. Should a customer be charged for the segments that have been successfully completed even if the call did not reach its final destination? Would it be counted as one unit toward the total local units allowed per month for a Message Rate Charge service?"

Discussion.   First of all, this is a problem of ambiguous requirements, but there can also be a mode confusion because of incorrect knowledge about the behaviour of the system. Because of the difficult billing questions, the user can easily have false expectations on the behaviour. A call segment may be in a charged connection mode earlier than expected. The user will notice this confusion only much later, when he receives the bill. For example, he might be charged for long distance call attempts that he knows were never completed. An overrun of the allowed Message Rate Charge units could in principle also be a surprise, but it is much less likely that the user really counts all his local calls.

3.9 *Example 20 – Calling from hotel rooms (Hotel)*

Description [2].   "Many hotels contract with independent vendors to collect access charges for calls originated from phones in their premises. Without being able to access to the status of call connections, some billing applications developed by these vendors use a fixed amount of time to determine if a call is complete or not – thus one can be billed for incomplete calls that rang a long time, or not billed for very short duration calls (even long distance)."

Discussion.   This is another mode confusion with a particularly long delay. The user will detect it several days later when paying the hotel bill. The cause is incorrect knowledge about the behaviour of the system. The user is usually not informed about the unusual way to determine the start of billing by a timer and incorrectly assumes that completing a connection starts billing. In case that the user knows the behaviour of the system correctly, a mode confusion due to the implicit state change can still occur, that is, due to incorrect observation. This happens when the system and the user perceive a call duration just below / just above the threshold due to imprecise time measurement.

3.10　Benchmark Feature Interactions Which are No Mode Confusions

The remaining examples present undesired feature interactions with a wide range of causes. Often, there are ambiguous, incomplete, or conflicting requirements. In some examples, the restrictions of the current implementation cause a problem, or the implementation is just deficient. In all of these examples, there are either no surprising modes, or the user is not actively involved.

3.11　A Non-Benchmark Example – Key Lock and Volume Adjust (Lock & Vol)

This example is not from the benchmark, but its causes are particularly interesting with respect to our classification. Our colleague Axel Lankenau experienced this problem, we report it here.

Description (by ourselves).　"Our colleague's mobile phone has the feature to lock its keys. This prevents unintended commands while carrying it in the pocket. The lock mode is indicated permanently by a small key symbol on the display. The lock can be released only by pressing the pound key for a long time. If any key is pressed, the lock mode is shown clearly on the display ("press '#' to unlock"), such that the user knows that he must unlock the phone before any further usage. There is one exception to the lock: when the phone rings, the user can press the hook button to accept the call. The phone remains locked otherwise. The phone also has two buttons at the side of its case which allow to adjust the volume of the speaker. It happened that our colleague carried his phone in the pocket in locked mode, and the phone rang. He took out the phone, pressed the hook button to accept the call, and held the phone to his ear. He then noticed that the volume level was not right and tried to adjust it. This did not work, and it surprised and annoyed him."

Discussion.　This problem has three causes: incorrect processing by the user, incorrect observation for psychological reasons, and incorrect observation for physical reasons. First, there was a slip of memory when our colleague did not remember that his phone was in locked mode after some time of non-use. This was incorrect processing. Second, he committed a lapse as he did not look at the display while accepting the call. He can do it without looking and therefore missed to check the mode. This was an incorrect observation for psychological reasons. Afterwards, he could not see the display of the phone while it was close to his ear. This was an incorrect observation for physical reasons.

　　A solution to the problem could be: when the user presses any key in the locked mid-call mode, the phone not only shows a textual warning message, but also generates an unambiguous warning beep tone. Another solution could be to redesign (and weaken) the lock feature such that it releases the lock entirely when the user accepts a call.

Table 2: the causes of the mode confusions in the benchmark and of our one extra example.

cause \ ID no.	CW&TWC 2	POTS&POTS 2b	911&TWC 3	CCC&VM 7	CW&PCS 10	CW&CW 14	CW&TWC/2 15	LDC&MRC 19	Hotel 20	Lock&Vol —
incorrect observation									•	••
incorrect knowledge				•	•	••	•	•	•	
incorrect abstraction	•	•	•	•						
incorrect processing										•

4 Does the Safety-Critical System's Notion of Mode Confusion Work in Telephony?

4.1   Is the Classification of Causes Useful For Telephony?

Our classification of causes for mode confusions in safety-critical systems shows a distribution of causes in telephony which is still reasonable. Table 2 classifies all mode confusions in the feature interaction benchmark of Cameron *et al.* [2] and also our one extra example according to their causes. The dominant cause is incorrect knowledge of the user about the system (7 cases). Also important is an incorrect abstraction of the user's knowledge to the relevant parts of it (4 cases). Rare is an incorrect observation by the user (1 case). One cause does not appear in the benchmark, but in our extra example: incorrect processing by the user (1 case). This extra example also shows more incorrect observations by the user (2 cases). One cause does not appear at all: incorrect knowledge of the user about his/her own senses.

Two of the four classes appear to be less important for telephony: incorrect observation by the user and incorrect processing by the user. One can suspect that this is only because the authors of the benchmark concentrated on "technical" problems and were not interested in human factors problems. But we would need more empirical data to support this.

If incorrect observations should not be relevant, this is not a problem for our approach. The observation relation SENSE just becomes a one-to-one mapping. It must remain nevertheless in the rigorous definition of mode confusion. It ensures "type correctness" for the events.

If incorrect processing by the user should occur only rarely, this is even an advantage for our approach. The user then sticks more closely to the mental model on which our mode confusion analysis is based.

4.2   What Does Not Fit?

Our definition of mode confusion for safety-critical systems does not fit nicely for telephone switching systems with one respect: the user does not abstract to safety-relevant aspects of the system, but to the set of telephony features which are relevant currently.

4.3   How Can We Adapt the Definition to Telephone Switching Systems?

We must use a different kind of abstraction. The telephone user does not abstract to the safety-relevant behaviour $\mathrm{REQ}^M_{\mathrm{SAFE}}$. Instead, the user abstracts his/her knowledge about the behaviour of the telephone switching system REQ^M to the behaviour $\mathrm{REQ}^\mathrm{M}_\mathrm{R}$ of those features which are "relevant".

The relevant features are those that are currently active or can become active. They must be active for the user considered and in the scope of time considered. A feature is active if it can contribute to the visible behaviour of the system. A feature can be activated either by the user considered or by another user in the telephone network. The relevant scope of time ends when all features involved become inactive. Often this is the end of the current call. Sometimes, the scope of time is not limited at all, if the effects of a feature's behaviour are permanent. An example is the billing of calls from hotel rooms. The money spent will never return.

We can abstract the actual behaviour of the entire telephone switching system REQ to the behaviour REQ_R of the relevant features in the same way as the user abstracts REQ^M to $\mathrm{REQ}^\mathrm{M}_\mathrm{R}$. The same holds for $\mathrm{SENSE}/\mathrm{SENSE}_\mathrm{R}$ and $\mathrm{SENSE}^\mathrm{M}/\mathrm{SENSE}^\mathrm{M}_\mathrm{R}$. We need all these for the adapted definition of mode confusion. The same criterion for the relevance of a feature applies. The adapted definition is:

Definition 2 (Mode confusion in telephony) A *mode confusion in telephony* between $\mathrm{SENSE}_\mathrm{R}(\mathrm{REQ}_\mathrm{R})$ and $\mathrm{SENSE}^\mathrm{M}_\mathrm{R}(\mathrm{REQ}^\mathrm{M}_\mathrm{R})$ occurs if and only if $\mathrm{SENSE}_\mathrm{R}(\mathrm{REQ}_\mathrm{R})$ is not a failure refinement of $\mathrm{SENSE}^\mathrm{M}_\mathrm{R}(\mathrm{REQ}^\mathrm{M}_\mathrm{R})$, i.e., iff $\mathrm{SENSE}^\mathrm{M}_\mathrm{R}(\mathrm{REQ}^\mathrm{M}_\mathrm{R}) \not\sqsubseteq_F \mathrm{SENSE}_\mathrm{R}(\mathrm{REQ}_\mathrm{R})$.

5 How Can the Notion of Mode Confusion Help Against Feature Interactions?

Attention to mode confusions helps to design features and sets of features with less undesired surprises.

The design should help the user to abstract his/her mental model. The correct abstraction to the relevant features is difficult for a user. Table 2 shows this. In particular, it is difficult to determine which features are active currently.

Enhanced feedback helps. The system must notify the user when a feature becomes active or inactive. For example remember the interaction between credit-card calling and voice mail in benchmark example 7. The announcement of the voice mail service must make clear the point of time from which on voice mail commands may be entered. The credit card feature must announce from which point of time on its command processing is suspended. Unfortunately, we cannot improve the feedback by a richer hardware interface, for example with many indicator lights. The hardware costs prevent us from installing better customer premises equipment everywhere. Improved switch-side feedback is possible, though. An example are announcements.

The designer of a new feature must always check if it is obvious to the user whether the feature is active or not. The designer must also check whether the user perceives the feature as a single entity. If necessary, the design must be changed.

Correct abstraction is easier if the active features are simple and if only a few are active.

The design should limit the complexity of the abstracted mental model. Two factors in particular increase this complexity: a long duration of feature activation and non-determinism.

The longer a feature is active, the higher is the chance that its period of activity overlaps the period of activity of other features. This increases the number of features that the user must include into his/her abstracted mental model. A feature should terminate its activity after the end of a call, if possible. The purpose of some features does not allow this. In this case the user should get appropriate feedback on the set of active features, at least.

It is difficult for a user to interpret an observed sequence of events correctly on a non-deterministic model. This requires to follow simultaneously several alternative paths in the model. It can exhaust a user's mental memory capacity soon. The system therefore must give the user an immediate feedback signal about any internal choice that changes its behaviour.

A distributed system such as the telephone network inherently exhibits a lot of non-determinism to the individual user. The user cannot perceive what other users do. The user's senses SENSE mask out all telephone usage events in the world except of the local events. Again, feedback for relevant events is necessary. For example, remember the interaction of the call waiting feature and the three-way calling feature in the benchmark example 2. It persists even after a precedence rule has been added. We can avoid the problematic race condition if the newly activated call waiting feature proceeds in two steps: first, it informs the user about its activation. Second, only after it has completed this, it accepts hook flashes. For this, the feature could either use two signals with, e. g., a delay of one second, or it could just make the system ignore hook flashes for one second, starting with the signal tone, until the user must have noticed the mode change.

The design should help the user to know the system correctly.   A prospective user must be able to learn the behaviour of a feature easily. Either it must be intuitive to use, or the user must be trained suitably. Our telephone provider offers us a few use-cases only as the description of a newly provisioned feature. These leave many questions about the behaviour open. Research on better teaching material is required here.

A feature must be redesigned if the user cannot learn its behaviour. An example is the activation of Call Waiting (CW) and Personal Communication Services (PCS) on the line of another PCS user not subscribed to CW (example 10 in the benchmark). CW-PCS must not "hi-jack" the line without telling its current user what is going on.

An improved development process.   The operator directed design process ODP for avionics by Vakil and Hansman, Jr. [23] produces the user training material even before the software specification. If the system appears to be difficult to handle for its users, it is redesigned immediately. The same process can be applied to telephone switching.

Rushby [3] proposes an iterated development process for shared-control systems which model-checks an abstracted system against an abstracted mental model. The mental model is derived from the user training material such that it matches the mental model of an average user. The goal is to detect potential for mode confusions early. Our rigorous definition of mode confusion is a suitable foundation for such a tool supported development process, both for shared-control systems and in telephony.

A potential obstacle for this model-checking approach is the complexity of a complete telephone switching system. The analysis might be infeasible because of the state space explosion problem. A potential solution is to analyse small sets of features at a time. We need statistical information for this about which features are used together most often. These combinations are most likely to annoy customers if they are prone to mode confusions.

Online mode confusion detection and resolution. Shared-control systems can be designed with an "intelligent" interface component. This component monitors the behaviour of both the rest of the system and of the user. If it detects potential for mode confusion in the current situation, it becomes active and resolves the problem. For example, it may give the user additional information about the mode in which the system currently is. When there is no problem ahead, it is silent and does not distract the user. It detects mode confusions online, at run-time. In each situation, the component model-checks the currently active features only. We do not need to model-check the entire system. This reduces the complexity of the analysis. A potential difficulty are the increased computational costs for the switch. A research project on this kind of intelligent interface started at the University of Bremen in the end of 2002. The application domain of this project is spacial cognition and robotics. But we expect that we can transfer its results to telephony directly.

6　What Remains to Be Done?

The aim of this paper is to present the new way in which one can view and tackle feature interactions. We demonstrate that many undesired telephony feature interactions are also shared-control mode confusions. Several measures for preventing mode confusions are known in the literature on human-computer interaction. We show that these measures can be applied to this kind of feature interaction. We sketch several more measures for the telephony domain. The next step should be to apply the existing ideas in the literature to feature interactions practically, and to work out the ideas sketched in the previous section.

The relation between ease of abstraction and the structure of the features deserves more research. Are there any additional feature design rules that help the user to abstract to the currently active set of feature behaviour correctly? Research in shared-control systems is interested in the minimal safe mental model [22, 25]. This model is the "smallest" abstraction that is failure equivalent to the safety-relevant part of the behaviour of the technical system [24]. We should find out how the user can have smaller abstracted mental models of the telephone switching system without experiencing mode confusions.

Acknowledgements

We thank Dr. Axel Lankenau for sharing his invaluable expertise on mode confusions with us.

References

[1] CHARLES E. BILLINGS. "Aviation automation: the search for a human-centered approach". Human factors in transportation. Lawrence Erlbaum Associates Publishers, Mahwah, N.J. (1997).

[2] E. JANE CAMERON, NANCY D. GRIFFETH, YOW-JIAN LIN, ET AL.. A feature interaction benchmark in IN and beyond. In L. G. BOUMA AND HUGO VELTHUIJSEN, editors, "Feature Interactions in Telecommunications Systems", pages 1–23, Amsterdam (1994). IOS Press.

[3] JOHN RUSHBY. Modeling the human in human factors. In UDO VOGES, editor, "Computer Safety, Reliability and Security – 20th Int'l Conf., SafeComp 2001, Proc.", volume 2187 of "LNCS", pages 86–91, Budapest, Hungary (September 2001). Springer.

[4] D. A. NORMAN. Some observations on mental models. In D. GENTNER AND A. L. STEVENS, editors, "Mental Models". Lawrence Erlbaum Associates Inc., Hillsdale, NJ, USA (1983).

[5] J. J. CAÑAS, A. ANTOLÍ, AND J. F. QUESADA. The role of working memory on measuring mental models of physical systems. *Psicológica* **22**, 25–42 (2001).

[6] J. CROW, D. JAVAUX, AND J. RUSHBY. Models and mechanized methods that integrate human factors into automation design. In K. ABBOTT, J.-J. SPEYER, AND G. BOY, editors, "Proc. of the Int'l Conf. on Human-Computer Interaction in Aeronautics: HCI-Aero 2000", Toulouse, France (September 2000).

[7] R. W. BUTLER, S. P. MILLER, J. N. POTT, AND V. A. CARREÑO. A formal methods approach to the analysis of mode confusion. In "Proc. of the 17th Digital Avionics Systems Conf.", Bellevue, Washington, USA (1998).

[8] JOHN RUSHBY. Analyzing cockpit interfaces using formal methods. In H. BOWMAN, editor, "Proc. of FM-Elsewhere", volume 43 of "Electronic Notes in Theoretical Computer Science", Pisa, Italy (October 2000). Elsevier.

[9] RACHID HOURIZI AND PETER JOHNSON. Beyond mode error: Supporting strategic knowledge structures to enhance cockpit safety. In "HCI 2001, People and Computers XIV", Lille, France (10–14 September 2001). Springer.

[10] N. SARTER AND D. WOODS. How in the world did we ever get into that mode? Mode error and awareness in supervisory control. *Human Factors* **37**(1), 5–19 (1995).

[11] A. DEGANI, M. SHAFTO, AND A. KIRLIK. Modes in human-machine systems: Constructs, representation and classification. *Int'l Journal of Aviation Psychology* **9**(2), 125–138 (1999).

[12] BETTINA BUTH. "Formal and Semi-Formal Methods for the Analysis of Industrial Control Systems". Habilitation thesis, University of Bremen, Germany (2001).

[13] N. G. LEVESON, L. D. PINNEL, S. D. SANDYS, S. KOGA, AND J. D. REESE. Analyzing software specifications for mode confusion potential. In "Workshop on Human Error and System Development", Glasgow, UK (1997).

[14] GAVIN DOHERTY. "A Pragmatic Approach to the Formal Specification of Interactive Systems". PhD thesis, University of of York, Dept. of Computer Science (October 1998).

[15] HAROLD THIMBLEBY. "User Interface Design". ACM Press, New York, USA (1990).

[16] PETER WRIGHT, BOB FIELDS, AND MICHAEL HARRISON. Deriving human-error tolerance requirements from tasks. In "Proc. of ICRE'94 – IEEE Int'l. Conf. on Requirements Engineering", Colorado, USA (1994).

[17] JAN BREDEREKE AND AXEL LANKENAU. A rigorous view of mode confusion. In STUART ANDERSON, SANDRO BOLOGNA, AND MASSIMO FELICI, editors, "Computer Safety, Reliability and Security – 21st Int'l Conf., SafeComp 2002, Proc.", volume 2434 of "LNCS", pages 19–31, Catania, Italy (September 2002). Springer.

[18] A. W. ROSCOE. "The Theory and Practice of Concurrency". Prentice-Hall (1997).

[19] C. A. R. HOARE. "Communicating Sequential Processes". Prentice-Hall (1985).

[20] M. A. SASSE. "Eliciting and Describing Users' Models of Computer Systems". PhD thesis, School of Computer Science, The University of Birmingham (April 1997). Available online at www.cs.ucl.ac.uk/staff/A.Sasse/thesis/LINK_ME.html.

[21] JAMES REASON. "Human error". Cambridge University Press (1990).

[22] D. JAVAUX. Explaining Sarter & Woods' classical results. The cognitive complexity of pilot-autopilot interaction on the Boeing 737-EFIS. In "Proc. of HESSD '98", pages 62–77 (1998).

[23] S. S. VAKIL AND R. J. HANSMAN, JR. Approaches to mitigating complexity-driven issues in commercial autoflight systems. *Reliability Engineering & System Safety* **72**(2), 133–145 (February 2002).

[24] AXEL LANKENAU. "The Bremen Autonomous Wheelchair 'Rolland': Self-Localization and Shared Control". PhD thesis, University of Bremen, Germany (August 2002).

[25] DENIS JAVAUX. A method for predicting errors when interacting with finite state systems. How implicit learning shapes the user's knowledge of a system. *Reliability Engineering & System Safety* **72**(2), 147–165 (February 2002).

*Feature Interactions in Telecommunications
and Software Systems VII*
D. Amyot and L. Logrippo (Eds.)
IOS Press, 2003

Context and Intent in Call Processing

Tom GRAY[1], Ramiro LISCANO[2], Barry WELLMAN[3],
Anabel QUAN-HAASE[3], T. RADHAKRISHNAN[4], Yongseok CHOI[4]

[1] Pinetel, Ottawa, Canada
[2] Carleton University, Ottawa, Canada
[3] University of Toronto, Toronto, Canada
[4] Concordia University, Montréal, Canada

Abstract. A new feature set suited for IP telephony is described. The user value of
this feature set is discussed in terms of social science results. This feature set supports
natural human collaboration within a human environment. The use by humans of
casual awareness to support collaboration is discussed and models taken from
architectural research are used to show how casual awareness may be facilitated
within the physical and communication environments. A tripartite policy-based
architecture which can support this new feature set is described.

1 Introduction

Mobility at both at the personal and device levels will be the hallmark of VoIP systems.
The network will be able to track user location through both automatic and manual
registration services [1]. Personal wireless mobile devices will supply the user with always-
on connectivity. These capabilities of VoIP and wireless technologies will force a sea
change in the types of features that are supplied by telephony systems. Current systems find
much of their customer value by being able to find the called user. Features such as voice
mail and the various forms of call forwards were designed to find the called user through
both space and time. Next generation systems will be able to locate the user at all times.
The current feature set will be a supplanted by a new one that is tailored to the specific
capabilities and limitations of VoIP.

This paper is a brief description of a major project that was sponsored by Mitel
Networks Corporation. Research was sponsored with a number of academic research
groups and undertaken internally to investigate the types of features that will be supplied by
VoIP systems. Work was undertaken to determine the basic customer value of IP telephony
features and feasible architectures that could provide such services.

2 Features in IP Telephony

2.1 *Location Systems and Their Effect on Feature Value*

Location awareness is an integral part of the mobility capability of the VoIP architectures
and protocols being designed today. This type of awareness raises fundamental issues about
the utility and value of the feature set supplied with current communication and specifically
telephony systems. If users can be found at all times, the implicit ability of current system
systems to filter calls by user location will no longer be available. Users' collaborators will

know that they are locatable by the system and will expect that they will be available for what they consider important calls at all times. However, the uncontrolled use of location awareness and mobility services can be disruptive, as when cell phones ringing and conversations bother others in the vicinity.

Despite such annoyances, the capability of always-on, always-locatable systems can have much merit and can open the possibility for a completely new feature set. These features will enable a new type of customer value whose effect and worth have been the object of many years of social science research. They find their value by the enabling of greater user visibility and thus greater capability for informal and ad hoc interactions. We are already witnessing some of this in the new presence-based Instant Messaging (IM) communication services like Microsoft Messenger, Yahoo Messenger, and ICQ. All of these services are based on the ability of users being able to detect when other users are logged onto their computing device.

2.2   Social Science Background of New Feature Types

In bureaucratic organizations, communication between employees is regulated by the formal structure of the organization [9,10]. Who talks to whom is to some extent prescribed in the organizational chart. The information revolution has brought about changes in organizations' communication structures leading to new forms of work, such as networked organizations, virtual organizations, and high-performance teams. In these new forms of work, organizations have moved from being bound up in hierarchically arranged, relatively homogeneous, densely knit, bounded groups ("little boxes") to become social networks [11,12]. In a networked society, boundaries are more permeable, interactions are with diverse others, links switch among multiple networks, and hierarchies are flatter and more complex. In networked organizations, informal communications become essential because they provide employees with important information, new ideas, work opportunities, and influence [13]. Besides leading to the exchange of information, social relationships are responsible for the provision of tacit knowledge: knowledge that is sticky, difficult to transfer and comprises insights and deep understandings [15].

Research showing the relevance of informal communication for the success of an organization was published in the 1960's with the pioneering work of Thomas J. Allen of MIT [2]. His study shows that the success of R&D projects is related to the amount of informal interactions both inside as well as outside of the design team. Rob Cross at IBM's Institute of Knowledge Management showed that tacit knowledge and ideas are provided by close working relationships [14]. It is not the informal communication per se, that seems to be responsible for the success of teams, but specifically the opportunities for problem-oriented and unplanned, spontaneous interactions that allow people to take advantage of the collective knowledge available in the team. In the architectural field, Bill Hillier provided evidence for the importance of office space in fostering unplanned interactions [6]. Hillier recognized that it is *casual visibility* that promotes and facilitates informal interactions. His comparison of high-tech labs shows that open spaces lead to more frequent and spontaneous conversations between colleagues working on different teams promoting cross-pollination of ideas and innovation.

AT IBM, Thomas Erickson, Wendy Kellogg, and Erin Bradner have studied the communication system, BABBLE, which promotes through visibility interactions among employees [15,16]. The study shows that visibility of employees' work activities promotes conversations leading to higher performance. They apply the notion of "social affordances" to describe how the interface and its ability to depict the social world facilitates information exchange. BABBLE has three important features: 1) a display of all the actors who are online, 2) a display of their current status of activity (active/inactive), and 3) a display of their involvement with other actors. In this way, the system creates an interface that allows its members to recognize each other's availability and current status of activity, enabling and inviting toward spontaneous interactions. Communication systems similar to BABBLE are especially valuable for colleagues, who work virtually and thus cannot monitor each other in the same way that colleagues working co-located can. While many companies still work co-located, the number of dispersed teams and virtual teams is increasing.

2.3   Features as Proactive Availability Filters

The results of Allen, Hillier and other social science researchers are useful for developing the always-on, always-locatable capabilities of next generation communication systems. These systems will move beyond the 'user-finding' value of current systems to attempt to engender the types of informal activities as Hillier is doing in his building designs. Hillier is accomplishing this by increasing the casual visibility within the office space. New communication systems could do the same thing by proactively projecting the availability of potential collaborators to other users. Instead of the typical telephone tag played with today's systems, users with a glance at their telephone displays will see who is available for conversation. They will then be able to make quick informal calls or set up informal conferences to discuss issues of the moment. Voice mail has turned POTS into an asynchronous messaging system. These new capabilities will return the telephone to being a synchronous device for spontaneous conversation.

Availability has been implicitly defined above as the willingness to talk to someone about a topic of the caller's interest. This definition reveals much on how these new VoIP systems will work. Current systems attempt to locate called users. Next generation systems will be always aware of where a user is [7]. They will thus act more as filters that allow users to be visible to their colleagues and yet protect them from unsuitable interaction. In the physical office that Hillier is providing, this can be provided by visible clues. Users can indicate their desire for momentary privacy by the position of their door, their posture, the volume of their voices etc. Users are in multiple physical locations, all of which will have rules that indicate the types of interaction that may take place there. Some of these physical clues have already been implemented into electronic communication systems like the telepresence project described by Buxton et al. [8].

These types of informal interaction can be replicated with the mobile. always-on capabilities of new systems. Users will be reachable not only at their desk as in today's systems but wherever they happen to be. The result of this is that user interaction in the new systems will take place in multiple locations, which will necessarily have multiple expectations for proper feature behavior. Privacy indicators and filters equivalent to those available in the physical office space will have to be provided. Feature operations that are suited to users' private offices will be entirely unsuitable when users are located in meeting rooms. Operations suited to users working alone in their offices will be unsuitable when they have important visitors. To be truly useful, an always-on network must supply means to adapt its operation to these differing circumstances in the environments of its users. The system will be required to take action that is consistent with the current situation of users, both the physical environment and the social context.

2.4 Roles and Groups in the Social Sciences

The social science work discussed above reveals that users often collaborate in groups and social networks. It is these groups and networks that provide users with the capability of informal interaction that brings the benefits that these new communication systems seek. As a result, communication systems will function best if they operate in cognizance of the actual structure and functioning of groups and social networks.

Groups contain participants, each of whom has a status that is equivalent, for example, to a job title. These statuses may be formal or informal. Examples could be: manager, Java guru, secretary, company director, friend, etc. In an enterprise situation, users will participate in one or more groups and social networks. As such, they will have one or more statuses that define their behavior.

Participants relate to each other in role relationships that join a pair of statuses. Most importantly to a call-processing discussion, participants' behaviors in groups and social networks are defined by the relationships between the roles they fulfill and the roles of their collaborators. Examples of roles are boss-secretary, boss–VIP customer, etc. Roles hold the policies that determine the preferred and suitable interactions between participants. For example, when people play the role of boss in an organization, their behavior as bosses will be determined by their status and the status of those with whom they relate. There will be entirely different behavior and acceptable feature operation in a boss–secretary interaction than in a boss–VIP customer interaction.

The functioning of roles is more complex in social networks than in groups. In groups, users usually have stable statuses and roles, and they usually interact routinely with the same set of group members. In social networks, users move between work teams, and they usually relate to different other users in each team. Moreover, their status and roles in one team may be different than in others. Hence, awareness of social context is necessary for acceptable feature operation.

In both groups and networks, a role is a directional relationship. For example, the boss-secretary role is very different from the secretary-boss role. The policies that direct proper behavior are very different in each direction. The required behavior and operation can be attached to the policies for each role. Call processing to function within the mobile framework of next generations must be sensitive to the identity of the roles that the users are currently playing. Next generation call processing could be improved if it had the means for extracting this role information automatically as well as specifying *a priori* specific rules for classes of statuses (for example all bosses) and specific statuses (for example the person who is the immediate boss).

3   Architecture and Functioning of Context-Aware Call Processing System

3.1 Next Generation Call Processing - CoC and PoA

The discussion above indicates the types of features that will be valuable in next generation systems can be divided into two categories. Existing systems act reactively to user action to provide services for incoming and outgoing telephone calls. This reactive service will be maintained in a modified form in next generation systems. However it will be supplemented by a set of proactive features that will encourage useful informal discussion within an organization. In our work we have classified the reactive feature type as CoC (control of communications) and the proactive feature type as PoA (projection of availability).

Firstly, CoC features will act as call filters that will be aware of users' current situations. Secondly, PoA features will be filters that analyze users' current situations and project their availability to potential collaborators based on that determination.

3.2   Context – Call Processing and Presence Combined

There is much current interest in the issue of presence and availability. Presence is defined as users' current contact information. This could be the directory number of the telephones in the rooms they are occupying, their wireless numbers, the directory numbers of the secretaries who can locate them, etc. Availability is an indication of users' willingness to converse given their current situations.

There are standards groups and industry forums dedicated to the development of standards for these services. Because of the obvious similarity between PoA and presence, we considered carefully the issue of interaction between presence and call processing systems. We were initially concerned about the potential for harmful interactions between these two services. However in working with these systems, we soon realized that this potential was only illusory. The availability indications provided by presence systems are an indication of the users current wish to be contacted. However users can be contacted only if their call processing policies allow it.

No matter what users' availability policies might say, if they have set their call processing features to Do Not Disturb, then they cannot be contacted. Unlike some other standards, there are no specific presence and availability policies. Rather, users' current availability can be determined as a result of a hypothetical call attempt. If call processing indicates that a call will be completed given the users' current policies, then they are available and not otherwise. Presence is really a form of hypothetical call processing. Without this realization call processing and presence would be very difficult to operate together as a system. There would be major problems in making their policies compatible.

We have argued that call-processing features must be aware of users' current situation. The discussion above has shown that users' real presence or availability are determined by their current call processing policies. This indicates to us that presence and call processing are really aspects of a single larger composite entity. It is this entity that will become the focus of next generation call processing. We have identified this entity with context. Next generation telephony systems will have their feature set defined beyond call processing and into this new area of context.

3.3   Physical Context and Human Intent

Context has a research literature of its own. Anind K. Dey, a leading researcher in this field, has defined context [3] as 'any information that can characterize the situation of an entity. An entity is a person, place or object that is relevant to the interaction between a user and an application including the user and applications themselves'. Note that this definition implicitly defines context to be a relationship that controls behavior much like the role relationship in sociology.

User context can be determined from sensing the environment. Context can determine users' location, who they are with, what they are doing on the computer system, when it is, etc. The important aspect of this is that surmises may be made about users' activity in the social context from the sensing of their physical context. That is, from sensing the physical environment, the system may make surmises about which statuses of users are currently active and which roles they are playing. This will allow the selection of features that are appropriate to the current social context. For example, if it is determined that a user is in a meeting room with VIP customers then it can be assumed that she is in the sales status and the current role is the sales-customer role. From this it could be surmised that she does not wish to be disturbed. If the company owner calls this user, then it can be assumed that she is available to take the call no matter when it is.

This is supported by Hillier's development of his model of human interaction based on building structure. Rules can be created which link the physical context to the intent of the user in the human environment. Just as Hillier's office space allows for physical clues to indicate a user's current intent, it is possible for rules to link clues determined by sensors to the user's intent. This linking of physical sensing to the determination of the user's human intent will allow for the selection of features that are appropriate to the user's current circumstances. It is an answer to the issues that wireless and VoIP create when they create the always-on, always-locatable communication systems.

The sensing of the who, where, when, and what in the physical environment will allow the system to surmise the why in the human environment. Next generation call processing will be context-aware so that it can function within the human environment.

3.4   *Tripartite Architecture for Context-Aware Call Processing*

Figure 1 shows a conceptual diagram of a policy-based, tripartite architecture for context-aware call processing. Policies of different forms will exist in each of the basic three sections shown in the illustration.

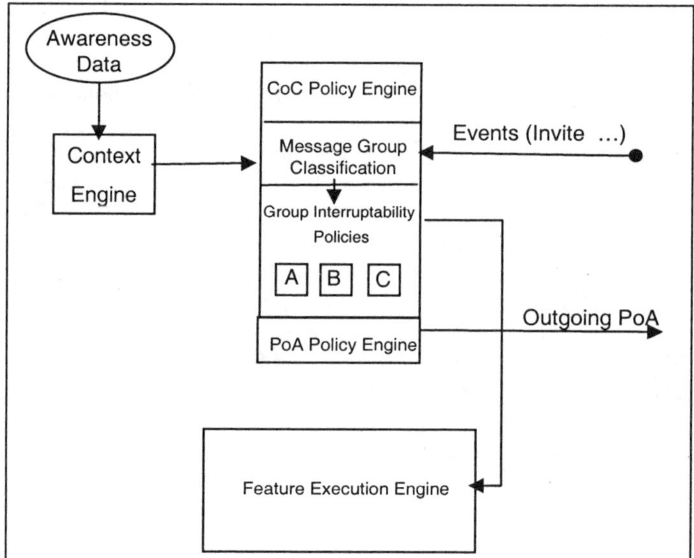

Figure 1: Policy-based, tripartite architecture for context-aware call processing.

Awareness data is raw information about the physical environment. It will be gathered by sensors in users' environments and by analysis of their interactions with the computer and telephony systems. This information will be analyzed by rules in a context engine that will make assertions about the users' current situation. For example, it can take into account that a user is in a meeting room with others and that these others have registered at reception as visitors from an important customer.

The second part of the architecture is concerned with feature selection. Given actions by the user or requests from other users the appropriate feature or action to take can be determined. This feature selection will be done by rules that link the physical context to actions appropriate at the human level. This will implement the CoC filters required.

Availability can be determined by hypothetical requests to this part of the architecture and projected as PoA indicators as shown. Amer et al's policy architecture [17] presented at FIW00 gives an indication as to how these first two sections can function.

Third, once the proper feature has been selected it can be sent to the feature execution engine. We have considered using Barbuceanu's et al [5] OPI policy system in this part of the architecture.

4   Current Work

4.1   Prototype Context-aware Call Processing System

A prototype context-aware call processing system was developed [4]. This system is based on a blackboard architecture implemented on top of a tuple space. This system is based on the tripartite architecture described in this paper. Novel methods of context determination and interaction resolution have been implemented within this system.

4.2   Policy Languages

We are partnering with several university research groups in the development of policy languages for the specification and the detection of feature interaction between policies. We expect that this work can be fitted into the tripartite architecture.

4.3   Context and User Activity Sensing

Work is going on to improve methods of sensing user environments and interpreting this as context. This includes means of sensing user activity on the computer system and physical location sensors to determine users' location and users' co-presence (whom they are with).

4.4   Test and Development Systems

A prototype system to simulate natural human interaction in an office environment has been developed. Users are represented by software agents that autonomously interact within a programmable office environment. These user agents are scripted with high-level goals to perform day-to-day activities such as working in a private office, attending meetings, looking for others to deliver messages, etc. The agents are provided with heuristics so that they can collaborate autonomously in performing their individual goals in typically human manners. This system is based on a blackboard architecture. It can thus be coupled directly to the prototype system described in this paper. Work to accomplish this is now going on.

4.5   Native Call Processing Engine

Questions are raised about the scalability of tuple space systems in which large amounts of information are interpreted in the handling of a single call. Work was done to improve the performance of software tuple spaces by the creation of high efficiency operators. To obviate this issue, a hardware-assisted tuple space was developed. This is a hardware device that is capable of matching tuples at hardware memory speed with the speed limitation being the memory's access time only. This work is now being supplemented to create a hardware policy accelerator in which the policy rules themselves will be matched and executed at memory speed as well. This will enable the development of a native call-processing engine that exploits the possibility of policies.

5   Acknowledgements

We gratefully acknowledge our colleagues at Mitel Networks who were vital in the development of these ideas. We acknowledge the contribution of Katherine Baker and Natalia Balaba for their strong effort in the development of the presence system. We acknowledge the leadership that Daisy Fung and Peter Perry provided us. We acknowledge the contributions of Daniel Amyot, Gunter Mussbacher, Kelvin Steeden, Tonis Kasvand, Serge Mankovskii, Michael Weiss, Babak Esfandiari, Tom Ware, Eliana Peres and all former members of Strategic Technology in the genesis of these ideas. We would like to thank Debbie Pinard for initiating this effort within Mitel. We would also like to acknowledge the academic partners of Mitel Networks who include Ray Buhr, Ahmed Karmouch, Jo Atlee, Luigi Logrippo, Murray Woodside, Ken Turner, Stephan Reiff-Marganiec, Eric Yu, Daniel Gross, Roger Impey, Aurora Diaz, Sue Abu Hakima, Mihai Barbuceanu, Trevor Rainey, Colin Bänger, Hugh McLaren, Dave Athersych, Bernie Pagurek and Nick Dawes. We would like to thank Ernst Munter, Harry Kennedy, the MNS and the RPE for demonstrating how systems really work.

References

[1]   Rosenberg, J. et al, *RFC 3261 SIP: Session Initiation Protocol*, http://www.ietf.org/rfc/rfc3261.txt, accessed on December 15, 2002

[2]   Allen,T., *Managing the Flow of Technology: Technology Transfer and the Dissemination of Technological Information Within the R&D Organization*, MIT Press, Cambridge Massachusetts, 1984

[3]   Dey, A., Abowd, G. *Towards a Better Understanding of Context and Context-Awareness*, GVU Technical Report GIT-GVU-00-18, Graphics, Visualization and Usability Center, Georgia Institute of Technology, 1999

[4]   Choi, Y. *Context Based Delivery of Voice Messages*, Master's thesis, Department of Computer Science, Concordia University, Montreal, 2001

[5]   Barbuceanu, M., Gray, T., Mankovski, S., *Coordinating with Obligations*, Second International Conference on Autonomous Agents, Minneapolis 1998.

[6]   Hillier, B. *Space is the Machine: A Configurational Theory of Architecture*, Cambridge University Press, 1996

[7]   Schilit,B. Adams,N. and Want.R. *Context-Aware Computing Application*. IEEE Workshop on Mobile Computing System and Application, December 1994.

[8]   Buxton, W. (1995). *Integrating the Periphery and Context: A New Model of Telematics* Proceedings of Graphics Interface '95, 239-246.

[9]   Monge, P. R., & Contractor, N. S. (1997). *Emergence of communication networks*. In F. M. Jablin & L. L. Putnam (Eds.), *Handbook of organizational communication* (2nd ed.). Thousand Oaks, CA: Sage Publications.

[10]   Monge, P. R., & Contractor, N. S. (2001). *Theories of communication networks*. Unpublished manuscript.

[11]   Hiltz, S. R., & Wellman, B. (1997). *Asynchronous learning networks as virtual communities*. Journal of the ACM, 40(9), 44-49.

[12]   Wellman, B. (2002). *Designing the Internet for a networked society. Communications of the ACM, 45(5)*, 91-96.

[13]   Uzzi, B. (1996). *The sources and consequences of embeddedness for the economic performance of organizations: The network effect*. American Sociological Review, *61*, 674-698.

[14]   Cross, R., Rice, R. E., & Parker, A. (forthcoming). *Information seeking in social context: Structural influences and receipt of information benefits*. Organization Science.

[15]   Erickson, T., & Kellogg, W. A. (2000). *Social translucence: An approach to designing systems that support social processes*. ACM Transactions on Computer-Human Interaction, *7*(1), 59-83.

[16]   Bradner, E., Kellogg, W. A., & Erickson, T. (1998). *Babble: Supporting conversation in the workplace*. SIGGROUP Bulletin, *19(3)*, 8-10.

[17]   Amer.M. Karmouch, A. Gray, T. Mankovskii, S. (2000) *A multi-agent architecture for the resolution of feature conflicts in telephony*. Proceedings of the International Conference on Practical Application of Intelligent Agents and Multi-Agents, 5th (PAAM 2000), April, Manchester, UK, pages 53-74.

Detection and Resolution
Methods I

*Feature Interactions in Telecommunications
and Software Systems VII*
D. Amyot and L. Logrippo (Eds.)
IOS Press, 2003

Generalising Feature Interactions in Email

Muffy CALDER and Alice MILLER
Department of Computing Science
University of Glasgow
Glasgow, Scotland.
`muffy,alice@dcs.gla.ac.uk`

Abstract. We report on a property-based approach to feature interaction analysis for a client-server email system. The model is based upon Hall's email model [12] presented at FIW'00 [3], but the implementation is at a lower level of abstraction, employing non-determinism and asynchronous communication; it is a challenge to avoid deadlock and race conditions. Our analysis differs in two ways: interaction analysis is fully automated, based on model-checking the entire state-space, and the results are scalable, that is they generalise to email systems consisting of any number of email clients.

Abstraction techniques are used to prove the general results. The key idea is to model-check a system consisting of a constant number (m) of client processes, in parallel with a mailer process and an "abstract" process which represents the product of any number of other (possibly featured) client processes. We give a lower bound for the value of m.

All of the models – for any specified set of client processes and selected features – are generated automatically using Perl scripts.

1 Introduction

We consider modelling features and analysing feature interactions in an *email* system. Our model is derived from Hall's email model [12] presented at FIW'00 [3], but our analysis differs in two significant ways:

- interaction analysis is *fully automated*, based on a model-checking approach,

- results *generalise* to email systems consisting of *any* number of email clients.

We adopt a *property-based* approach to interaction analysis [4], that is we develop an explict model of the basic service and features which is checked against a set of more abstract, temporal properties. Interactions are uncovered through the analysis of property violations. The (parameterised) model is developed in Promela [13], a high-level, state-based, language for modelling (asynchronously) communicating, concurrent processes. Spin is the bespoke model-checker for Promela. Individual models and model-checking runs are generated using Perl scripts.

Our first goal is faithful modelling of an email system as client-server with explicit concurreny and asynchronous communication; this is challenging for a property based approach because of the high degree of concurrency and consequent state-space explosion. Nevertheless feature interaction analysis is comprehensive.

Our second goal is generalisation of interaction results. Model-checking alone is limited to reasoning about a *given* number of processes. This aspect is often overlooked, and proof for a fixed number, say m, processes, is informally assumed to scale up to imply proof for an *arbitrary* number of processes, i.e. for n processes, for any n. In this paper we address the problem explicitly and show how to generalise results without resorting to explicit induction (which is difficult in this case). Our approach is based upon a combination of abstraction and model-checking.

The paper is divided into two parts, in the first part we consider feature interaction analysis for a fixed number of client processes, in the second part, we consider how to generalise these results to an arbitrary number of clients.

In section 2, we give a brief overview of Promela and Spin. In section 3 we give an overview of the basic email service and feature behaviour, the Promela implementation, the properties for the basic service and features and the corresponding LTL formulae. In section 4 we define feature validation and interaction analysis, and give corresponding results for systems of 3 or 4 client processes. We also discuss how we use Perl scripts and the model-checker Spin for analysis. In section 5 we outline the abstraction technique and give results. We conclude in section 6.

2 Reasoning in Spin

Promela is an imperative, C-like language with additional constructs for non determinism, asynchronous and synchronous communication, dynamic process creation, and mobile connections, i.e. communication channels can be passed along other communication channels. Spin is the bespoke model-checker for Promela and provides several reasoning mechanisms: assertion checking, acceptance and progress states and cycle detection, and satisfaction of temporal properties.

In order to perform verification on a model, Spin translates each process template into a finite automaton and then computes an asynchronous interleaving product of these automata to obtain the global behaviour of the concurrent system. This interleaving product is referred to as the *state-space*.

As well as enabling a search of the state-space to check for deadlock, assertion violations etc., Spin allows the checking of the satisfaction of an LTL formula over all execution paths. The mechanism for doing this is via *never claims* – processes which describe *undesirable* behaviour, and Büchi automata – automata that accept a system execution if and only if that execution forces it to pass through one or more of its accepting states infinitely often [13, 11]. Checking satisfaction of a formula involves the depth-first search of the synchronous product of the automaton corresponding to the concurrent system (model) and the Büchi automaton corresponding to the never-claim.

If the original LTL formula f does not hold, the depth-first search will "catch" at least one execution sequence for which $\neg f$ is true. If f has the form $[]p$, (that is f is a *safety* property), this sequence will contain an *acceptance state* at which $\neg p$ is true. Alternatively, if f has the form $\langle\rangle p$, (that is f is a *liveness* property), the sequence will contain a cycle which can be repeated infinitely often, throughout which $\neg p$ is true. In this case the never-claim is said to contain an *acceptance cycle*. In either case the never claim is said to be *matched*.

When using Spin's LTL converter (a feature of XSpin – Spin's graphical interface) it is possible to check whether a given property holds for *All Executions* or for *No Executions*. A universal quantifier is implicit in the beginning of all LTL formulas and so, to check an LTL property it is natural, therefore, to choose the *All Executions* option. However, we sometimes wish to check that a given property (p say) holds for *some state* along *some execution path*. This is not possible using LTL alone. However, Spin can be used to show that "p holds for *No Executions*" is **not** true (via a never-claim violation), which is equivalent. Therefore, when listing our properties (section 3.4), we use the shorthand $\exists p$, meaning *for some path p*, i.e. for No Executions p is not true.

2.1 Parameters and further options used in Spin verification

When performing verification with Spin, three numeric parameters must be set. These are *Physical Memory Available*, *Estimated State-Space Size* and *Maximum Search Depth*. The meaning of the first of these is clear, and the second controls the size of the state-storage hash table. The *Maximum Search Depth* parameter determines the size of the *search-stack*, where the states in the current search are stored. If comparisons are to be made with other model-checkers, then the value of the Maximum Search Depth should be taken into account.

Partial order reduction (POR) [17] is based on the observation that execution sequences can be divided into equivalence classes whose members are indistinguishable with respect to a property that is to be checked. We apply POR in most cases.

Compression (COM) is a method by which each individual state is encoded in a more efficient way. We apply compression in all cases.

3 Basic email service and features

The email system consists of a number of *clients* and one server, in this case the *mailer*. Each client has a unique mail address. Clients send mail messages, addressed to other clients (or themselves) to the mailer; the mailer delivers mail messages to clients. Communication between client and server is asynchronous. Therefore, mail messages are not necessarily received by clients in the (global) order in which they were sent, but local temporal ordering is maintained, i.e. if client 1 sends messages A and B to client 2, in that order, then client 2 will receive message A before message B. We assume (like Hall) that the system does not lose or corrupt messages, because our motivation is feature interaction analysis, not error detection and/or recovery.

We assume (weak) fairness, i.e. an enabled process cannot be ignored infinitely often, when verifying liveness properties (e.g. 3 and 8, see section 3.4). In all other cases, it is not relevant (and just increases the state-space).

The overall system is illustrated in Figure 1. High level, abstract automata for the client and mailer processes are given in Figures 2 and 3, respectively. Note that in these figures, transitions are labelled by conditions, e.g. in Figure 2 a transition from *initial* to *sendmail* is only possible if the channel *mbox* is empty and the channel *network* is not full. Local and global variables are updated at various points; variable assignments omitted from the diagrams. We refer to states in these abstract automata as *abstract states*, these are not to be confused with states in the Promela model.

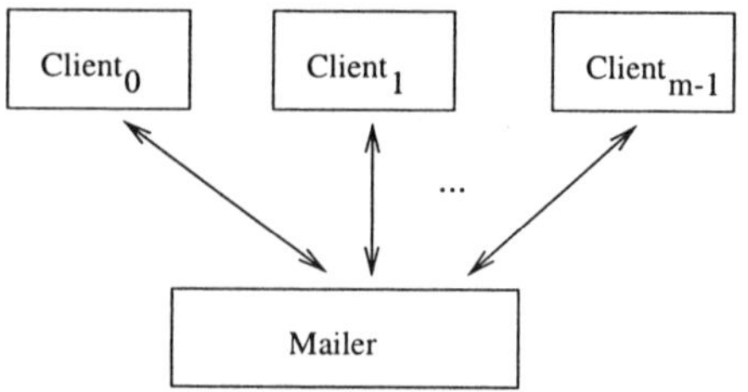

Figure 1: Email service with m clients

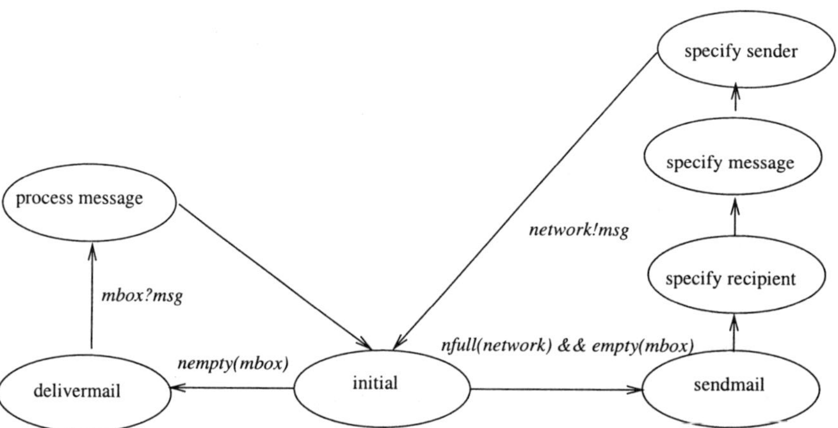

Figure 2: Client process with mailbox *mbox*

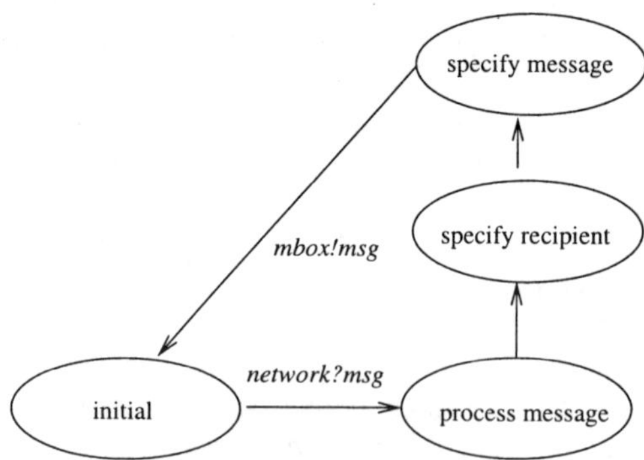

Figure 3: Mailer process

3.1   Basic email service in Promela

The model for m client processes consists of m instantiations of the parameterised proctype *Client*, all in parallel with one instance of the proctype *Mailer*.

A mail message consists of a *sender, receiver, message body*, and *key*. Mail messages may be sent from clients to the mailer, and delivered to clients from the mailer (via a *mailbox* belonging to the receiver). All communication between clients and the mailer is *asynchronous*. We chose to adopt this position because it is closer to actual system behaviour, however we have to take care that it does not result in state explosion. Delivery of mail takes precedence over sending, i.e. a client has to take delivery of any mail which has been delivered by the network, before sending mail. A client can otherwise send mail at any time, provided the channel $network$ has capacity.

The parameter associated with a client process is the identification (a byte) of that process. A client process can either send mail, or have mail delivered, the latter taking precedence over the former. The former can only occur if the network is not full. If neither is possible, the client is, in effect, in *busy waiting*.

Communication between clients and the mailer is via asynchronous, global channels, one for each client and one for the mailer. The sizes of the channels are set using the constants k for the client channels and N for the mailer. During most of this investigation $k = 2$ and $N = 4$. M is a constant denoting the number of clients in the given system and is often used to denote a default, or unassigned, value.

The client channels are, in effect, *mailboxes*. The role of the mailer is therefore to deliver mail messages to the appropriate mailboxes; clients take delivery of message by reading from the appropriate mailbox/channel. In Figure 3, note that *mbox* is a free variable which must be correctly instantiated during the *specify recipient* abstract state.

Mail addresses are simple integers, used to index the client processes. Initially, we implemented a more sohpisticated addressing mechanism, with login names and hierarchical domains. However, this resulted in a very large state vector and state-space (due in part to Promela's poor handling of structured types). Since we found no additional analysis benefit to this approach (apart from the aesthetic one), we have implemented a simpler, more abstract addressing mechanism.

Mail messages themselves are of no consequence, save to observe whether or not they are encrypted. We denote encrypted text by the value 1 and plain text by the value 0. Keys are simple mail addresses, i.e. simple bytes.

An important issue for any distributed system is that of *atomicity*. This is especially important from a model-checking perspective as it provides a means of controlling state-space explosion and resolving race conditions. Promela provides a facility for grouping together statements as atomic, provided only the initial statement has the potential to block. Our model employs as much atomicity as possible within each consituent process. Specifically, in the *Client* process, each iteration from the *initial* state, back to the *initial* state, is a single atomic step (i.e. Figure 2 encapsulates a single atomic step) – with suitable guards which block if the process can neither read nor write. This is crucial in order to avoid deadlock – since all channels are of finite size. On the other hand, the *Mailer* process consists of two atomic steps: one for reading a message and the other for sending the appropriate message. Any variable about which we may intend to reason should not be updated more than once within any atomic statement (so that each change to the variables is visible to the never-claim), other variables may of course be updated as required.

Another implementation issue is the size of the state vector, i.e. the number and nature of global and local variables. We must be careful not to introduce any extraneous variables (see also 3.3) nor introduce extraneous state values. To avoid the latter, we must be careful to reset variables when returning to so-called initial, abstract states, to ensure that we are indeed representing the same abstract state.

The interplay between atomicity, the number and nature of variables, and faithful modelling/levels of abstraction is very subtle and a challenge in this domain, particularly due to the asynchronous communication. It took considerable time and expertise to develop a tractable, deadlock free model. Fortunately, we were able to employ some lessons learned from modelling *POTS* [6].

3.2   The Features

We consider here a set of five features.

- **encrypt** a message, using a (private) key, the *intended* recipient.

- **decrypt** a message, using a (private) key, the *actual* recipient.

- **filter** all messages from a given mail address.

- **forward** all messages to another mail address.

- **autorespond** to incoming messages. The automatic response is only sent in response to the first message from a given client. Any subsequent message from that client is received, but no automatic response is issued as a result.

The features encrypt, decrypt and autorespond reside at the client side, the remaining reside at the server side. Note that only the features encrypt and decrypt alter the actual mail message, forwarding does not affect the message.

We have considered all the features proposed in [12]; but for brevity, we omit them here because they do not reveal any further "interesting" behaviour paradigms for our analysis. That is not to say that they do not reveal further interactions, and interesting aspects of email, but that they do not reveal any further aspects with respect to generalisation.

3.3   Features in Promela

Features are implemented within the Promela model via a number of *inline* functions (procedures with dynamic bindings). Most features are relatively straightforward to implement, simply involving additional transitions or steps during one or more of the abstract states of the client or mailer processes.

The exception is autorespond, because this feature involves both *reading* – a message from a client channel, and *writing* – a message to the network channel. Both events are potentially blocking, hence cannot take place within one atomic step. Therefore, to implement this feature we add an additional data structure to indicate whether or not a client requires to send an autoresponse. We enhance the *initial* state to include the possibility that an autoresponse message needs to be sent, and give priority to this over any other event. This means that it is possible for an autoresponse to *never* be sent, if the network channel is continuously full. We cannot avoid this situation[1].

In order to reason about feature behaviour (see section 3.4) we introduce a number of "observation" variables. These are not integral to the behaviour of the service and/or features, but exist solely for the purposes of analysis. For this reason, these variables are included/excluded on a per model/property basis.

Examples of "observation" (process indexed) variables include:

- $last\_del\_to_i\_to$ the intended receiver of the mail message last delivered to $Client[i]$

- $last\_del\_to_i\_from$ the intended sender of the mail message last delivered to $Client[i]$

- $last\_del\_to_i\_body$ the body of the mail message last delivered to $Client[i]$.

- $last\_sent\_from_i\_to$ the intended recipient of the mail message last sent from $Client[i]$.

A further variable required both for correct function and for reasoning is the array of bit vectors $autoarray$. The function

- $IS\_0(autoarray[i], j)$ indicates whether $Client[i]$ has already sent an autoresponse to $Client[j]$.

3.4　Feature Properties

We give a small number of illustrative properties for both the basic service and the features. The properties are linear temporal logic (LTL) formulae over propositions about states. Temporal operators include [] (always), <>(eventually) and X (next). Propositional connectives are || (disjunction), && (conjunction), → (implication) and ¬(negation). The path quantifier is (implicitly) ∀, except when explicitly given as ∃. Feature properties are properties that are expected to hold when *one* such feature is present.

Property 1 – Basic Messages are delivered only to intented recipients.

If $Client[i]$ receives a message from $Mailer$, then the (intended) recipient of that message is $Client[i]$. Alternatively, $Client[i]$ has not yet received any messages (in which case the $last\_del\_to_i\_to$ variable will remain set to the default value M).

LTL: $[](p||q)$
$p = (last\_del\_to_i\_to == M)$, $q = (last\_del\_to_i\_to == i)$

Property 2 – Basic Messages can be sent between any two clients.

It is possible for $Client[i]$ to recieve a message such that the sender of that message is $Client[j]$.

---

[1]In any operational email system, buffers can become full. Although this is unlikely in reality, as model-checking involves the exploration of all *feasible* behaviours, this possibility must be considered.

LTL: $\exists <> (p)$
$p = (last\_del\_to_i\_from == j)$

Property 3 – Basic Messages are eventually delivered correctly.

If Client[i] sends a message to Client[j], then Client[j] will eventually receive a message from Client[i].

LTL: $[](((\neg p)\&\& X(p)) \rightarrow X(<> q))$
$p = (last\_sent\_from_i\_to == j)$, $q = (last\_del\_to_j\_from == i)$

Property 4 – Encryption Messages are properly encrypted.

If Client[i] has encryption on, then if Client[j] receives a message whose sender is Client[i], then the message will be encrypted.

LTL: $[](p \rightarrow q)$
$p = (last\_del\_to_j\_from == i)$, $q = (last\_del\_to_j\_body == 1)$

Property 5 – Decryption Messages are properly decrypted.

If Client[i] has decryption on, then all messages received by Client[i] will have been decrypted.

LTL: $[](p)$
$p = (last\_del\_to_i\_body == 0)$

Property 6 – Filtering Messages are discarded by a filter.

If Mailer filters messages from Client[i] to Client[j] then it is not possible for Client[j] to receive a message from Client[i].

LTL: $[](\neg p)$
$p == (last\_del\_to_i\_from == j)$

Property 7 – Forwarding Messages are forwarded.

If Client[i] forwards messages to Client [j], then it is possible for Client[j] to receive messages not addressed to Client[j] (or to the default value M).

LTL: $\exists <> (\neg(p||q))$
$p == last\_del\_to_j\_to == M)$, $q == (last\_del\_to_j\_to == j)$

Property 8 – Autorespond Single automatic response messages are sent out.

If Client[i] has autorespond on, then if Client[j] sends a message to Client[i], and Client[j] hasn't already received an automatic response from Client [i], then Client[j] will eventually receive a reply from Client[i]. Alternatively, Client[i] eventually stops sending messages because network can't be accessed.

LTL: $[](p- > (<> q||(<> ([]\neg r))))$
$p = ((last\_sent\_from_j\_to == i)\&\&((autoarray[i,j] == 0)))$
$q = (last\_del\_to_j\_from == i), r = (network??[i,x,y,z])$
 (The function $network??[i,x,y,z]$ determines whether there is a message *at any position* on the network channel in which the sender field is i.)

4   Analysis for a constant number of clients

The basic idea of feature interaction analysis is to detect when features behave as expected in isolation, but not in the presence of each other. So, interaction analysis involves feature *validation* (checking a feature in isolation) and then analysis of *tuples* of features (checking for violation of expected behaviour). Fortunately, we need only restrict our attention to pairwise analysis, as empirical evidence shows that it is extremely rare to have a 3-way interaction which is not detected as 2-way interaction [14]. In each case we consider a model consisting of either 3 or 4 client processes and 1 mailer. An example Promela model of a system of 3 Client processes and a Mailer process in which $Client[0]$ has encryption, $Client[1]$ filters messages from $Client[2]$ and property 4 is to be verified for $i = 0, j = 2$ can be found on our website at [5].

For all verification runs we used a PC with a 2.4GHZ Intel Xenon processor, 3GB of available main memory, running Linux (2.4.18).

An overview of the reasoning process is given in Figure 4.

4.1   Use of Perl Scripts

For each pair of features, set of feature parameters, associated property and set of property parameters, a relevant model needs to be individually constructed, to ensure that only relevant variables are included and set. We have developed two Perl scripts, *mailchange.pl* and *auto_mailchange.pl* for automatically configuring the model and for generating model-checking runs. These scripts greatly reduce the time to prepare each model and the scope for errors.

During initial investigations, *mailchange.pl* is used to generate a model for a given set of features, feature parameters, property and property parameters. The resulting model is then loaded into SPIN with an appropriate set of search parameters (MSD, POR, WF for example) and results interpreted manually. Once confidence has been gained in the model, suitable values assertained for the value of MSD and the applicability of POR and WF determined for successful verification in each case, *auto_mailchange.pl* is used to iteratively select pairs of features and parameters, set up model checking runs and interpret results. An overnight run is required to collect all results from all pairs of features and suitable parameter sets. It is important to note that a certain amount of simple symmetry reduction is incorporated within the Perl script to avoid repeating runs of configurations which are identical up to renaming of processes.

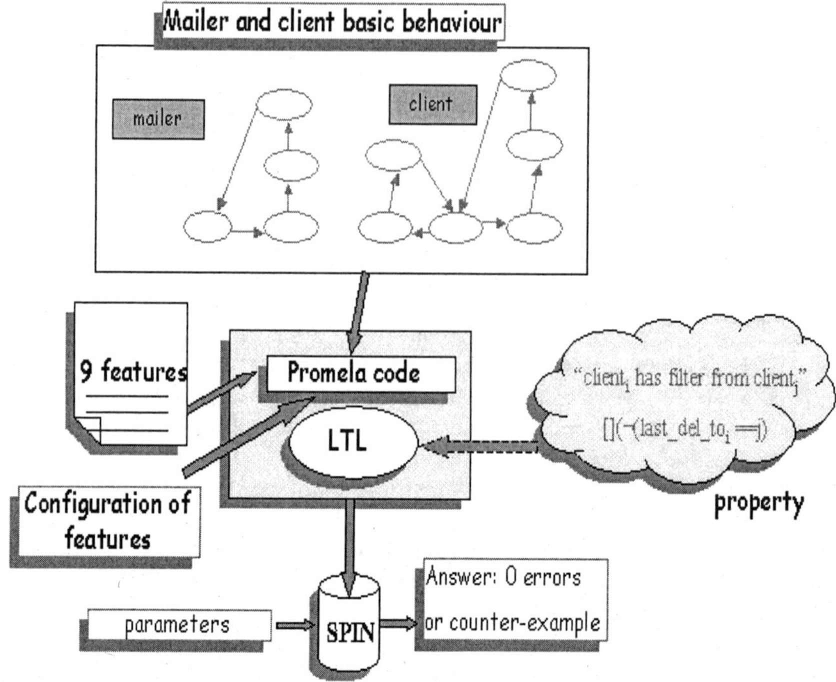

Figure 4: The reasoning set up

Table 1: Results of verification of the properties

Feature	Prop	WF?	POR?	MSD ($\times 10^4$)	States ($\times 10^4$)	Depth (Mb)	Mem (s)	Time
basic	1	$\times$	$\checkmark$	6	8.3	52989	4.4	2
basic	2	$\times$	$\checkmark$	0.01	0.001	85	2.3	0.1
basic	3	$\checkmark$	$\times$	17	98.2	162286	38.2	44
encryption	4	$\times$	$\checkmark$	13	9.3	121792	4.8	3
decryption	5	$\times$	$\checkmark$	5	10.2	45365	4.2	3
filtering	6	$\times$	$\checkmark$	6	7.1	52069	3.7	2
forwarding	7	$\times$	$\checkmark$	0.01	0.001	66	2.3	0.1
autorespond	8	$\checkmark$	$\checkmark$	17	250	163176	111.5	334

The interpretation stage of *auto_mailchange.pl* involves reading the output file from the SPIN verification run. Firstly it is checked that the maximum search depth (MSD) has not been reached and that the total memory available has not been exhausted. If neither of these is true, the second phase of the interpretation phase involves checking if there are any errors – and as such, whether the associated property is true. Finally, the interpretation stage involves determining whether a feature interaction has occurred and, if so, what type (SU or MU). All parameter sets and corresponding results are output to a results file.

4.2　Results - single property validation

Table 1 gives the results of verification of properties 1 – 8. In each case the feature (if any) associated with the property is given in the column labelled 'feature'. When there is no feature present, (during the verification of properties 1–3) the term 'basic' is written in this column. When a feature is present, the verification corresponds to checking the associated property for a model consisting of a Mailer process, two basic Client processes and a Client processes for which the given feature is 'turned on'. (When no feature is present the associated model simply consists of a *Mailer* process and three *Client* processes.) In all cases we give results for verification of the model in which $Client[0]$ has the given feature (in relation to $Client[1]$ if appropriate) and i (and j) is (are) assigned the value(s) 0 (and 1). The Prop column contains the property being checked and a $\checkmark$ or a $\times$ in the 'WF?' and 'POR?' columns indicate whether weak-fairness and partial order reduction are selected respectively. Note that property 3 is the only property for which POR is not applied. This is due to the presence of the next operator (X) in the property. Also, WF is only applied during the verification of *liveness* properties – properties that contain the *eventually* operator $<>$. The entries in the MSD column show the value to which the maximum search depth is set prior to verification. The remaining columns contain the number of stored states, the depth reached, the memory required for state storage (in Mbyte) and the time taken (in seconds) for each verification.

4.3　Results - feature interactions

Now we turn our attention to consideration of *pairs* of features. For each pair of features we generate a model for each distinct set of parameters (the union of the sets of parameters for each feature) and for each appropriate set of property parameters. This may mean that up to 5 client processes are required. For example, if $Client[i]$ has filtering from $Client[j]$ and

$Client[k]$ has filtering from $Client[l]$, $(i, j, k, l$ distinct$)$, then 4 client processes are required. For each suitable pair of features, f_i, f_j, an interaction is said to occur if the feature property associated with f_i does not hold for the model in which features are f_i and f_j are present. Note that we do not consider the basic service in our analysis, as all other features interact with it in some way. This can be determined without the need for model-checking.

We enumerate the interactions found below. In each case, we indicate whether the interaction is single user (SU), i.e. both features reside at the same network component, or multiple user (MU), i.e. the features reside at different network components. We also give a *witness* for the interaction. We do not give details of timing or memory requirements etc. as these vary depending on the parameter set under consideration. (There are 111 feasible parameter sets after symmetry reduction. It would be impractical to give details of such requirements for each case.) In some cases more $Client$ processes are required to fully check for interaction. Clearly when more $Clients$ are required, verification takes longer and more memory is required. In addition, when an error is reported during verification (in most cases, excluding the verification of property 2, indicating an interaction) a full search of the state-space is not required. This again results in far smaller time and memory requirements.

1. encryption and decryption (SU)
   witness i=j=0 – $Client[i]$ has encryption and decryption

2. encryption and decryption (MU)
   witness i=0, j=1 – $Client[i]$ has encryption, $Client[j]$ has decryption

3. filter and forward (MU)
   witness $i = 0, j = 1$ – $Client[i]$ has filter from j, $Client[j]$ has forwarding to i

4. forward and forward (MU)
   witness i=0,j=1,k=2 – $Client[i]$ has forwarding to j, $Client[j]$ has forwarding to i

5. autoresponse and filter (SU)
   witness $i = 0, j = 1$ –$Client[i]$ has autoresponse, $Client[i]$ has filter from j

6. autoresponse and filter (MU)
   witness i=0,j=1 – $Client[i]$ has autoresponse, $Client[j]$ has filter from i

7. autoresponse and forward (SU)
   witness $i = 0, j = 2$ – $Client[i]$ has autoresponse, $Client[i]$ has forwarding to j

8. autoresponse and forward (MU)
   witness $i = 0, j = 1$ – $Client[i]$ has autoresponse, $Client[j]$ has forwarding to i

Each of these pairs of features are listed in Hall's results [12] but in all cases, he only reports a MU example. While Hall explicitly states that his method is not complete, it is not clear if the SU interactions would be found in his approach, or he stopped after the MU interaction was found. Our method is combinatorially complete. We note that in the MU cases, our witnesses are identical to Hall's (modulo translation).

5　Generalisation

We have shown above that a property holds (or does not hold) for a fixed number of clients, i.e. for
$Client[0]||Client[1]||\ldots Client[m]||Mailer$, where $||$ denotes parallel composition. But how can we deduce that (if at all) a property holds for
$Client[0]||Client[1]||\ldots ||Client[n]||Mailer$, for an arbitrary n? It is not possible to demonstrate this with straight-forward model-checking [1].

More formally, the generalisation problem is how to prove (disprove)

$$M(Client[0]||Client[1]||\ldots ||Client[n]||Mailer) \models \phi[0,1,\ldots,t]$$

where the left hand side is the finite-state model of the parallel composition of client and mailer processes (the former are instances of the parameterised process $Client$) and $\phi[0,1,\ldots,t]$ is a temporal logic formula containing free variables indexed by $0,1,\ldots,t$. The indices refer to instances of $Client$ (e.g. the variables i and j in section 3.4). In general, the $Client[i]$ are not isomorphic because they have different sets of features enabled.

We offer a solution based on abstraction and model-checking. The technique and theoretical justification are described in more detail in [8, 7], here we apply the results. Briefly, the technique involves choosing a fixed m, such that t is constrained by $0 \leq t \leq m-1$. We refer to $Client[0]||\ldots ||Client[m-1]$ as *concrete* processes and the $Client[m]||\ldots ||Client[n-1]$ as *abstract* processes. We represent the behaviour of $Client[m]||\ldots ||Client[n-1]$, by a new abstract process, *Abstractclient*. We do not assume that the (original) abstract processes, i.e. $Client[m]||\ldots ||Client[n-1]$ are isomorphic: they might have different combinations of features enabled. However, we do assume that the features are all drawn from our given set and we know how they can communicate with each other and more importantly, how they communicate with the concrete processes.

A model of the m *concrete* processes, together with the abstract process, is generated automatically from a model of the concrete processes together with a single (parameterised) $Client$. This is summarised by Figure 5. The value of m depends upon the particular feature set considered. Here, because each feature involves at most two parameters, a worst case analysis suggests that 5 concrete $Clients$ are required, however further detailed analysis shows that in this case, we require only a maximum of 4. For some combinations, 3 will suffice.

Our approach is based upon our result:

$$M(Client[0]||Client[1]||\ldots Client[m]||Abstractclient||Mailer') \models \phi[0,1,\ldots,t]$$
$$\Rightarrow$$
$$M(Client[0]||Client[1]||\ldots Client[n]||Mailer) \models \phi[0,1,\ldots,t].$$

The process $Mailer'$ is a slightly modified version of $Mailer$, modified to take into account communication with $Abstractclient$ (instead of communication with the original abstract processes).

Thus to generalise interaction analysis results, we need only consider interaction analysis of the *finite* (model of) $Client[0]||Client[1]||\ldots Client[m]||Abstractclient||Mailer'$.

In the next section we outline the form of $Abstractclient$ and $Mailer'$ and give our interaction analysis results.

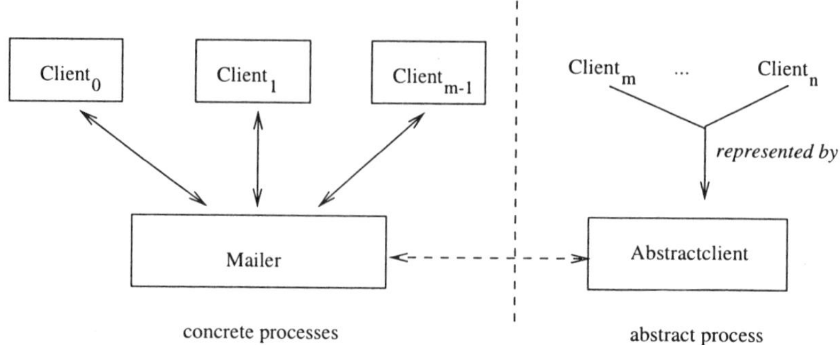

Figure 5: Generalised Email service

5.1 Abstractclient specification and analysis results

Abstractclient is defined as follows. *Abstractclient* can only affect the behaviour of the m concrete processes indirectly via $Mailer'$. Therefore, communication to/from a concrete process from/to $Mailer'$ takes place via a *virtual* channel. Rather than concrete processes reading/writing to this (virtual) channel and behaving accordingly, each possible read is replaced by a non-deterministic choice over the possible contents of the channel. In this way, all possible behaviours are explored (a write to the virtual channel is not relevant).

As an example, when $m = 3$ *Abstractclient* is as follows:

```
proctype Abstractclient(byte id)
{Mail msg;
 atomic
 {
 msg.receiver=M;
 msg.sender=M;
 msg.key=M;
 msg.body=0};
 do
 ::blocked==1->blocked=0
 ::atomic{nfull(network)->
   if
   ::msg.receiver=0
   ::msg.receiver=1
   ::msg.receiver=2
   ::msg.receiver=3
   /*another client within Abs process */
   fi;
   msg.sender=id;
   network!msg;
   msg.receiver=M;
   msg.sender=M}
 od
 }
```

Note that the *Abstractclient* process can send messages to $Mailer'$ (via the *network* channel), but does not receive messages. *Abstractclient* is also able to set the *blocked* variable to 0. This simulates "unblocking" $Mailer'$ when it is unable to "send" a message to a particular process within the abstract process. Note that *Abstractclient* is always able to unblock $Mailer'$ but can only send messages when the *network* channel is not full. This reflects the finite model. When $Mailer'$ wants to "deliver" a message to *Abstractclient*, it first checks whether the relevant channel is blocked (via non-deterministic choice). If so, $Mailer'$

waits until the channel becomes unblocked (when the *blocked* variable is reset to 0 by *Abstractclient*) before delivering the message. (In fact no message is actually sent, but *Mailer'* continues as if it has been.) Here we give the *Mailer'* proctype:

```
proctype Mailer'()
{
Mail msg;
chan deliverbox=null;
atomic{
bit myanswer=0;
msg.sender = M;
msg.receiver= M;
        msg.key=M;
        msg.body=0;
}
loop:
atomic{
        network?msg;
        filter_message(msg.receiver,msg.sender,myanswer);
        if
        :: myanswer -> /* throw away message from this sender*/
           myanswer=0;
           msg.sender = M;
           msg.receiver= M;
           msg.body = 0; msg.key = M;
           deliverbox = null;goto loop
        :: else -> skip
        fi;
        if
        ::msg.receiver==3->/* abstract process */
          if
          :: blocked=0
          :: blocked=1
          fi
        ::else->
          mailbox_lookup(msg.receiver,deliverbox)
        fi;
/* now pass on message */
}

atomic{

if
::((msg.receiver!=3)&&(nfull(deliverbox)))->deliverbox!msg
::((msg.receiver==3)&&(blocked==0))->skip /*delivered virtual message*/
fi;

/*reset variables to initial values*/
msg.sender = M; msg.receiver= M; msg.body = 0; msg.key = M;
deliverbox = null;
goto loop
     }}
```

The concrete *Client* processes are declared in the usual way and communication between them and *Mailer* is unchanged.

It is important to note that this model is not, strictly, a conservative extension, because *Abstractclient* allows additional behaviour. Namely, an (abstract) client can send mail even when there is mail to be delivered (to that client). This is not possible in any concrete model. However, the constraint that mail delivery takes priority is in the concrete model only to prevent deadlock (when mail buffers are full), not for any reason of *functional* behaviour. Relaxing this constraint in the general model neither allows deadlock nor affects the observational behaviour of the system. Thus the constraint is safely relaxed and our approach is sound.

The interaction analysis results reveal no new interactions, nor new witnesses. We therefore do not give details, save to indicate that time and space complexity lie in between those for the system with m (concrete) $Clients$ and $m + 1$ (concrete) $Clients$. The value of m depends on the parameter set and the property to be verified. Again, all analysis was automated through the use of Perl scripts.

An (abridged) example Promela model of a system of $m = 3$ $Client$ processes, a $Mailer'$ process and an $Abstractclient$ process in which $Client[0]$ has encryption, $Client[1]$ filters messages from $Client[2]$ and property 4 is to be verified for $i = 0$, $j = 2$ can be found in the appendix below. The full model can be found on our website at [5].

6 Conclusions

We have developed a property-based approach to feature interaction analysis for a client-server email system. The feature set described here is not as extensive as Hall's [12], but it is sufficient to reveal most of the interesting behaviour (from a modelling point of view) and to validate our approach. On the other hand, our analysis is complete and fully automated. We note that a difficult feature for our implementation was autoresponder, this is because unlike most other features, it initiated the sending of a completely new message. This was a difficulty because we chose to use asynchronous communication, with fixed size communication channels (thus leading to more interleavings and potentially, more interactions).

Additionally, our results are scalable, that is they generalise to email systems consisting of any number of $Client$ processes.

Abstraction techniques are used to prove the general results. The key idea is to model-check a system consisting of a constant number (m) of client processes, in parallel with an "abstract" process which represents the product of any number of other client processes. We give a lower bound for the value of m, for our given feature set.

While the general results did not reveal any new interactions (and we admit it is difficult to think of situations where they would, for *fixed* feature sets), it is nevertheless important to prove rigorously that results scale up. We have achieved this.

Our results demonstrate the feasibility of the abstraction technique for this application domain – the model-checking requirements are well within the capability of our machine. Also, the transformation to a general model is relatively straightforward: we need only consider the *communication* between the abstract process and the concrete process(es). (In this case, we need only consider communication with $Mailer$ process, as there is no peer-peer communication.) An alternative, an induction approach [9, 15], requires the construction of an inductive invariant. This involves incorporating the behaviour of the entire system within the invariant; moreover, it requires that both the concrete and abstract m $Clients$ are isomorphic. Our abstraction approach offers a more suitable and tractable alternative. However, at some level they are similar, future work aims to establish this.

References

[1] Krzysztof R. Apt and Dexter C. Kozen. Limits for automatic verification of finite-state concurrent systems. *Information Processing Letters*, 22:307–309, 1986.

[2] M. Calder and E. Magill, editors. *Feature Interactions in Telecommunications and Software Systems*, volume VI. IOS Press, Amsterdam, 2000.

[3] M. Calder and E. Magill, editors. *Feature Interactions in Telecommunications and Software Systems VI*. IOS Press (Amsterdam), 2000.

[4] M. Calder, E. Magill, and S. Kolberg, Reiff-Marganiec. Feature interaction: A critical review and considered forecast. *Computer Networks*, 41/1:115 – 141, 2003.

[5] M. Calder and A. Miller. Veriscope publications website: `http://www.dcs.gla.ac.uk/research/veriscope/publications.html`.

[6] M. Calder and A. Miller. Using SPIN for feature interaction analysis - a case study. In *[10]*, pages 143–162, 2001.

[7] M. Calder and A. Miller. Feature validation for any number of processes. Technical Report TR2002-110, University of Glasgow, Department of Computing Science, 2002.

[8] Muffy Calder and Alice Miller. Automatic verification of any number of concurrent, communicating processes. In *Proceedings of the 17th IEEE International Conference on Automated Software Engineering (ASE 2002)*, pages 227–230, Edinburgh, UK, September 2002. IEEE Computer Society Press.

[9] E.M. Clarke, O. Grumberg, and S. Jha. Verifying parameterized networks using abstraction and regular languages. In *[16]*, pages 395–407, 1995.

[10] M.B. Dwyer, editor. *Proceedings of the 8th International SPIN Workshop (SPIN 2001)*, volume 2057 of *Lecture Notes in Computer Science*, Toronto, Canada, May 2001. Springer-Verlag.

[11] R. Gerth, D. Peled, M.Y. Vardi, and P. Wolper. Simple on-the-fly automatic verification of linear temporal logic. In *Proceedings of the 15th international Conference on Protocol Specification Testing and Verification (PSTV '95)*, pages 3–18. Chapman & Hall, Warsaw, Poland, 1995.

[12] R.J. Hall. Feature interactions in electronic mail. In *[3]*, pages 67–82, 2000.

[13] Gerard J. Holzmann. The model checker Spin. *IEEE Transactions on Software Engineering*, 23(5):279–295, May 1997.

[14] M. Kolberg, E. H. Magill, D. Marples, and S. Reiff. Results of the second feature interaction contest. In *[2]*, pages 311–325, May 2000.

[15] R. P. Kurshan and K.L. McMillan. A structural induction theorem for processes. In *Proceedings of the eighth Annual ACM Symposium on Principles of Distrubuted Computing*, pages 239–247. ACM Press, 1989.

[16] Insup Lee and Scott A. Smolka, editors. *Proceedings of the 6th International Conference on Concurrency Theory (CONCUR '95)*, volume 962 of *Lecture Notes in Computer Science*, Philadelphia, PA., August 1995. Springer-Verlag.

[17] Doron Peled. Partial order reduction: Linear and branching temporal logics and process algebras. In *[18]*, pages 233–257, 1996.

[18] Doron A. Peled, Vaughan R. Pratt, and Gerard J. Holzmann, editors. *Proceedings of the DIMACS Workshop on Partial-Order Methods in Verification (POMIV '96)*, volume 29 of *DIMACS Series in Discrete Mathematics and Theoretical Computer Science*. American Mathematical Society, 1996.

Appendix: The email service with features

The following model is an abridged version of one generated (by a Perl script from template) for features: encryption[0] and filter[1]=2 to verify property 4 with i=0 and j=2. Note that 'etc.' indicates that some lines of code have been omitted, for space reasons. The full model (including the appropriate never-claim and bit vector definitions) can be found on our website at [5].

Example Promela model

```
typedef Mail
{byte sender;byte receiver;byte key;bit body};
/*need bit vector definitions in here*/
#define k 1 /*size of Mailbox */
#define N 2  /* size of network channel */
#define M 4 /*default value of variables */
bit blocked=0;
chan null=[k] of {Mail};
chan zero=[k] of {Mail};
(etc.)
chan network = [N] of {Mail};
BITV_8 Encrypt=0; byte Filter[M]=M;
byte last_del_to2_from=M;
bit last_del_to2_body=0;
inline mailbox_lookup(login,box)
{
if
:: (login==0) -> box = zero
   (etc.)
fi}
inline encrypt_message(login,answer)
{if
::(IS_1(Encrypt,login))->answer=1
::else->answer=0
fi
/*if encryption is on, answer=1,
  otherwise answer=0*/}
inline filt_mess(to,from,answer)
{if
   ::(Filter[to]==from)->answer=1
   ::else->answer=0
fi
/*if appropriate filter is on, answer=1,
  otherwise answer=0*/
}
inline reset_vars(i)
{(etc.)}
inline set_deliv_vars(i,from,to)
{(etc.)}
inline set_body(i,text)
{if
::i==2->last_del_to2_body=text
::else->skip
fi}

proctype Client(byte id)
{chan mybox=null; Mail msg;
atomic{msg.sender=M; msg.receiver=M;
msg.key=M; msg.body=0;bool myanswer=0;
/*get appropriate mailbox*/
mailbox_lookup(id,mybox)};
initial:atomic{
(nempty(mybox)||nfull(network));
/* wait here if cannot send or deliver */
 reset_vars(id);
  if
  :: nempty(mybox) -> goto delivermail
  :: empty(mybox)&&nfull(network)->goto sendmail
  fi;
delivermail:
mybox?msg; set_body(id,msg.body);
set_deliv_vars(id,msg.sender,msg.receiver);
goto endClient;
sendmail:/*specify recipient */
if
:: msg.receiver= 0
   (etc.)
fi;
encrypt_message(id,myanswer);
```

```
if
:: myanswer -> /*encryption on */
   myanswer=0;
   /* use  reciever id as  key */
   msg.body = 1; msg.key = msg.receiver
:: else -> msg.body = 0; /*no encryp*/
fi;
msg.sender = id; /* specify sender */
network!msg; /* send mail */
endClient:
/* reset other variables */
msg.sender = M; (etc.) goto initial}}

proctype Network_Mailer()
{
Mail msg; chan deliverbox=null;
atomic{
bit myanswer=0; (etc.)}
loop:
atomic{network?msg;
filt_mess(msg.receiver,msg.sender,myanswer);
if
:: myanswer ->/* throw away message*/
   myanswer=0;msg.sender = M;
   msg.receiver= M; (etc.)
   deliverbox = null;goto loop
:: else -> skip
fi;
if
::msg.receiver==3->
  if
  :: blocked=0
  :: blocked=1
  fi
::else->
  mailbox_lookup(msg.receiver,deliverbox)
fi;
/* now pass on message */
}
atomic{
if
::((msg.receiver!=3)&&(nfull(deliverbox)))->
  deliverbox!msg
::((msg.receiver==3)&&(blocked==0))->skip
  /*delivered virtual message*/
fi;
/*reset variables to initial values*/
msg.sender = M; (etc.)
deliverbox = null; goto loop}};

proctype Abstractclient(byte id)
{Mail msg;
atomic
{msg.receiver=M;
msg.sender=M; msg.key=M; msg.body=0};
do
::blocked==1->blocked=0
::atomic{nfull(network)->
  if
  ::msg.receiver=0
    (etc.)
  ::msg.receiver=3 /*other client in Abs proc */
  fi;
  msg.sender=id;network!msg;
  msg.receiver=M;msg.sender=M}
od}

init
{atomic{SET_1(Encrypt,0);Filter[1]=2;
run Abstract(3);run Network_Mailer();
run Client(0); run Client(1);run Client(2);}}
```

*Feature Interactions in Telecommunications
and Software Systems VII*
D. Amyot and L. Logrippo (Eds.)
IOS Press, 2003

Formal Approaches for Detecting Feature Interactions, Their Experimental Results, and Application to VoIP

T. YONEDA, S. KAWAUCHI, J. YOSHIDA, and T. OHTA
SOKA University, 1-236, Tangi-cho, Hachioji-shi 192-8577 Japan
ohta@t.soka.ac.jp

Abstract. Services that operate normally independently will behave differently when initiated simultaneously with another service. This behavior is called feature interaction. This paper describes the results of experiments with formal approaches for detecting feature interactions in telecommunication services, evaluation of these results, and application to VoIP. One of the most difficult problems in this field is terminal assignment, which was discussed at the FIREWORKS workshop which was a collocated conference just before FIW2000, and nobody could resolve the problem. This paper describes briefly how to make terminal assignments. The results of experiments were evaluated by comparing them to a benchmark based on FIW98. The detection system was applied to a validation server, which was designed for an active VoIP gatekeeper.

Keywords: Feature interaction detection, filtering, specification validation, VoIP

1 Introduction

This paper describes the results of experiments using formal methods for detecting feature interactions in telecommunication services, their evaluation, and application of the detection system to a validation server for an active VoIP gatekeeper, which was designed based on an active network architecture [1][2]. These formal methods were proposed by the authors at FIW98 [3], at another international conference [4] and in papers [5][6] submitted by them.

As is well known, telecommunication service specifications can be described as state transition diagrams. There are two types of feature interactions [7]. One is called a logical interaction and the other is called a semantic interaction. The authors have already proposed formal methods for detecting semantic interactions and one of the logical interactions, a non-determinacy interaction [3][5]. This paper describes a detection system based on the proposed methods and evaluation results. A case study involving the application of the proposed detection system to a validation server for the active VoIP gatekeeper is also described.

In implementing the detection system, the following problems were solved.

(a) How to reduce the number of terminal assignments.

(b) How to reduce computation time required for a reachability test.

(c) How to elicit knowledge to be used for detecting semantic interactions.

To solve the above problems, new filtering methods and elicitation method were developed. But, in this paper, only terminal assignments is discussed.

For evaluations, the newest international benchmark published in 2000 [8][9] was used; detection coverage, redundancy and the number of feature interactions, which were not shown in the benchmark, were evaluated. All interactions described in the benchmark were detected. We could not compare with the results of FIW2000 contest, because of lack of detailed information. No redundancy was discovered. There were 2,571 interactions which were not shown in the benchmark; this is more than 20 times the number shown in the benchmark. Thus, it was confirmed that the proposed detection system is reasonable.

As a case study in the application of the detection system to an active VoIP network, an architecture for an active VoIP network and the results of experiments are reported on.

In section 2, the proposed detection algorithm is briefly reviewed. In section 3, the problem of terminal assignment and its solution are described. In section 4, the results of experiments are shown. In section 5, a case study of the application of the detection system to an active VoIP gatekeeper is described.

2 Detection Algorithm

In general, feature interactions are detected by investigating the compound state of two states for two services. But, investigating all of the compound states means that detection takes an extraordinarily long time. The authors proposed a formal detection algorithm whereby feature interactions are detected solely by analyzing each service specification described, using a rule based language, instead of generating all of the states.

First of all, a specification description language used in the proposed method is explained, and then, a brief explanation of the algorithm for detecting semantic interactions is described.

2.1 Specification Description Language

A rule based language, called STR [10][11], was used in our method. A system state is represented as a set of primitives, which are states of terminals or relationships between terminals connected to the system. The primitive consists of a primitive name, which represents a state, and arguments which indicate real terminals (e.g. A,B,C).

Using STR, telecommunication service specifications are described as a set of rules. The syntax of the rule is as follows:

Pre-condition   event: Post-condition

Pre-condition and Post-condition are each represented as a set of primitives. In the rule, the arguments of each primitive are described as terminal variables (e.g. x,y,z) so that the rule can be applied to any terminals.

Next, rule application and change of system state when the rule is applied are described. The appropriate states whereby Pre-condition or Post-condition terminal variables are replaced by real terminals is called the state corresponding to Pre-condition or Post-condition, respectively. When the state corresponding to Pre-condition exists in the system state, it is said that Pre-condition is included in the system state.

Rule 1) Basic rule application: When an event occurs in a system a rule that has the same event and whose Pre-condition is included in the system state is applicable.

Rule 2) Precedent rule application: When more than one rule are applicable the rule whose Pre-condition includes any other rules' Pre-conditions is applied.

Rule 3) Change of system state: When a rule is applied the next state of the system state is obtained as follows. The state corresponding to Pre-condition of the applied rule is deleted from the current system state and the state corresponding to Post-condition of the applied rule is added.

2.2 Detection Algorithm

A brief explanation of the detection algorithm for semantic interactions will be described. For detailed descriptions, please refer to paper [3][5].

One type of semantic interaction is the appearance of an abnormal state. Abnormal states can be classified into two categories [3]. In one case, the compound system state for both services can not be represented as the union of individual state for both services. In the other case, though the compound system state for both services can be represented as the union of individual state for both services, the compound system state violates service constraints for either or both services. For the latter case, knowledge is used to identify whether the compound system state violates the constraints or not.

(1) Interaction detection without using knowledge

Except for using knowledge to detect interaction [3], feature interactions are detected as follows: To make a rule pair, select a rule from each service, which is applicable to the same system state. Apply either rule to the system state. Check if the state transition by the applied rule causes an abnormal state transition from the viewpoint of the other service whose rule is not applied.

In some conventional methods, all possible states must be generated in one way or another and all state transitions must be checked to detect feature interactions. This causes an explosion in the number of states, resulting in a huge increase in computation time for detecting feature interactions.

With our method, on the other hand, interactions are detected solely by analyzing Pre-conditions, events, and Post-conditions of selected rules as follows:

step 1) If both rules of the rule pair can be applied to the same system state, go to step 2. Otherwise, a feature interaction is not detected.

step 2) If both rules have the same event, go to step 4, otherwise, go to step 3.

step 3) If the Pre-condition of the rule, which is not applied, is not preserved in the next system state, a feature interaction is detected. Otherwise, a feature interaction is not detected.

step 4) If the Post-condition of the rule, which is not applied, is not realized in the next system state, a feature interaction is detected. Otherwise, a feature interaction is not detected.

Suppose, r_a and r_b denote selected rules from service a and service b, respectively. r_{ac}, r_{an}, r_{bc}, and r_{bn} denote Pre-condition of r_a, Post-condition of r_a, Pre-condition of r_b, and Post-condition of r_b, respectively. Formal descriptions of conditions used to detect interactions in step3 and step 4, respectively, are given as follows:

$$\text{For step 3: } (r_{bc} - r_{ac}) \cup r_{an} \not\supseteq r_{bc} \qquad \text{----------(1)}$$

$$\text{For step 4: } \{(r_{bc} - r_{ac}) \cup r_{an} \not\supseteq r_{bn}\} \vee \{(r_{ac} - r_{an}) \not\supseteq (r_{bc} - r_{bn})\} \qquad \text{----------(2)}$$

(2) Interaction detection using knowledge

Constraints can be represented as two kinds of primitive set. One is an *inhibit* primitive set, the other is an *intention* primitive set. The inhibit primitive set for service a is a primitive set which should not appear in any system states of service a. The intention primitive set for service a is a primitive set which should appear in the next state after the specific state transition of service a. For a more detailed explanation, please refer to a paper [5].

3 Terminal assignment

To evaluate the proposed algorithm, a detection system based on the algorithm was implemented. In this section, problems in implementing the system are described.

(1) Problems in terminal assignment

Terminals in service specifications are written as terminal variables so that the specifications can be applied to any terminals. To assign real terminals to terminal variables is called 'terminal assignment'. Feature interactions may happen or may not happen according to terminal assignment. In general, to detect all feature interactions, all terminal assignments should be considered. This calls for the testing of extremely large numbers of system states. So, in the detection system, the number of system states to be tested should be reduced by filtering.

The following problems exist with conventional detection systems:

problem 1) The minimum number of terminals to detect all interactions.

problem 2) How to assign real terminals to terminal variables to detect all interactions.

These questions were discussed at the FIREWORKS workshop which was a collocated conference just before FIW2000. The authors asked the above problems as open questions to FIW2000 participants. But nobody could answer them. For these problems, considering equivalent states (permutation symmetry [12]), which represent the same states except for the name of the terminals, the authors proposed realistic solutions. Namely, only one system state among equivalent states has to be tested. A brief explanation is given. For more detailed explanation please refer to [4].

(2) Basic idea

When the rule of service a, r_a, is applied to the current state of the combined system state, s_c, the next state of the combined system state can be obtained by deleting r_{ac} from s_c and adding r_{an}. See Figure 1. s_{bc} in Figure 1 represents the current state for service b. Part P in Figure 1 represents a common part for r_{ac} and s_{bc}; part P belongs to both services.

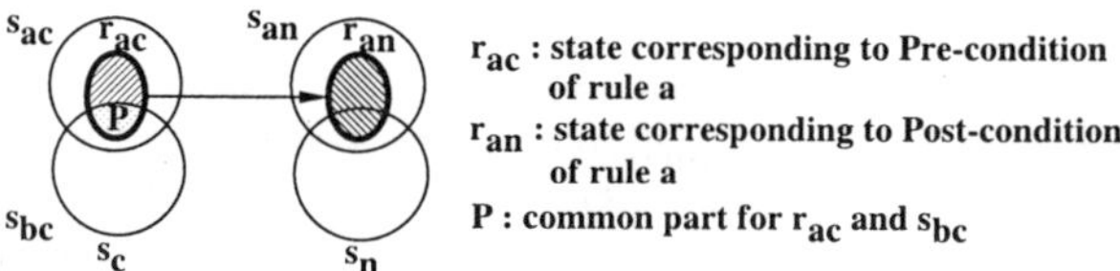

Figure 1. State change of combined system

Therefore, if part P is not null, the combined system state may transit to an abnormal state from the service b's view point. Considering that the definition of primitive is 'state of a terminal or relationship between terminals', if primitive p belongs to service a, the terminal t, which is an argument of p, also belongs to service a. Thus, if t does not belong to service a, p does not belong to service a. This suggests that to detect feature interactions it is sufficient to consider such terminal assignments where at least one terminal variable belongs to both services. This condition is satisfied by means of making combinations of terminal variables to which the same real terminals are assigned; pairs of terminal variables, each of which belong to different rules of a given rule pair, respectively.

Suppose a system state s_i has arguments A, B, C and D, and is denoted by $s_i(A,B,C,D)$. A system state s_j has arguments P, Q, R and S, and is denoted by $s_j(P,Q,R,S)$. Equivalent states are defined as follows:

Definition: For $s_i(A,B,C,D)$ and $s_j(P,Q,R,S)$, if $s_i(P,Q,R,S)$, which is obtained by replacing the arguments of s_i, A, B, C and D, by P, Q, R and S (A->P, B->Q, C->R, D->S), respectively, equals $s_j(P,Q,R,S)$, then $s_i(A,B,C,D)$ and $s_j(P,Q,R,S)$ are called equivalent states.

For equivalent states, the following Lemma is obtained.

Lemma: Suppose $s_i(A,B,C,D)$ and $s_j(P,Q,R,S)$ are equivalent states. For detecting feature interactions, if the transition from $s_i(A,B,C,D)$ to $s_{in}(A,B,C,D)$ by an event ei is checked, then the transition from $s_j(P,Q,R,S)$ to $s_{jn}(P,Q,R,S)$ by the same event ei does not need to be checked.

Consequently, for a given combination of terminal variables to which the same real terminals are assigned, one terminal assignment is sufficient since other terminal assignments give equivalent states. Thus, the following theorem can be obtained.

Theorem: For all different combinations of terminal variables to which the same terminals are assigned, only one terminal assignment should be considered for detection of feature interactions.

A terminal assignment that produces an equivalent state is called an equivalent terminal assignment.

(3) The number of terminal assignments
Select rule a and rule b from rules for service a and service b, respectively. Let the number of terminal variables in each rule be k_1 and k_2, respectively. Suppose a terminal variable in rule a and a terminal variable in rule b are assigned the same real terminal. Let the number of terminal variables, to which the same real terminals are assigned, be t. The total number of combinations of terminal variables, to which the same real terminals are assigned, is obtained in the following steps. Suppose $k_2 \geq k_1$.

step 1) Select t terminal variables from k_1 terminal variables of a rule whose number of terminal variables is equal to or less than that of the other rule. The number of ways of selecting t terminal variables out of k_1 terminal variables is shown as k_1Ct. k_1Ct denotes the following formula:
$$k_1Ct = k_1(k_1 - 1)(k_1 - 2)...(k_1 - t + 1)/t!$$

step 2) Select t terminal variables from k_2 terminal variables of the other rule and make pairs of terminal variables selected in step 1) and selected here. The number of ways to do this is shown as k_2Pt. k_2Pt denotes the following formula.
$$k_2Pt = k_2(k_2 - 1)(k_2 - 2)...(k_2 - t + 1)$$

step 3) Thus, when the number of terminal variables, to which the same terminals are assigned, is t, the number of terminal assignments is shown as $k_1Ct \times k_2Pt$.

step 4) For a pair of rules, a total number of terminal assignments is obtained by adding the numbers obtained in step 3) from t=1 to t=k_1, as follows in formula (3)

$$\sum_{t=1}^{t=k_1} k_1Ct \times k_2Pt \qquad \text{-----------------------------------} \quad (3)$$

(4) Solutions for the problems
solution 1) The minimum number of terminals is the sum of the minimum number of terminals which are required to describe each service specification, respectively.
solution 2) One terminal assignment is enough for each combination of terminal variables to which the same terminals are assigned.

4 Results of Experiments

As a benchmark, we used the papers summarizing the results of a feature interaction detection contest at FIW98 (published in 2000). From another contest held at FIW00, only the number of interactions detected was reported and detailed information about what kind of interactions were detected were not published. Therefore, the authors could not use the contest results of FIW00 as a benchmark.

In the benchmark, specifications for 12 *services* are described using Chisel. In our experiment, first of all we describe the specifications for 12 services using STR, and verified the STR descriptions using an STR simulator comparing with all state transitions described in Chisel.

In the benchmark, for 78 combinations out of 12 services, 98 interactions are described. But, among them, 19 interactions are actually not interactions, or at least there are ambiguities. Out of 79 interactions, there were 13 non-determinacy interactions and 66 semantic interactions. Out of 66 semantic interactions, 32 interactions required knowledge to be identified as interactions.

Table 1: The number of feature interactions detected by the proposed system.

	CFBL	CND	INFB	INFR	INTL	TCS	TWC	INCF	CW	INCC	RC	CELL	**TOTAL**
CFBL	6	5	5	18	0	14	95	13	95	0	2	4	257
CND	-	0	3	15	0	9	28	13	13	2	18	27	128
INFB	-	-	0	16	0	10	72	13	60	2	8	9	190
INFR	-	-	-	17	1	42	102	48	107	3	33	33	386
INTL	-	-	-	-	0	1	4	1	5	12	4	0	27
TCS	-	-	-	-	-	7	13	35	24	2	17	18	116
TWC	-	-	-	-	-	-	137	103	317	54	40	278	929
INCF	-	-	-	-	-	-	-	17	95	3	27	27	169
CW	-	-	-	-	-	-	-	-	122	42	9	207	380
INCC	-	-	-	-	-	-	-	-	-	0	2	2	4
RC	-	-	-	-	-	-	-	-	-	-	8	49	57
CELL	-	-	-	-	-	-	-	-	-	-	-	7	7
TOTAL	6	5	8	66	1	83	451	243	838	120	168	661	2,650

Using the proposed detection system, 78 combinations out of 12 services were tested and the results were compared with the benchmark. The total number of feature interactions detected in our system was 2,650. The number of non-determinacy interactions and semantic interactions were 697 and 1,953, respectively. For semantic interactions, the number of interactions detected using knowledge was 234. All the interactions described in the benchmark were detected, there was no redundancy, no miss-detection, and 2,571 new interactions, which were not described in the benchmark, were detected, this is more than 20 times the shown in the benchmark. For each combination, the number of interactions detected is shown in Table 1.

To evaluate the effect of filtering, and the effect on computation time reduction, the number of subjects to be tested is discussed. The number of real terminals, to be used for terminal assignment, is the sum of the number of terminal variables in each rule of a rule pair. The number of subjects to be tested in a case where equivalent terminal assignments are eliminated is compared with the number of subjects to be tested where equivalent terminal assignments are included. By eliminating equivalent terminal assignments, the number of terminal assignments was reduced to one 250th, in the mean.

Computation time is discussed. Total time to detect all interactions in 78 service pairs was 22 minutes, 52 seconds, and 675 milliseconds. The mean time to detect all interactions of one service pair is 17.6 seconds. Outstanding cases are for pairs of (TWC and TWC) and (TWC and CW). Even in these cases, computation times are around 170 seconds and 130 seconds, respectively, and there is no problem.

Evaluation results for detecting interactions are as follows:

1. A large number of interactions were detected, more than 20 times those shown in the benchmark. There was no redundancy or miss-detection. Thus it was confirmed that the proposed algorithm, including the detection system, is reasonable.

2. The reason so many interactions were detected is that, with the conventional detection methods, limited terminal assignments were tested to suppress detection time because the correct way of making terminal assignments had not been made clear, but with the proposed methods, the way of making terminal assignments was clarified and other filtering methods were developed. As the filtering methods reduced detection time considerably, all terminal assignments were considered and far more interactions were detected.

5 Application to VoIP

5.1 Proposed Active Network Architecture

Much research into Active Networks has been done all over the world. In the IWAN2000, we proposed an architecture for the Active Network (STAR; Software archiTecture for Active network using Rule based language) for VoIP where the up-loaded program, ('service program'), is described using a declarative language ESTR (Enhanced STR) [1]. We have also implemented an experimental VoIP active gatekeeper based on the proposed architecture (Fig.2).

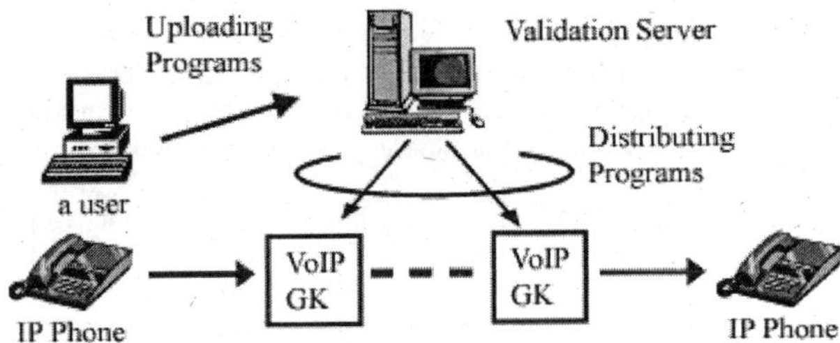

Figure 2. Proposed Active Network Architecture for VoIP.

In the validation server, feature interactions between service programs are detected. Characteristics of STAR are as follows:

1) A declarative language, ESTR [1], is developed by enhancing STR and is adopted to describe service programs instead of a procedural language, such as Java.

2) One common Execution Environment (EE) is used for all service programs, instead of using an individual EE for each service program.

3) A Validation Server is used to detect feature interactions between service programs described by unspecified users, before the service programs are installed to network nodes.

5.2 Results of Experiment

11 services programs: OCS (Originating Call Screening Service), TCS (Terminating Call Screening Service), CFV (Call Forwarding Service), FB (Free phone Billing), FR (Free phone Routing), CC (Charge Call service), CFBL (Call Forwarding Busy Line), TL (Teen Line service), TWC (Three Way Call Service), and two game service programs using specific numbers, a memory test game and a personality analysis game, were described using ESTR.

These services were tested in the proposed validation server and 2,339 feature interactions were detected. The mean time to detect all interactions between two services was about 20 seconds, mostly less than 10 seconds.

Thus, the proposed detection methods were confirmed to be effective for VoIP services.

6 Conclusion

In this paper, evaluation results of the formal approach for detecting feature interactions in telecommunication services were reported. Application of the proposed detection method to an active gatekeeper for VoIP was also described.

For future work, the algorithm for resolving or avoiding feature interactions detected should be studied. Though we applied the proposed detection method to VoIP services, application to non-voice services should also be studied.

References

[1]	S. Komatsu and T. Ohta, "Active Networks using Declarative Language", Proc. of IWAN2000, pp. 33-44, Oct. 2000.

[2]	K.L.Calvert et al., "Directions in Active Networks", IEEE Communications Magazine, Vol. 36, No. 10, pp. 72-78, Oct. 1998.

[3]	T. Yoneda and T. Ohta, "A Formal Approach for Definition and Detection of Feature Interactions", Proc. of FIW'98, pp. 202-216, Sep. 1998.

[4]	T. Yoneda and T. Ohta, "Reduction of the number of Terminal Assignments for Detecting Feature Interactions in Telecommunication Services", Proc. of ICECCS, pp. 202-209, Sep. 2000.

[5]	T. Yoneda and T. Ohta, "Automatic Elicitation of Knowledge for Detecting Feature Interactions in Telecommunication Services", IEICE Transactions on Information and Systems, vol. E83-D No.4, pp. 640-647, April 2000.

[6]	J. Kobayashi, T. Yoneda and T. Ohta, "An Effective Method for Testing Reachability Using Knowledge in Detecting Non-Determinacy Feature Interactions", IEICE Transactions on Information and Systems, vol. E85-D No.4, pp. 607-614, April 2002.

[7]	T. Ohta and F. Cristian, "Formal Definitions of Feature Interactions in Telecommunications Software", IEICE transactions on Fundamentals of Electronics, Communications and Computer Science, vol.E81-A No.4, pp. 635-638, April 1998.

[8]	N. Griffeth et al., "Feature Interaction Detection Contest of the Fifth International Workshop on Feature Interactions," The International Journal of Computer and Telecommunications Networking, Computer Networks 32 (2000) pp. 487-510, April 2000.

[9]	N. Griffeth et al., "A Feature Interaction Benchmark for the First Feature Interaction Detection Contest", The International Journal of Computer and Telecommunications Networking, Computer Networks 32 (2000) pp.389-418, April 2000.

[10] Y. Hirakawa, et al., "Telecommunication Service Description Using State Transition Rules", Intl. Workshop on Software Specification and Design, Oct. 1991.

[11] T. Yoneda and T. Ohta, "The declarative language STR", Language Constructs for Describing Features, pp. 197-212, Aug. 2000, Springer.

[12] M. Nakamura et al., "Feature Interaction Detection Using Permutation Symmetry", Proc. of FIW '98, pp. 187-201, Sep. 1998.

Emerging Architectures

*Feature Interactions in Telecommunications
and Software Systems VII*
D. Amyot and L. Logrippo (Eds.)
IOS Press, 2003

Detecting Script-to-Script Interactions in Call Processing Language

Masahide NAKAMURA[1], Pattara LEELAPRUTE[2], Ken-ichi MATSUMOTO[1],
and Tohru KIKUNO[2]

[1] *Graduate School of Information Science, Nara Institute of Science and Technology, Japan*
{masa-n, matumoto}@is.aist-nara.ac.jp
[2]*Graduate School of Information Science and Technology, Osaka University, Japan*
{pattara, kikuno}@ist.osaka-u.ac.jp

Abstract. This paper addresses a problem to detect feature interactions in a CPL (Call Processing Language) programmable service environment on Internet telephony. In the CPL environment, the previous works cannot be directly applied, because of new complications introduced:(a) features created by non-experts and (b) distributed feature provision. To cope with the problem (a), we propose eight types of *semantic warnings* which guarantee some aspects of semantic correctness in each individual CPL script. Then, as for (b), we present an alternative definition of feature interactions, and propose a method to implement run-time feature interaction detection. The key idea is to define feature interactions as the semantic warnings over multiple CPL scripts, each of which is semantically safe. We also demonstrate tools, called CPL checker and FI simulator, to help users to construct reliable CPL scripts.

1 Introduction

Internet telephony [8] is expected to enable a new generation of telecommunication services. It facilitates integration with other Internet services, which allows rich and sophisticated telephony services. Internet telephony has been widely studied at the *network protocol level* (i.e., SIP [9] and H.323 [14]). Some companies have already started commercial services, and its protocol stacks are also released by open source communities (e.g., VOCAL [16]).

The concern is now shifting to the *service level*, that is, how to provide value-added features in Internet telephony. There are two complementary approaches to achieving the feature provision.

The first one is based on *network convergence*, which integrates IP and the traditional IN/PSTN networks [13]. The idea is to activate the IN services from Internet telephony through APIs (e.g., JAIN API [15]). Many telecom industries are conducting research and development for it, in order to make full reuse of their legacy services. Though this approach is quite challenging, we do not discuss it in this paper.

Another approach, which is interesting for us here, is *programmable service* [6] [10]. The service and feature creation is opened to end users and third parties. The service definitions can be deployed in the local (and distributed) servers over the Internet. The users can create, delete and modify their own services at any time.

As in the IN/PSTN networks, feature interactions occur as well in Internet telephony. The previous research works might be helpful to understand a part of the interactions. However, feature interactions in Internet telephony are a more serious problem than the conventional ones, because of various new complications introduced [1][7]. Especially in the context of the programmable service, the following are essential issues that make the problem more difficult.

(a) Features created by non-expert users: In conventional telephony, quality of individual features is guaranteed by the telecom companies, and end users just *subscribe* to the ready-made features. On the other hand, programmable services allow end users or third parties to freely create and define their own custom-made features. Most of the end users are not as expert as telecom engineers. Each user is very likely to create a feature, without the greatest care of logical consistency and correctness in the (single) feature, much still less feature interactions with others.

(b) Distributed feature provision: The created features can be easily installed in local signaling servers. The features are completely distributed over the Internet and there is no centralized feature server. This fact means that it is impossible to enumerate all possible features. Thus, we cannot conduct off-line feature interaction detection, nor prepare resolution schemes in advance, such as feature priorities.

In this paper, we tackle the problem of feature interaction in Internet telephony with *Call Processing Language* (CPL, in short) [5][6]. The CPL is an XML-based language for the programmable service in Internet telephony, and is proposed as RFC2824 in IETF. The CPL is gaining popularity, and major VoIP systems (e.g., [16][17]) adopt it as feature description language. The goal of this paper is to establish a definition and a detection method of feature interactions within the CPL programmable service environment. To achieve this, we propose two new methods corresponding to the above problems (a) and (b).

Firstly, we propose *semantic warnings* for the CPL scripts to address the problem (a) feature created by non-experts. In a CPL environment, each user describes his/her own feature in a (single) *CPL script* at a time. The syntax of the CPL is strictly defined by DTD (Document Type Definition). However, compliance with the DTD is not a sufficient condition for correctness of a CPL script. As far as we know, there exist no guidelines for users to assure semantical correctness of individual CPL scripts. The proposed warnings are not necessarily errors, but they identify the source of ambiguity, redundancy and inconsistency for a given CPL script.

Secondly, to address the problem (b) distributed feature provision, we propose an alternative definition of feature interaction, and its detection method in the CPL environment. In [5][6], a brief categorization of feature interactions in the CPL environment is presented [1]. However, no concrete method to detect feature interactions in the CPL environment has been proposed yet.

In general, feature interactions can be defined (informally) as violation of user's requirement that is caused by combination of multiple features. Here, a CPL script described by a user can be considered as an exact requirement of the user. So, the violation of the requirement occurs when the script is not executed as described, under the influence of CPL scripts

---

[1]According to the categorization in [5], feature interactions discussed in this paper are *script-script* and/or *server-to-server* interactions

by other users within the call. Note that the violation can be observed only at run time, and can never be predicted by off-line analysis. Thus, the new definition of feature interactions must be dependent on a call scenario at run time, which is significantly different from definitions in the literatures [4] [12].

Our key idea is to define feature interactions (i.e., the violation) as the semantic warnings over multiple CPL scripts, each of which is semantically valid. For this, we propose a combine operator and some new notions for CPL scripts (i.e., complete, safe). Then, we present a procedure to implement run-time feature interaction detection. We also present tools to detect the semantic warnings in a single CPL script, and to feature interactions over multiple scripts.

The rest of the paper is organized as follows. Section 2 describes a brief review of CPL. In Section 3, we propose the semantic warnings for a single CPL script. Then, Section 4 presents definition and detection method of feature interactions among multiple CPL scripts. Section 5 describes the tool support for the proposed method. Finally we conclude the paper with discussion and future work in Section 6.

2 Call Processing Language (CPL)

2.1 Overview

Internet telephony basically consists of two types of components: end systems and signaling servers. The CPL is meant to describe *network-based features* which process calls on the signaling servers in a network. Terminal-based features, like camp-on, call waiting and voicemail, that heavily depend on end-system states and devices should be implemented on the end systems, and thus are out of scope of the CPL.

First of all, we review the CPL definition briefly. The full specification can be found in [5][6]. A CPL script is composed of mainly four types of constructors: *top-level actions*, *switches*, *location modifiers* and *signaling operations*.

Top-level actions: Top-level actions are firstly invoked when a CPL script is executed: `outgoing` (or `incoming`) specifies a tree of actions taken on the user's outgoing call (or incoming call, respectively). `subaction` describes a sub routine to increase re-usability and modularity.

Switches: Switches represent conditional branches in CPL scripts. Depending on types of conditions specified, there are five types: `address-switch`, `string-switch`, `language-switch`, `time-switch` and `priority-switch`.

Location modifiers: The CPL has an abstract model, called *location set*, for locations to which a call is to be directed. The set of the locations is stored as an implicit global variable during call processing action by the CPL. For the outgoing call processing, the location is initialized to the destination address of the call. For the incoming call processing, the location set is initialized to the empty set. During the execution, the location set can be modified by three types of modifiers: `location` adds an explicit location to the current location set; `lookup` obtains locations from outside; `remove-location` removes some locations from the current location set.

Signaling operations: Signaling operations trigger signaling events in the underlying signaling protocol for the current location set. There are three operations: `proxy` forwards

```
<?xml version="1.0" ?>
<!DOCTYPE cpl PUBLIC "-//IETF//DTD RFCxxxx
                       CPL 1.0//EN" "cpl.dtd">
<cpl>
  <outgoing>
    <address-switch field="destination" >
      <address is="sip:bob@home.org">
        <reject status="reject"
          reason="No call to Bob is permitted" />
      </address>
    </address-switch>
  </outgoing>
</cpl>
```

Figure 1: A CPL script s_a of OCS

```
<?xml version="1.0" ?>
<!DOCTYPE cpl PUBLIC "-//IETF//DTD RFCxxxx
                       CPL 1.0//EN" "cpl.dtd">
<cpl>
  <subaction id="voicemail">
    <location url="sip:chris@voicemail.example.com">
      <proxy />
    </location>
  </subaction>

  <incoming>
    <address-switch field="origin" subfield="host">

      <address subdomain-of="example.com">
        <location url="sip:chris@office.example.com">
          <proxy />
        </location>
      </address>

      <address subdomain-of="crackers.org">
        <reject status="reject"
          reason="No call from this domain allowed" />
      </address>

      <address subdomain-of="instance.net">
        <location url="sip:bob@home.org">
          <redirect />
        </location>
      </address>

      <otherwise>
        <sub ref="voicemail" />
      </otherwise>

    </address-switch>
  </incoming>
</cpl>
```

Figure 2: A CPL script s_c of DCF

the call to the location set currently specified; `redirect` prompts the calling party to make another call to the current location set, then terminates the call processing; `reject` causes the server to reject the call attempt and then terminates the call processing.

2.2 Feature examples

We start with a simple feature, namely Originating Call Screening (OCS, in short). Suppose the following requirement: Alice (`alice@instance.net`) wants to block any outgoing calls to Bob (`bob@home.org`) from her end system. Figure 1 shows an implementation of Alice's script s_a. In Figure 1, the first three lines represent declaration of XML and DTD. The tag `<cpl>` means the start of a body of the CPL script. The top-level action `<outgoing>` describes actions activated when Alice makes a call. Next, `<address-switch>` specifies a conditional branch. In this example, the condition is extracted from the destination address of the call (`field="destination"`). If the destination address matches `bob@home.org` (`<address is="bob@home.org">`), the call is rejected (`<reject status... />`). If it does not match, the call is be proxied to the destination address (This is done by *default behavior* of the CPL, although the proxy operation is not explicitly specified. See Section 4.2.1).

The next example is a bit complicated. A user Chris (`chris@example.com`) wants to:

- receive calls from domain `example.com` at office `chris@office.example.com`.

- reject any call from malicious crackers belonging to `crackers.org`.

- redirect any call from clients within `instance.net` to Bob's home at `bob@home.org`.

- proxy any other calls to his voicemail at `chris@voicemail.example.com`.

Figure 2 shows an implementation of Chris's script s_c. Let us call this feature Domain Call Filtering (DCF). The portion surrounded by `<subaction> </subaction>` defines a subaction called from the main-routine. `<incoming>` specifies actions activated when Chris receives an incoming call.

Next, in `<address-switch>`, a condition for the switch is extracted from the host address of the caller (`field="origin"` `subfield="host"`). If the domain matches example.com (`<address subdomain-of="example.com">`), then the location is set to chris@office.example.com, and the call is proxied to his office (`<proxy />`). If the domain matches crackers.org, the call is rejected by `<reject />`. Else if the domain matches instance.net, the location is set to bob@home.org. Then, the call is redirected to Bob and the caller places a new call to Bob. Otherwise, the subaction voicemail is called. In the subaction voicemail, the location is set to the voicemail at chris@voicemail.exam ple.com, and then the call is proxied there.

3　CPL semantic warnings

The definition of CPL with DTD [5][6] guarantees the syntactical structure of CPL, but does not cover any semantical aspects of each individual script. Hence, there is enough room, especially for non-expert users, to make semantical flaws in the script. The proposed warnings aim to help each user find such semantical flaws in his/her script. Focusing on constraints of CPL and semantic aspects of telephony features, we have identified eight kinds of warnings so far.

3.1　Multiple forwarding addresses (MFAD)

Definition: The execution reaches `<proxy>` or `<redirect>` while multiple addresses are contained in the location set.

Effects: The CPL allows calls to be proxied (or redirected) to multiple address locations by cascading `<location>` tags. However, if the call is redirected to multiple locations, then the caller would be confused to which address the next call should be placed. Or, if the call is proxied, a race condition might occur depending on the configuration of the proxied end systems. As a typical example, if a user simultaneously sets the forwarding address to his handy phone and voicemail that immediately answers the call. Then the call never reaches his handy phone.

Example CPL: Figure 3(a) shows an example. The user is setting the forwarding address to his handy phone pattara@mobile.example.com and voicemail pattara@voicemail .example.com, simultaneously. If the user configures the voicemail to immediately answer the call, then no call reaches the mobile phone.

3.2　Unused subactions (USUB)

Definition: Subaction `<subaction id= "foo" >` exists, but `<subaction ref= "foo" >` does not.

Effects: The subaction is defined but not used. The defined subaction is completely redundant, and should be removed to decrease server's overhead for parsing the CPL script.

Example CPL: Figure 3(b) shows an example. In this script, a subaction mobile that is declared in the subsection part is not used in the body of the script. So, the unused subaction mobile is redundant and should be removed.

3.3   Call rejection in all paths (CRAP)

Definition: All execution paths terminate at `<reject>`.

Effects: No matter which path is selected, the call is rejected. No call processing is performed, and all executed actions and evaluated conditions are nullified. This is not a problem only when the user wants to reject all calls explicitly. However, complex conditional branches and deeply nested tags make this problem difficult to be found, on the contrary to the user's intention.

Example CPL: Figure 3(c) shows an example. By this script, any incoming call is rejected, no matter who the originator is. All actions and evaluated conditions are meaningless after all.

3.4   Address set after address switch (ASAS)

Definition: When `<address>` and `<otherwise>` tags are specified as outputs of `<address-switch>`, the same address evaluated in the `<address>` is set in the `<otherwise>` block.

Effects: The `<otherwise>` block is executed when the current address does not match the one specified in `<address>`. If the address is set as a new current address in `<otherwise>` block, then a violation of the conditional branch might occur. A typical example is that, after screening a specific address by `<address-switch>`, the call is proxied to the address, although any call to the address must have been filtered.

Example CPL: Figure 3(d) shows an example. When the user make an outgoing call, this script checks the destination of the call. The call should be rejected if the destination address is `pattara@example.com`, according to the condition specified in `<address>`. However, in the `<otherwise>` block, the call is proxied to `pattara@example. com`, which must have been rejected.

3.5   Overlapped conditions in single switch (OCSS)

Definition: Let A be a switch, and let $cond_{A1}$ and $cond_{A2}$ (arranged in this order) be conditions specified as output tags of A. Then, $cond_{A1}$ is implied by $cond_{A2}$.

Effects: According to the CPL specification, if there exist multiple output tags for a switch, then the condition is evaluated in the order the tags are presented, and the first tag to match is taken. By the above definition, whenever $cond_{A2}$ becomes true, $cond_{A1}$ is true. So, the former tag is always taken and the latter tag is never executed, which is a redundant description.

Example CPL: Figure 3(e) shows an example. This script is supposed to do a typical call processing in a support center. Calls for general help (with a subject containing `help`) are meant to be redirected to `general-support`. Emergency calls (with a subject matching `emergency help`) are to be proxied to an attendant `staff`. However, in fact, all calls are redirected to `general-support`, and no emergency call reaches to the attendant. Since `"help"` is a substring of `"emergency help"`, two conditions are overlapped.

3.6   Identical actions in single switch (IASS)

Definition: The same actions are specified for all conditions of a switch.

Effects: No matter which condition holds, the same action is executed. Therefore, the conditional branch specified in the switch is meaningless. In such case, this switch should be eliminated to reduce complexity of the logic.

Example CPL: Figure 3(f) shows an example. This script has a `language-switch` to check the language preference of the call. This switch specifies a conditional branch depending on whether the preference is Japanese (`jp`) or not. However, the same action `<sub ref="voicemail">` occurs independently of the language preference. So, the switch is completely meaningless.

3.7   Overlapped conditions in nested switches (OCNS)

Definition: Let A and B be switches of the same type, and let $cond_A$ and $cond_B$ be the conditions of A and B, respectively. Then, [A is nested in B's condition block] and [$cond_B$ implies $cond_A$].

Effects: This warning is derived by the fact that CPL has no variable assignment. So, any condition that is evaluated to be true (or false) remains true (or false, respectively) during the execution. Assume that $cond_B$ implies $cond_A$. B's condition block, in which A is specified, is executed only when $cond_B$ is true. So, by the assumption, $cond_A$ always becomes true when evaluated. Thus, A's condition block is unconditionally executed. Also, if A has an otherwise block, then the block cannot be executed. As a result, the switch A is completely redundant and should be removed.

Example CPL: Figure 3(g) shows an example. When an incoming call arrives, the script first checks a domain of the caller's host. If the domain matches `home.org`, then the second switch does the same checking again. However, since the condition for the second switch is the same as the first one, which have already been shown to be true, it is redundant description. Also, `<reject />` in `<otherwise>` is unreachable.

3.8   Incompatible conditions in nested switches (ICNS)

Definition: Let A and B be switches of the same type, and let $cond_A$ and $cond_B$ be the conditions of A and B, respectively. Then,

(α) [A is nested in B's condition block] and [$cond_A$ and $cond_B$ are mutually exclusive], or

(β) [A is nested in B's otherwise block] and [$cond_A$ implies $cond_B$].

Effects: Let us consider (α) first. B's condition block, in which A is specified, is executed only when $cond_B$ is true. However, $cond_A$ and $cond_B$ are exclusive, so $cond_A$ cannot be true at this time. Therefore, A's condition block is unexecutable. (β) is the complemental case of (α). B's otherwise block is executed only when $cond_B$ is false. Now that $cond_A$ must be false, which is implied by $\neg cond_B$. Consequently, A's condition block is unexecutable also.

```
<?xml version="1.0" ?>
<!DOCTYPE cpl PUBLIC "-//IETF//DTD RFCxxxx
                  CPL 1.0//EN" "cpl.dtd">
<cpl>
 <incoming>
   <location
     url="sip:pattara@mobile.example.com">
   <location
     url="sip:pattara@voicemail.example.com">
     <proxy />
   </location>
   </location>
 </incoming>
</cpl>
```
(a) Example of MFAD

```
<?xml version="1.0" ?>
<!DOCTYPE cpl PUBLIC "-//IETF//DTD RFCxxxx
                  CPL 1.0//EN" "cpl.dtd">
<cpl>
   <subaction id="mobile">
     <location url="sip:jones@mobile.example.com"
       <proxy />
     </location>
   </subaction>
   <incoming>
     <location url="sip:jones@example.com">
       <proxy />
     </location>
   </incoming>
</cpl>
```
(b) Example of USUB

```
<?xml version="1.0" ?>
<!DOCTYPE cpl PUBLIC "-//IETF//DTD RFCxxxx
                  CPL 1.0//EN" "cpl.dtd">
<cpl>
   <incoming>
     <address-switch field="origin">
       <address is="anonymous">
        <reject status="reject" reason=
        "I don't accept anonymous calls" />
       </address>
       <otherwise>
        <reject status="reject" reason=
        "I don't accept call from anyone" />
       </otherwise>
     </address-switch>
   </incoming>
</cpl>
```
(c) Example of CRAP

```
<?xml version="1.0" ?>
<!DOCTYPE cpl PUBLIC "-//IETF//DTD RFCxxxx
                  CPL 1.0//EN" "cpl.dtd">
<cpl>
   <outgoing>
     <address-switch field="destination">
       <address is="sip:pattara@example.com">
         <reject status="reject"
             reason="I don't call Pattara" />
       </address>
       <otherwise>
         <location url="sip:pattara@example.com">
           <proxy/>
         </location>
       </otherwise>
     </address-switch>
   </outgoing>
</cpl>
```
(d) Example of ASAS

```
<?xml version="1.0" ?>
<!DOCTYPE cpl PUBLIC "-//IETF//DTD RFCxxxx
                  CPL 1.0//EN" "cpl.dtd">
<cpl>
   <incoming>
     <string-switch field="subject">
      <string contains="help">
       <location url=
         "sip:general-support@example.com">
        <redirect />
       </location>
      </string>
      <string is="emergency help">
       <location url="sip:staff@example.com">
        <proxy />
       </location>
      </string>
     </string-switch>
   </incoming>
</cpl>
```
(e) Example of OCSS

```
<?xml version="1.0" ?>
<!DOCTYPE cpl PUBLIC "-//IETF//DTD RFCxxxx
                  CPL 1.0//EN" "cpl.dtd">
<cpl>
   <subaction id="voicemail">
     <location url="sip:nakamura@voicemail.org" >
       <proxy />
     </location>
   </subaction>
   <incoming>
     <language-switch>
       <language matches="jp">
         <sub ref="voicemail" />
       </language>
       <otherwise>
         <sub ref="voicemail" />
       </otherwise>
     </language-switch>
   </incoming>
</cpl>
```
(f) Example of IASS

```
<?xml version="1.0" ?>
<!DOCTYPE cpl PUBLIC "-//IETF//DTD RFCxxxx
                  CPL 1.0//EN" "cpl.dtd">
<cpl>
 <incoming>
   <address-switch field="origin"
                    subfield="host">
    <address subdomain-of="home.org">
     <address-switch field="origin"
                      subfield="host">
      <address subdomain-of="home.org">
       <location url="sip:pattara@mobile.net">
        <proxy />
       </location>
      </address>
      <otherwise>
        <reject status="reject"
         reason="Some reason" />
      </otherwise>
     </address-switch>
    </address>
   </address-switch>
 </incoming>
</cpl>
```
(g) Example of OCNS

```
<?xml version="1.0" ?>
<!DOCTYPE cpl PUBLIC "-//IETF//DTD RFCxxxx
                  CPL 1.0//EN" "cpl.dtd">
<cpl>
 <incoming>
   <address-switch field="origin"
                    subfield="host">
    <address subdomain-of="home.org">
     <address-switch field="origin"
                      subfield="host">
      <address subdomain-of="example.com">
       <location url="sip:pattara@mobile.net">
        <proxy />
       </location>
      </address>
     </address-switch>
    </address>
   </address-switch>
 </incoming>
</cpl>
```
(h) Example of ICNS

Figure 3: Example CPL scripts

Example CPL: Figure 3(h) shows an example for the above (α). When an incoming call arrives, the script first checks a domain of the caller's host. If the domain matches `home.org`, the second address switch evaluates the domain again. If the domain matches `example.com`, the call is proxied to `pattara@mobile.net`. However, this proxy operation never occurs, since conditions for the two switches are mutually exclusive. That is, it is impossible that the domain matches both `home.org` and `example.com`, simultaneously.

The above eight warnings can occur even if a given CPL script is syntactically well-formed and valid in the sense of XML. Note that the semantic warnings in a single script can be detected by a simple *static* (thus, off-line) analysis.

Definition　(Semantically Safe): We say that a CPL script is *semantically safe* iff the script is free from the semantic warnings.

4　Feature interaction detection in CPL scripts

4.1　Key idea

Even if each user creates a safe script by means of the proposed semantic warnings, feature interactions may occur when multiple scripts are executed simultaneously. In the CPL environment, each user cannot have more than one script at a time. Hence, interactions between features allocated in the same user (e.g, CW v.s. TWC) cannot occur [7]. Instead, interactions may occur between scripts owned by different users.

Interaction between OCS & DCF: Let us recall two features OCS and DCF in Section 2.2, implemented as s_a in Figure 1 and s_c in Figure 2, respectively. Now, consider a call scenario where Alice (`alice@instance.net`) calls Chris (`chris@example.com`). First, Alice's script s_a is executed. Since Chris is not screened in s_a, the call is proxied to Chris. Next, Chris's script s_c is executed. Since Alice belongs to a domain `instance.net`, the call is redirected to Bob (`bob@home.org`). As a result, Alice makes a call to Bob, although this call must have been blocked in s_a. Thus we can say that s_a and s_c *interact*.

The situation in the above example is quite similar to the semantic warning ASAS (See Section 3.4), although it occurs within the combination of multiple scripts s_a and s_c. The key idea of our approach is to define feature interactions as *the semantic warnings over multiple CPL scripts*.

4.2　Preliminaries

Before formalizing feature interactions in the CPL environment, we define some new notions with respect to CPL scripts.

4.2.1　Complete CPL scripts

When an execution of a CPL in a signaling server reaches an unspecified condition or an empty signaling operation, the execution follows the *default behaviors* (See Section 11 of [5] for more details). Here are some examples:

```
<?xml version="1.0" ?>
<!DOCTYPE cpl PUBLIC "-//IETF//DTD
                         RFCxxxx CPL 1.0//EN" "cpl.dtd">
<cpl>
  <outgoing>
    <address-switch field="destination" >
      <address is="sip:bob@home.org">
        <reject status="reject"
              reason="No call to Bob is permitted" />
      </address>
      <otherwise>
        <proxy />
      </otherwise>
    </address-switch>
  </outgoing>
  <incoming>

  </incoming>
</cpl>
```

Figure 4: A complete CPL script s_a of OCS

D1: In an outgoing action, if there is no location modifier and no signaling operation is reached, then proxy to the destination of the call.

D2: In an incoming action, if there is no location modifier and no signaling operation is reached, then treat as if there is no CPL script (i.e., the server tries to connect the call to an end system of the owner of the script).

D3: If location modifier exists but no signaling operation is specified, proxy or redirect to the location, based on the server's standard policy.

These default behaviors are usually taken *implicitly* from user's point of view, based on the server's policy and/or the underlying protocol [2], and may sometimes contradict to the user's intension. Hopefully, the implicitness caused by the default behaviors should be eliminated from every script. For this purpose, we define a new class of CPL scripts:

Definition (Complete Script): We say that a CPL script is *complete* iff no default behavior is taken in any possible execution path.

The default behaviors must be simulated deterministically by using auxiliary information on the signaling server. Hence, we assume that every CPL script on a signaling server can be transformed into a completed script without changing logics of the original script. The followings are guidelines to achieve the transformation.

(a) Make all conditional branches complementary. For instance, `<otherwise>` block must be added to every switch, if it is not present.

(b) Based on the server's standard policy, specify an appropriate signaling operation in every terminating node (i.e., leaf of XML's tree structure) that has location modifier.

(c) Add empty `<incoming>` or `<outgoing>` blocks if either of them is not present.

As an example, consider again the CPL script in Figure 1. This script is not complete, since there is no action specified when the destination address is not `bob@home.org`. Based on the default behavior and the guidelines above, the script can be transformed into a complete one as shown in Figure 4.

---

[2]For example, the VOCAL system [16] adopts redirect for the above D3.

4.2.2   Successor functions

Suppose that we have a complete CPL script s, and that we want to examine feature interactions between s and other related scripts. Then, we need to know at least which script should be executed after s is terminated. Since s is complete, the execution of s must exit on an empty tag or a certain signaling operation (proxy, redirect or reject), with a location set containing the next address(es) the call is directed to.

Note that the above information dynamically varies depending on given call scenarios. More specifically, we assume that the following functions are available at run time for a given CPL script s and a call scenario c.

Definition (Functions): For a complete CPL script s and a call scenario c, we define the following functions.

$exit(s, c)$: returns a pointer to a signaling operation in s executed at the end under c.

$next(s, c)$: returns the next CPL script executed following s under c, obtained based on the next address.

$type(s, c)$: returns a type of the signaling operation: *proxy*, *redirect*, *reject* or *end* (for empty signaling operation).

For example, consider again the example in Section 2.2 and the scripts s_a in Figure 4 and s_c in Figure 2. Table 1 summarizes values of the functions, with respect to two instances c_1 and c_2 of call scenarios.

Table 1: Example of successor functions for s_a

Call scenarios	$exit(s_a, c_i)$	$next(s_a, c_i)$	$type(s_a, c_i)$
c_1 (Alice calls Bob)	`<reject .../>` line 8–9	*none*	reject
c_2 (Alice calls Chris)	`<proxy .../>` line 12	s_c	proxy

4.3   Feature interactions among two scripts

Firstly, let us consider two complete scripts s and t only. In order to define feature interactions between s and t, we need to capture a combined behavior of s and t. For this purpose, we propose a *combine operator*.

Intuitively, the combine operator merges two scripts s and t such that t *is executed after s*. This partial order is defined only when the call is *proxied* from s to t. In the case that s *redirects* a call to t, the call is once reverted to the caller, and s terminates. Then, a new call is originated from the caller to t without passing through s. Note that the combine operator depends on a given call scenario, because t depends on the scenario.

Definition (Combine operator): Let c be a given call scenario, and let s and t be complete scripts such that $type(s, c) = proxy$ and $next(s, c) = t$. Then, a *combined script $r = s \triangleright_c t$* is a CPL script obtained from s and t by the following procedures:

Step1: If any subaction (let it be `<subaction id="foo">`) is defined in s (or t), eliminate it by replacing `<sub ref="foo" />` with the body of the subaction.

```
<?xml version="1.0" ?>
<!DOCTYPE cpl PUBLIC "-//IETF//DTD RFCxxxx CPL 1.0//EN" "cpl.dtd">
<cpl>
  <outgoing>
    <address-switch field="destination" >
      <address is="sip:bob@home.org">
        <reject status="reject"
            reason="No call to Bob is permitted" />
      </address>
      <otherwise>
        <remove-location>
        <address-switch field="origin" subfield="host">

          <address subdomain-of="example.com">
            <location url="sip:chris@office.example.com">
              <proxy />
            </location>
          </address>

          <address subdomain-of="crackers.org">
            <reject status="reject"
                reason="No call from this domain is permitted" />
          </address>

          <address subdomain-of="instance.net">
            <location url="sip:bob@home.org">
              <redirect />
            </location>
          </address>

          <otherwise>
            <location url="sip:chris@voicemail.example.com">
              <proxy />
            </location>
          </otherwise>

        </address-switch>
        </remove-location>
      </otherwise>
    </address-switch>
  </outgoing>
  <incoming>
  </incoming>
</cpl>
```

Figure 5: A combined CPL script $s_a \triangleright_{c_2} s_c$

Step2: Let $In(t)$ be the body of incoming action of t (i.e., the portion surrounded by `<in-coming>` $\cdots$ `</incoming>`). In s, replace `<proxy />` pointed by $exit(s,c)$ with `<remove-location>` $In(t)$ `</remove-location>`. Let the resulting script be r.

The combine operator $\triangleright_c$ makes a chain between s and t, by merging the `<proxy>` operation executed at the end of s, with the `<incoming>` action of t executed next. The `<remove-location>` inserted in Step 2, is to simulate that the location set is initialized to empty when the incoming action occurs (see Section 2.1). Note that the combine operation does not ruin the syntax structure, since both `<remove-location>` $In(t)$ `</remove-location>` and `<proxy />` are defined as *nodes* in the DTD of CPL. So, if both s and t are syntactically valid, then $s \triangleright_c t$ is also valid.

Now we are ready to define feature interactions among two scripts.

Definition   (Feature interaction among two scripts): Let s and t be given complete scripts, and let c be a given call scenario. Then, we say that s *interacts* with t with respect to c, iff both s and t are semantically safe, but $s \triangleright_c t$ is not safe.

Let us consider two scripts s_a and s_c in Figures 4 and 2, respectively. Also, take a scenario c_2 in Table 1, where Alice calls Chris. Figure 5 shows a combined script $s_a \triangleright_{c_2} s_c$. Now, both s_a and s_c are semantically safe, but $s_a \triangleright_{c_2} s_c$ is not safe. It contains a semantic warning ASAS, since address `bob@home.org` evaluated in `<address>` is set in `<otherwise>` block. This is just the interaction explained in Section 4.1.

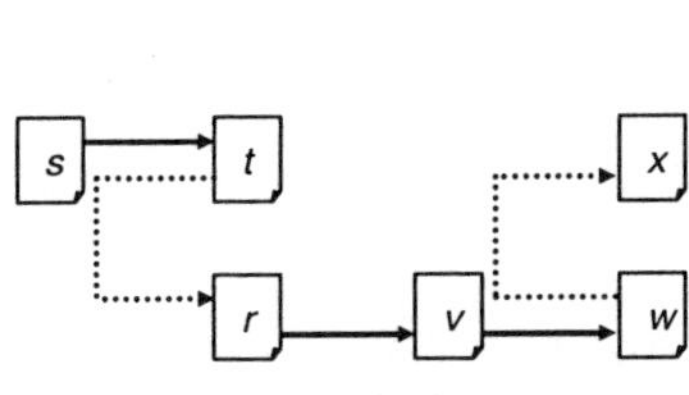

Figure 6: Multiple scripts involved in a call scenario

```
scripts Succ(script s, scenario c) {
        R = s;
        if (type(s, c)) == 'proxy') {
                check_loop(next(s, c), c);
                foreach t ∈ Succ(next(s, c), c)
                        R = R ∪ (s ▷_c t);
        } else if (type(s, c) == 'redirect') {
                R = R ∪ Succ(next(s, c), c);
        }
        return R;
}
```

Figure 7: Algorithm Succ(s,c) for computing a set of scripts to be checked

4.4 Feature interaction detection

In the previous subsection, we have defined feature interactions between two scripts. However, a call could involve more than two scripts in general, because of successive redirect and proxy operations. So, the definition of feature interactions is generalized as follows:

Definition (Feature interactions): Let s_0 be a given script of the call originator, and let c be a given call scenario. Also, let s_1, s_2, $\cdots$, s_n be scripts, where s_i proxies the call to s_{i+1} under a call scenario c. Then, we say that *feature interactions occur with respect to s_0 and c*, iff all of $s_i (0 \leq i \leq n)$ are semantically safe, but there exists some $k(1 \leq k \leq n)$ such that $s_0 \triangleright_c s_1 \triangleright_c \cdots \triangleright_c s_k$ is not semantically safe.

Figure 6 shows an example of a call scenario where multiple scripts are successively executed. In the figure, a box represents a CPL script. A solid arrow represents a proxy operation between scripts, while a dotted arrow describes a redirect operation. To identify feature interactions in this call scenario c, we must check the semantic warnings for the following six scripts (1) s, (2) $s \triangleright_c t$, (3) $s \triangleright_c r$, (4) $s \triangleright_c r \triangleright_c v$, (5) $s \triangleright_c r \triangleright_c v \triangleright_c w$ and (6) $s \triangleright_c r \triangleright_c v \triangleright_c x$.

We present an algorithm to compute a set of combined scripts that must be checked in feature interaction detection. Figure 7 shows a C-like pseudo code to compute the set R of the scripts for a given originating script s and a call scenario c. In the algorithm, we define a procedure `check_loop(t,c)`. This abstracted procedure checks if script t forms a *forwarding loop* in the call scenario c, by using a loop detection mechanism in the underlying protocol [5]. If a loop is detected, the procedure terminates the algorithm with some error reports.

The algorithm Succ first puts the given script s itself in the set R. Next, if the processing type is proxy, Succ first checks a forwarding loop by `check_loop`. If no loop is detected, it combines s with its successive scripts, which are recursively computed by setting the proxied script as the initial script, and put them in R. If the processing type is redirect, Succ recursively obtains a set of scripts starting with the redirected script, and then puts them in R. Finally, Succ returns the set R. For example, consider again a call scenario c in Figure 6. Succ(s, c) computes the six combined scripts: (1) s, (2) $s \triangleright_c t$, (3) $s \triangleright_c r$, (4) $s \triangleright_c r \triangleright_c v$, (5) $s \triangleright_c r \triangleright_c v \triangleright_c w$ and (6) $s \triangleright_c r \triangleright_c v \triangleright_c x$.

We are ready to present a feature interaction detection algorithm. We assume that each individual script is semantically safe.

Feature interaction detection algorithm :

> **Input:** A CPL script s of a call originator, and a call scenario c.
>
> **Output:** Feature interactions occur or not.
>
> **Procedure:** Compute Succ(s,c), and check if each script in Succ(s,c) is semantically safe. If all of the scripts are semantically safe, return "feature interaction does not occur". Otherwise, return "feature interaction occurs" with the corresponding (combined) scripts.

5   Tool support

We are currently implementing a set of tools for the proposed framework. Here we introduce two of them: CPL checker and FI simulator.

CPL checker: For a given CPL script, CPL checker detects the proposed semantic warnings. It also performs syntax checking to validate the conformance to the XML syntax and the DTD of CPL. Thus, it can be used for debugging CPL scripts as well. Figure 8(a) shows a screenshot, where semantic warning OCSS is detected. Every validated script can be registered in the system with the (virtual) owner's address of the script. The registered scripts can be used by FI simulator to perform off-line simulation.

FI simulator: This tool simulates execution of CPL scripts registered through CPL checker. Then, for a given call scenario, it tries to identify feature interactions by combining appropriate scripts. A user of the tool firstly chooses some of the registered scripts, then configures a call scenario [3]. Finally, the tool computes a combined script by algorithm Succ, and detects feature interactions as semantic warnings. Figure 8(b) shows a screenshot, where interaction ASAS presented in Section 4.1 is detected. The call scenario and simulated call processing are also seen in the figure.

The tools are implemented as a collection of CGI programs written in Perl open-source modules [2][3]. Since all operations to the tools are performed through a simple Web interface, a user can easily conducts validation and simulation of his/her CPL script. A prototype of the tools can be freely used at http://www-kiku.ics.es.osaka-u.ac.jp/~pattara/CPL/, though it is still experimental.

We are also planning to develop modules and libraries that can be used for on-line feature interaction detection (See Section 6.2).

6   Discussion

In this paper, we have presented two major issues of the CPL programmable service in Internet telephony: semantic warnings and feature interactions among CPL scripts. Finally, we summarize some important issues to be discussed further.

---

[3]Currently, a call scenario involves selection of caller and callee only.

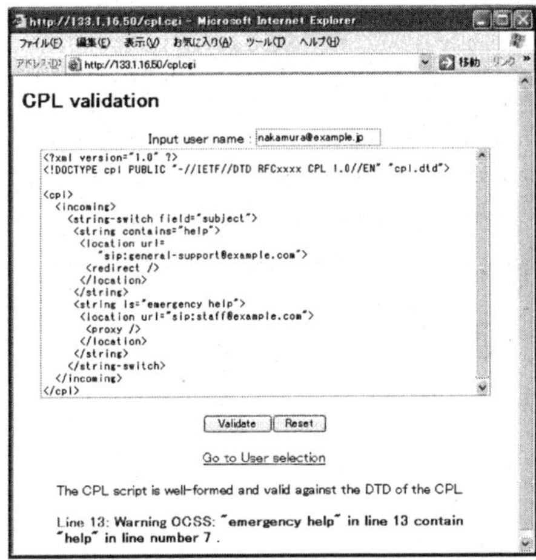

(a) CPL checker

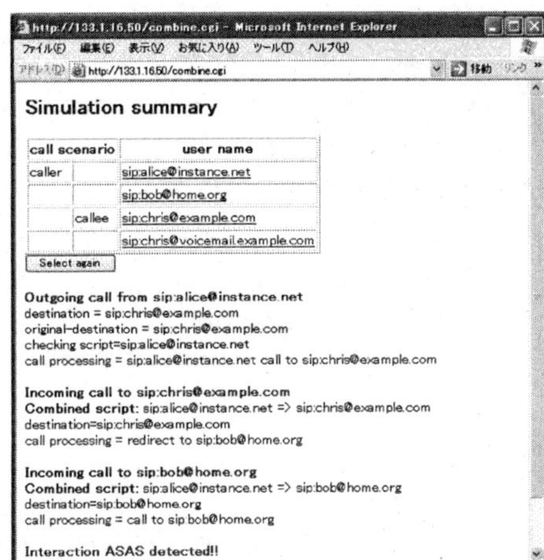

(b) FI simulator

Figure 8: Screenshots of the developed tools (prototype)

6.1 Other semantic warnings

We have proposed eight classes of the semantic warnings in this paper. However, there could exist other types of semantic warnings. We need to investigate more case studies and some quantitative evaluation to make it clear how much feature interactions can be covered by the proposed eight classes.

Also, we have discussed the semantic warnings and feature interaction in the context of the CPL programmable service only. However, feature interactions can occur between programmable services and the conventional telephony services provided by network convergence framework [15]. This is a very challenging issue and our future work.

6.2 Architecture for run-time detection

In order to perform a run-time detection of feature interaction in the CPL environment, we would need a special architecture to compute $Succ(s, c)$ and detect semantic warnings. A possible solution is to deploy an *Feature Interaction server* in the global network. Upon every call setup, signaling servers involving the call upload the relevant CPL scripts to the FI server. Then, the FI server performs appropriate combine operations and then detects feature interactions in the call. The overhead of the script uploading can be reduced if users voluntarily registers their own scripts in a *global service repository* of the FI server beforehand. To implement the architecture, we have to, of course, tackle related issues such as security, privacy and authentication.

6.3 Resolution of feature interactions

In the conventional telephony network, once an feature interaction is detected, some resolution schemes, such as feature priorities, are applied. However, in the CPL environment, it is impossible to prepare in advance appropriate resolution schemes. This is the point that the conventional run-time approaches (e.g., [11]) cannot be applied directly.

The Internet basically adopts *"use at your own risk"* policy. So, it would be natural to prompt users to make decision on how the call should be processed by themselves. However, if the programmable service environment is operated on the commercial basis, a certain guideline for users to resolve feature interactions must be prepared. The examination of the resolution schemes is also our feature research.

Acknowledgment

This research was partially supported by: the Ministry of Education, Science, Sports and Culture, Grant-in-Aid for Young Scientists (B) (No.13780234, 2002), and Grant-in-Aid for 21st century COE Research (NAIST-IS — Ubiquitous Networked Media Computing, 2003).

References

[1] L. Blair, J. Pang, "Feature Interactions - Life Beyond Traditional Telephony", Proc. of Sixth Int'l. Workshop on Feature Interactions in Telecommunication Networks and Distributed Systems (FIW'00), pp.83-93, May. 2000.

[2] C. Cooper, "XML::Parser - A perl module for parsing XML documents", http://search.cpan.org/author/COOPERCL/XML-Parser-2.31/Parser.pm

[3] E. Derksen, "Overview of libxml-enno", http://www.socsci.umn.edu/ssrf/doc/xml/enno-xml-docs/users.erols.com/enno/xml/index.html

[4] D. Keck and P. Kuehn, "The feature interaction problem in telecommunications systems: A survey," *IEEE Trans. on Software Engineering*, Vol.24, No.10, pp.779-796, 1998.

[5] J. Lennox and H. Schulzrinne, "Call processing language framework and requirements," Request for Comments 2824, Internet Engineering Task Force,May 2000, http://www.ietf.org/rfc/rfc2824.txt?number=2824

[6] J. Lennox and H. Schulzrinne, "CPL:A Language for User Control of Internet Telephony Service", Internet Engineering Task Force, Jan 2002, http://www.ietf.org/internet-drafts/draft-ietf-iptel-cpl-06.txt

[7] J. Lennox and H. Schulzrinne, "Feature Interaction in Internet Telephony", Proc. of Sixth Int'l. Workshop on Feature Interactions in Telecommunication Networks and Distributed Systems (FIW'00), pp.38-50, May. 2000.

[8] H. Schulzrinne and J. Rosenberg, "Internet Telephony: Architecture and protocols - an IETF perspective," Computer Networks and ISDN Systems, vol.31, pp.237-255, Feb 1999.

[9] M. Handley, H. Schulzrinne, E. Schooler, and J. Rosenberg, "SIP:session initiation protocol", Request for Comments 2543, Internet Engineering Task Force, Feb 2002, http://www.ietf.org/internet-drafts/draft-ietf-sip-rfc2543bis-09.txt

[10] M. Smirnov, "Programming Middle Boxes with Group Event Notification Protocol", Proceedings of the Seventh IEEE International Workshop on Object-oriented Real-time Dependable Systems (WORDS2002), pp. 198-205, Jan 2002.

[11] S. Tsang and E. H. Magill, "Learning to Detect and Avoid Run-Time Feature Interactions in the Intelligent Network", IEEE Transactions on Software Engineering, Volume 24, Number 10, Oct 1998.

[12] "Feature Interaction in Telecommunications", Vol. I-VI, IOS Press (1992-2000)

[13] ITU-T Recommendations Q.1200 Series: Intelligent Network Capability Set 1, ITU-T (1990)

[14] ITU-T Recommendation H.323, "Packet-Based Multimedia Communications Systems", February 1998.

[15] JAIN initiative, "The $JAIN^{TM}$ APIs: Integrated Network APIs for the Java Platform", http://java.sun.com/products/jain/

[16] "VOCAL: The Vovida Open Communication Application Library", http://www.vovida.org/

[17] $NetCentrex^{TM}$, "Application Execution and Service Creation Environment", http://www.netcentrex.net/products/application_server.shtml

Feature Interactions in Policy-Driven Privacy Management[1]

George YEE and Larry KORBA
Institute for Information Technology
National Research Council Canada
Montreal Road, Building M-50
Ottawa, Ontario, Canada K1A 0R6
Email: {George.Yee, Larry.Korba}@nrc-cnrc.gc.ca

Abstract. The growth of the Internet is increasing the deployment of e-services in such areas as e-business, e-learning, and e-health. In parallel, the providers and consumers of such services are realizing the need for privacy. The widespread use of P3P privacy policies for web sites is an example of this growing concern for privacy. However, while the privacy policy approach may seem to be a reasonable solution to privacy management, we show in this paper that it can lead to unexpected feature interaction outcomes such as unexpected costs, the lost of privacy, and even cause serious injury. We propose a negotiations approach for eliminating or mitigating the unexpected bad outcomes.

1 Introduction

The growth of the Internet has been accompanied by a growth in the number of e-services available to consumers. E-services for banking, shopping, learning, and even Government Online abound. Each of these services requires a consumer's personal information in one form or another. This leads to concerns over privacy. Indeed, the public's awareness of potential violations of privacy by online service providers has been growing. Evidence affirming this situation include a) the use of P3P privacy policies [1] by web server sites to disclose their treatment of users' private information, and b) the enactment of privacy legislation in the form of the Privacy Principles [2] as a sort of owners' "bill of rights" concerning their private information. We take a policy-based approach in privacy management. We believe this offers both effectiveness and flexibility. Both providers and consumers have privacy policies stating what private information they are willing to share, with whom it may be shared, and under what circumstances it may be shared. Privacy policies are attached to software agents that act as proxies for service consumers or providers. Prior to the activation of a particular service, the agents for the consumer and provider undergo a privacy policy exchange, in which the policies are examined for compatibility. The service is only activated if the policies are compatible (i.e. there are no conflicts). Figure 1 illustrates our policy exchange model. For the purposes of this paper, it is not necessary to consider the details of service operation.

Given the above scenarios, we show how the privacy policies of consumers and providers can interact with unexpected negative consequences. We then propose an approach to prevent or mitigate the occurrences of such consequences. Traditionally, feature interactions have been considered mainly in the telephony or communication

---

[1] NRC Paper Number: NRC 45785

services domains [3]. More recent papers, however, have focused on other domains such as the Internet, multimedia systems, mobile systems [4], and Internet personal appliances [5].

Section 2 looks at the content of privacy policies by identifying some attributes of private information collection, using the Privacy Principles as a guide. Section 3 presents a categorization of privacy policy interactions with their outcomes. Section 4 proposes an approach to prevent or mitigate the occurrence of negative consequences from privacy policy interactions. Section 5 gives conclusions.

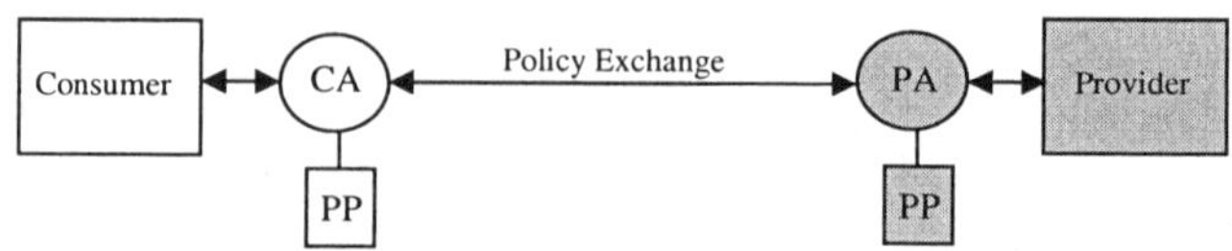

Figure 1. Exchange of Privacy Policies (PP) Between Consumer Agent (CA) and
Provider Agent (PA)

2   Privacy Policies

We identify some attributes of private information collection using the Privacy Principles [2] as a guide. We will apply these attributes to the specification of privacy policy contents.

Table 1. The Ten Privacy Principles Used in Canada

Principle	*Description*
1. Accountability	An organization is responsible for personal information under its control and shall designate an individual or individuals accountable for the organization's compliance with the privacy principles.
2. Identifying Purposes	The purposes for which personal information is collected shall be identified by the organization at or before the time the information is collected.
3. Consent	The knowledge and consent of the individual are required for the collection, use or disclosure of personal information, except when inappropriate.
4. Limiting Collection	The collection of personal information shall be limited to that which is necessary for the purposes identified by the organization. Information shall be collected by fair and lawful means.
5. Limiting Use, Disclosure, and Retention	Personal information shall not be used or disclosed for purposes other than those for which it was collected, except with the consent of the individual or as required by the law. In addition, personal information shall be retained only as long as necessary for fulfillment of those purposes.
6. Accuracy	Personal information shall be as accurate, complete, and up-to-date as is necessary for the purposes for which it is to be used.
7. Safeguards	Security safeguards appropriate to the sensitivity of the information shall be used to protect personal information.
8. Openness	An organization shall make readily available to individuals specific information about its policies and practices relating to the management of personal information.
9. Individual Access	Upon request, an individual shall be informed of the existence, use and disclosure of his or her personal information and shall be given access to that information. An individual shall be able to challenge the accuracy and completeness of the information and have it amended as appropriate.
10. Challenging Compliance	An individual shall be able to address a challenge concerning compliance with the above principles to the designated individual or individuals accountable for the organization's compliance.

We interpret "organization" as "provider" and "individual" as "consumer". Principle 2 implies that there could be different providers requesting the information, thus implying a *who*. Principle 4 implies that there is a *what*, i.e. what personal information is being collected. Principles 2, 4, and 5 state that there is a *purpose* for which the private information is being collected. Finally, Principle 5 implies a *time* element to the collection of personal information, i.e. the provider's retention time of the private information. We thus arrive at 4 attributes of private information collection, namely *who, what, purpose,* and *time.*

The Privacy Principles also prescribe certain operational requirements that must be satisfied between provider and consumer, such as identifying purpose and consent. Our use of proxy agents and their exchange of privacy policies automatically satisfy some of these requirements, namely Principles 2, 3, and 8. The satisfaction of the remaining operational requirements depends on compliance mechanisms (Principles 1, 4, 5, 6, 9, and 10) and security mechanisms (Principle 7), which are outside the scope of this paper.

Based on the above exploration, the contents of a privacy policy should, for each item of private information, identify a) who wishes to collect the information, b) the nature of the information, c) the purpose for which the information is being collected, and d) the retention time for the provider to keep the information. Figure 2 (top) gives examples of provider privacy policies from 3 types of providers: an e-learning provider, an e-commerce provider, and a nursing practitioner who uses the Internet to obtain referrals. Figure 2 (bottom) gives corresponding example consumer privacy policies. These policies need to be expressed in a

Privacy Policy: E-learning *Owner: E-learning Unlimited*	*Privacy Policy: Book Seller* *Owner: All Books Online*	*Privacy Policy: Medical Help* *Owner: Nursing Online*
Who: Any *What:* name, address, tel *Purpose:* identification *Time:* As long as needed *Who:* Any *What:* Course Marks *Purpose:* Records *Time:* 1 year	*Who:* Any *What:* name, address, tel *Purpose:* identification *Time:* As long as needed *Who:* Any *What:* credit card *Purpose:* payment *Time:* until payment complete	*Who:* Any *What:* name, address, tel *Purpose:* contact *Time:* As long as needed *Who:* Any *What:* medical condition *Purpose:* treatment *Time:* 1 year

Privacy Policy: E-learning *Owner: Alice Consumer*	*Privacy Policy: Book Seller* *Owner: Alice Consumer*	*Privacy Policy: Medical Help* *Owner: Alice Consumer*
Who: Any *What:* name, address, tel *Purpose:* identification *Time:* As long as needed *Who:* Any *What:* Course Marks *Purpose:* Records *Time:* 2 years	*Who:* Any *What:* name, address, tel *Purpose:* identification *Time:* As long as needed	*Who:* Any *What:* name, address, tel *Purpose:* contact *Time:* As long as needed *Who:* Dr. Alexander Smith *What:* medical condition *Purpose:* treatment *Time:* As long as needed

Figure 2. Example Provider Privacy Policies (top) and Corresponding Consumer Privacy Policies (bottom)

machine-readable policy language such as APPEL [6] (XML implementation). The authors are presently experimenting with a prototype privacy policy creation system, which will be reported in a future paper.

3 Privacy Policy Interactions

Once consumer and provider agents exchange privacy policies, each agent examines the other's policy to determine if there is a match between the two policies. If each agent finds a match, the agents signal each other that a match has been found, and service is initiated. If either agent fails to find a match, that agent would signal a mismatch to the other agent and service would then not be initiated. In this case, the consumer (provider) is free to exchange policies with another provider (consumer). In our model, the provider always tries to obtain more private information from the consumer; the consumer, on the other hand, always tries to give up less private information. We say that there is a *match* between a consumer's privacy policy and the corresponding provider's policy where the consumer's policy is giving up less than or equal to the amount of private information required by the provider's policy. Otherwise, we say that there is a *mismatch*. Where time is involved, a private item held for less time is considered less private information. Thus in the policies above, there is a match for e-learning, since the time required by the provider (1 year) is less than the time the consumer is willing to give up (2 years), i.e. the provider requires less private information than the consumer is willing to give up. There is a mismatch for book seller (consumer not willing to provide credit card data) and a mismatch for medical help (consumer only willing to tell medical condition to Dr. Smith). A privacy policy is considered *upgraded* if the new version represents more privacy than the prior version. Similarly, a privacy policy is considered *downgraded* if the new version represents less privacy than the prior version.

In telecom, the individual features work as designed, but the combination of features working together interact and produce unexpected outcomes. In the case of consumers and providers, each privacy policy is a statement of how private information is to be handled, much like how a telecom feature is a statement of how telecom traffic is to be handled. A privacy policy is therefore analogous to a telecom feature. Given any consumer-provider pair, the execution of their privacy policies is analogous to the simultaneous execution of two or more telecom features, and can also produce unexpected outcomes. Here, "execution" includes the examination by the respective agents to determine if there is a match. As for telecom, each privacy policy or feature is correct by itself (in the sense that it reflects the wishes of its owner), but can produce unexpected outcomes when executed in combination.

There are also differences with telecom feature interactions. Firstly, telecom feature interactions are regarded as side effects of the features. Policy interactions, on the other hand, are part of the normal workings of privacy management, i.e. consumer and provider privacy policies must interact or work together. There is thus no special mechanism needed to detect policy interactions – they occur normally. Secondly, there is a difference in the degree of certainty of unexpected outcomes. In telecom interactions, the unexpected outcomes are the results of physics, and are certain outcomes. In policy interactions, some unexpected outcomes are less certain to occur because they are based on predictions of social behaviour, which can be far more difficult to predict than the outcome of physical laws.

We categorize privacy policy interactions according to how many providers and consumers are exchanging policies at the same time and give examples of outcomes under each category. These policy interactions represent what we consider would be typical occurrences. Example outcomes for simpler exchange structures may also apply for more complex structures. This is determined by the degree to which the simpler structure is part of the more complex structure. We will indicate when this is the case. Our categorization follows:

A. <u>One Consumer to One Provider Interactions</u>
This is the case depicted in Figure 1, with one consumer exchanging policy with one provider.
Case 1: Policies match. Provider may begin service. Possible outcomes:

a) If this match occurred after a series of mismatches with other providers, then the newly found provider is probably less attractive according to criteria such as reputation and cost (this consumer probably started with more attractive providers).

b) If this match occurred after the provider or the consumer downgraded their privacy policies, the provider or consumer may not realize the extra costs that may result from not having access to the private information item or items that were eliminated through downgrading. For example, leaving out the social insurance number may lead to more costly means of consumer identification for the provider. As another example, suppose All Books Online in section 2 downgraded its privacy policy by eliminating the credit card requirement. This would lead to a match with Alice's privacy policy, but may cost Alice longer waiting time to get her order, as she may be forced into an alternate slower means of making payment (e.g. mail a cheque), if payment is required prior to shipping.

c) The provider now has the responsibility to safeguard the consumer's private data (Privacy Principle 7). The provider may not realize that the cost of the safeguard may be very high.

d) Unexpected outcomes that derive from the specifics of the privacy policies. For example, suppose the Nursing Online provider above modifies its policy to match Alice's policy because Dr. Smith is on staff. Alice is able to subscribe to Nursing Online. Then if Dr. Smith becomes unavailable due to an accident, just at the time Alice needs medical attention, Nursing Online would not be able to help Alice – an unexpected outcome.

Case 2: Policies mismatch. Provider may not begin service. Possible outcomes:

a) Consumer may decide to downgrade his privacy policy to try to get a match.

b) Provider may decide to downgrade its privacy policy to try to get a match.

c) The mismatch may result in a denial of service that has serious consequences, where the service is one that is required for safety reasons. For example, if the sought after provider is a health services provider, the consumer may suffer serious injury. In the example policies for medical help above, there is a mismatch due to Alice's requirement to only reveal her medical condition to Dr. Smith. However, if Dr. Smith is not available, a nursing service might still be better than no service.

B. <u>One Consumer to Many Providers Interactions</u>

This is the case shown in Figure 3, where agents for the same consumer exchange the same policy with many provider agents at the same time, with each provider agent representing a different provider that provides the same service.

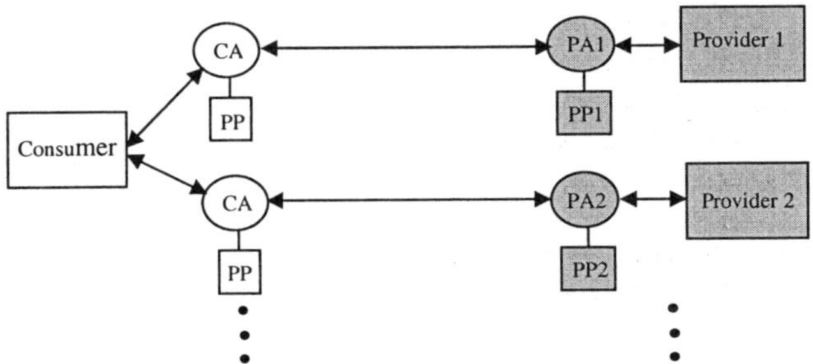

Figure 3. Exchange of Privacy Policies Between One Consumer and Many Providers

Case 1: Policies match for at least one provider. Provider may begin service. Possible outcomes:
   a) Same outcomes as in A, Case 1.
   b) Where more than one match is found, the consumer has the opportunity to select the best provider based on other criteria such as reputation and cost.
Case 2: Policies mismatch. Provider may not begin service. Possible outcomes:
   a) Same outcomes as in A, Case 2.
   b) The consumer may be influenced by the many provider policies he has seen to adjust his privacy policy to get a match with the provider that requests the least amount of private information.

C. <u>Many Consumers to One Provider Interactions</u>
   This is the case shown in Figure 4, where agents for different consumers exchange different policies at the same time with agents for the same provider.

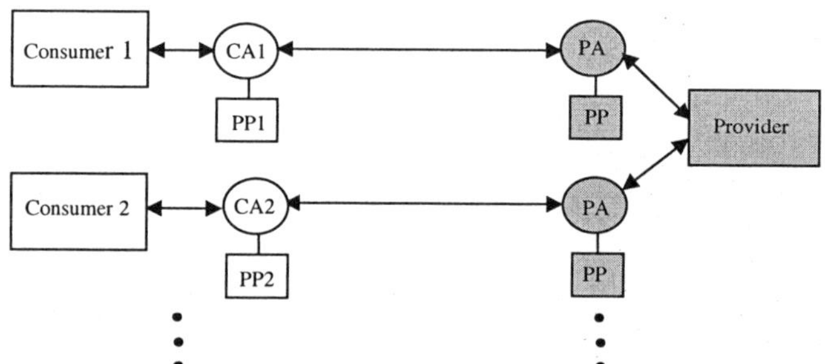

Figure 4. Exchange of Privacy Policies Between Many Consumers and One Provider

Case 1: Policies match for at least one consumer. Provider may begin service to that consumer. Possible outcomes:
   a) Same outcomes as in A, Case 1.
   b) Where more than one match is found, the provider has the opportunity to select the best consumer based on other criteria such as reputation and credit history.
Case 2: Policies mismatch. Provider may not begin service. Possible outcomes:
   a) Same outcomes as in A, Case 2.
   b) The provider may be influenced by the many consumer polices it has seen to adjust its privacy policy to get a match with the consumer that offers the most amount of private information.

D. <u>Many Consumers to Many Providers Interactions</u>
   This situation is a combination of A, B, and C with the same outcomes.

4   Preventing Unexpected Bad Outcomes

The problem at hand is how to detect and prevent the unexpected outcomes that are bad or dangerous. Not all the possible outcomes in section 3 are unexpected or bad. Solutions to telecom feature interaction problems are varied, ranging from formal analysis [7] to negotiating agents [8] and architectural approaches [9]. In this work, we propose the use of privacy policy negotiation between consumer and provider agents to mitigate or eliminate the unexpected outcomes that are bad. Consider section 3, category A. In case 1, part a) would be less serious since negotiation would reduce the number of mismatches. Case 1, parts b), c), and d) may also be less likely to happen since negotiation may force the provider or consumer to consider all implications. Case 2 c) would likely not occur since negotiations would reveal that Alice's policy is overly restrictive (see example negotiation below). The new outcomes in categories B and C are ones that favour either the consumer or the provider, based on the possibility of several matches, or the examination of many policies. These additional outcomes are neither good nor bad. It suffices to observe that they are still possible with negotiation, i.e. by negotiating several matches or examining many policies concurrently. We have proposed methods for agent-based privacy policy negotiation in [10,11]. In these papers, we provide formal descriptions of the negotiations process and methods for negotiating in cases where there is uncertainty of what offers and counter-offers to make. Table 2 illustrates how negotiation can detect and prevent the unexpected bad outcome of having no access to medical service when it is needed (read from left to right and down):

Table 2. Preventing Unexpected Bad Outcomes

Nursing Online (Provider)	*Alice (Consumer)*
OK if a nurse on our staff sees your medical condition?	*No, only Dr. Alexander Smith can see my medical condition.*
We cannot provide you with any nursing service unless we know your medical condition.	*OK, I'll see Dr. Smith instead.*
You are putting yourself at risk. What if you need emergency medical help for your condition and Dr. Smith is not available?	*You are right. Do you have any doctors on staff?*
Yes, we always have doctors on call. OK to allow them to know your medical condition?	*That is acceptable.*

The result of this negotiation is that Nursing Online will be able to provide Alice with nursing service whenever Alice requires it. If this negotiation had failed (Alice did not agree), Alice will at least be alerted to the possibility of a bad outcome, and may take other measures to avoid it. We have assumed that the provider will want to inform the consumer about bad policy implications that it knows about. We believe this is a reasonable assumption given that it is in their mutual interest to avoid unexpected bad outcomes.

5   Conclusions

The Privacy Principles impose legislative conditions on the rights of individuals (consumers) to privacy. They imply that the collection of private information may be done under the headings of *who, what, purpose,* and *time.* Privacy policies may be constructed using these headings to specify each private informational item to be shared. In an online community, consumers and providers of electronic services specify their privacy preferences using privacy policies. Agent proxies for consumers and providers exchange and compare these polices in an attempt to match up a consumer of an electronic service with the provider of that service. However, such exchanges can lead to unexpected feature interaction outcomes that have serious negative consequences. Rather than a simple matching process, privacy policies should be negotiated between consumer and provider to develop a mutually agreed upon policy for operation [12]. Such negotiation reduces or eliminates the harmful feature interaction outcomes. It does, however, lead to questions regarding revisiting mutually agreed policies, as consumer or provider policies change over time. As future work, we plan to continue experimenting with a prototype we have built for agent-based privacy negotiation, to identify issues and their resolution. We also plan to investigate ways of avoiding harmful outcomes that may be used in conjunction with negotiation, forming a multi-pronged approach to preventing unexpected bad outcomes.

6   References

[1] W3C, "The Platform for Privacy Preferences", http://www.w3.org/P3P/

[2] Department of Justice, Privacy Provisions Highlights,
http://canada.justice.gc.ca/en/news/nr/1998/attback2.html

[3] D. Keck and P. Kuehn, "The Feature and Service Interaction Problem in Telecommunications Systems: A Survey", IEEE Transactions on Software Engineering, Vol. 24, No. 10, October 1998.

[4] L. Blair, J. Pang, "Feature Interactions – Life Beyond Traditional Telephony", Distributed Multimedia Research Group, Computing Dept., Lancaster University, UK.

[5] M. Kolberg et al, "Feature Interactions in Services for Internet Personal Appliances", University of Stirling, UK, Telcordia Technologies, USA.

[6] W3C, "A P3P Preference Exchange Language 1.0 (APPEL1.0)", W3C Working Draft 15 April 2002, http://www.w3.org/TR/P3P-preferences/

[7] Amyot, D., Charfi, L., Gorse, N., Gray, T., Logrippo, L., Sincennes, J., Stepien, B. and Ware T, "Feature Description and Feature Interaction Analysis with Use Case Maps and LOTOS", Sixth International Workshop on Feature Interactions in Telecommunications and Software Systems (FIW'00), Glasgow, Scotland, UK, May 2000.

[8] N. D. Griffeth and H. Velthuijsen, "The Negotiating Agents Approach to Runtime Feature Interaction Resolution", Proc. of 2nd Int. Workshop on Feature Interactions in Telecommunications Systems, pp. 217-235, IOS Press, 1994.

[9] P. Zave, "Architectural Solutions to Feature-Interaction Problems in Telecommunications", Proc. 5th. Feature Interactions in Telecommunications and Software Systems, pages 10-22, IOS Press, Amsterdam, Netherlands, Sept. 1998.

[10] G. Yee, L. Korba, "Bilateral E-services Negotiation Under Uncertainty", accepted for publication, The 2003 International Symposium on Applications and the Internet (SAINT2003), Orlando, Florida, Jan. 27-31, 2003.

[11] G. Yee, L. Korba, "The Negotiation of Privacy Policies in Distance Education", accepted for publication, 14th IRMA International Conference, Philadelphia, Pennsylvania, May 18-21, 2003.

[12] L. Korba, "Privacy in Distributed Electronic Commerce", Proc. of the 35th Hawaii International Conference on System Science (HICSS), Hawaii, January 7-11, 2002.

*Feature Interactions in Telecommunications
and Software Systems VII*
D. Amyot and L. Logrippo (Eds.)
IOS Press, 2003

A Policy Architecture for Enhancing and Controlling Features

Stephan REIFF-MARGANIEC and Kenneth J. TURNER
Department of Computing Science and Mathematics
University of Stirling, FK9 4LA Stirling, Scotland, UK
{srm,kjt}@cs.stir.ac.uk

Abstract. Features provide extensions to a basic service, but in new systems users require much greater flexibility oriented towards their needs. Traditional features do not easily allow for this. We propose policies as the features of the future. Policies can be defined by the end-user, and allow for the use of rich context information when controlling calls. This paper introduces an architecture for policy definition and call control by policies. We discuss the operation of systems based on such an architecture. An important aspect of the architecture is integral feature interaction handling.

1 Motivation

Telecommunications has a central role in daily life, be it private or within the enterprise. We have left behind times when two users were simply connected for a verbal communication and now encounter a trend towards merging different communications technologies such as video conferencing, email, Voice over IP as well as new technologies like home automation and device control. This is combined with a move to greater mobility, e.g. wireless communications, mobile telephony and ad hoc networking. Services such as conference calling and voice mail have been added to help deal with situations beyond simple telephony.

Currently such services only support communication, that is they allow the user to simplify and more closely integrate telecommunications in their activities. In the future services will make use of the merged technologies on any device. We believe that services then will enable communications by allowing the user to achieve particular goals.

Increasing numbers of mobile users with multiple communication devices lead to a situation where users are always reachable. However, users might not always wish to be disturbed, or at least not for everyone or for any type of enquiry. Future services need to provide support for users to control their *availability*.

Availability will be highly dependent on the *context* of the user. Services must, amongst others, take into account where users are, what devices they currently use and who they might be with, as well as simple concepts such as time of day.

We conclude that the end-user must have a central place in communications systems. Services must be highly customizable by lay end-users, as only the individual is aware of his requirements. It might even be desirable for end-users to define their own services. However, any customization or service development must be simple and intuitive to suit lay users.

We discuss the advantages of policies over features, and define an architecture in which policies can be used to control calls. The operation of a system based on such an architecture is discussed. We will find that policies do not remove the feature interaction problem, but provide different angles on possible solutions. Interaction detection and resolution mechanisms will form an essential part of the proposed system. The system is user oriented and addresses the fact that future call control needs to enable individuals to achieve their goals.

2 Background

Policies and Services. Policies are defined as *information which can be used to modify the behaviour of a system* [6]. Considerable interest has been aroused by policies in the context of multimedia and distributed systems. Policies have also been applied to express the enterprise viewpoint [10] of ODP (Open Distributed Processing) and agent based systems [1].

In telecommunications systems, customized functionality for the user is traditionally achieved by providing services, i.e. capabilities offered on top of a basic service. Services are supplied by the network provider and thus do not offer completely customized functionality. Consider a call forwarding service: the user chooses whether the service is available or not, and which number the call gets forwarded to. There is no possibility for the user to forward only some calls or to treat certain calls differently, e.g. forward private calls to the mobile and others to a voicemail facility. It is exactly the flexibility, adaptability and end-user definition of policies that makes them an ideal candidate technology for services of the future.

Policies and Feature Interaction. One might hope that policies remove the feature interaction problem, simply by being higher-level. The policy community has recognised that there are issues, referred to as *policy conflict*, but has not considered any general solutions, assuming that this is not a crucial problem.

On the one hand, policies are defined by end-users, so policies will be larger in number and more diverse than features. Also, the lay nature of the user adds to the problem. On the other hand, policies can contain preferences. The context can also provide priorities usable to resolve conflicts in conjunction with richer protocols of new communications architectures.

3 The Policy Architecture

We propose a three layer architecture consisting of (1) the communications layer, (2) the policy layer and (3) the user interface layer. A three-tier architecture is used in completely different ways for other applications. However, a three-tier policy architecture emerges naturally.

When we consider existing call control architectures, similar architectures have been in use some time. For example, in the IN the three layers are given by the SSP (Service Switching Point), SCP (Service Control Point) and SCE (Service Creation Environment). Similarly in a SIP environment the layers are provided by the SIP proxies, CPL or CGI scripts and tools for creation of such scripts.

However our proposed architecture (Figure 1) differs in several key aspects, from similar existing architectures (e.g. the IN or SIP). Some of these aspects are:

- The Policy Servers can negotiate goals or solutions to detected problems.

- The User Interface Layer provides end-users with a mechanism to define functionality.

- The User Interface Layer and the Policy Servers make use of context information.

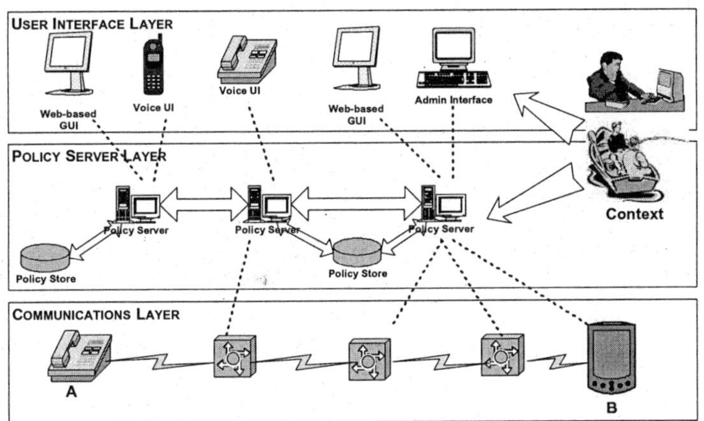

Figure 1: Overview of the Proposed Architecture

- The Policy Server layer is independent of the underlying call architecture.

The architecture makes the underlying communications network transparent to the higher layers. It enables end-user configurability and provides communication between policy servers which can be used for interaction handling. We will briefly discuss the details of each layer.

The Communications Layer. This represents the chosen call architecture. For this paper, we assume a general structure consisting of end-devices and a number of switching points. We impose two crucial requirements on the communications layer. (1) The policy servers must be provided with any message that arrives at a switching point; routing is suspended until the policy server has dealt with the message. (2) A mapping of low level messages of the communications layer into more abstract policy events must defined. Our investigations have shown that these are realistic assumption.

The Policy Server Layer. This contains a number of policy servers that interact with the underlying call architecture. It also contains a number of policy stores (database or tuple space servers) where policies are maintained by the policy servers as required. We assume that several policy servers might share a policy store, and also that each policy server might control more than one switching point or apply to more than one end device. The policy servers interact with the user interfaces in the policy creation process discussed in section 4. They also interact with the communications layer where policies are enforced; discussed in detail in section 5. The policy servers have access to up-to-date information about the user's context details which are used to influence call functionality.

The User Interface Layer. This allows users to create new policies and deploy these in the network. A number of interfaces can be expected here. We would assume the normal user to use a web-based interface for most functions. For mobile users, voice controlled interfaces are more appropriate. A voice interface is essential for disabled or partially sighted users. Both web and voice interfaces should guide the user in an intuitive way, preferably in natural language or in a graphical fashion. We also envision libraries of policies that users can simply adapt to their requirements and combine to obtain the functionality required, in a similar way that for example clip-art libraries are common today. System-oriented administrative interfaces exist for system administrators to manage more complex functionality.

4   Defining Policies

Policies should provide the end-user with capabilities to get the most out of their communications systems. End-users usually use their communications devices in a social or commercial context that imposes further policies. For example an employee is often subject to company policies. Hence policies will be defined at different domain levels (users, groups, companies, social clubs, customer premises, etc.) by differently skilled users. Any policy definition process needs to take this into account.

A Policy Description Language. In previous work [9] we have introduced initial ideas for a policy description language (PDL) to express call control policies. Here we present detail relevant for this paper. We have analysed a set of more than 100 policies before defining this language. Further, any traditional feature can also be expressed. The policy definition language is defined as an XML schema and hence policies will be stored as XML.

Complex policies are combinations of policy rules. There are a number of ways in which simple rules can be combined: sequencing, parallel composition (simultaneous application of rules), guarded and unguarded choice (conditional choice of one rule or indifference as to which rule is applied). Each policy is uniquely identified by an id and can be activated or deactivated. Also, each policy states who it applies to; this can be the person who defined the policy or everyone within a certain domain. This mechanism allows policies to be defined at different levels, e.g. individual, group and enterprise.

Policy rules provide a means to express simple facts about what a user wishes to happen with her communications. Each policy rule is composed of a modalities block, an action block, a condition block and and an trigger event block. All parts but the actions part are optional.

We consider a number of modalities such as obligation, permission and interdiction. 'Never' and 'always' are further simple modalities. Preferences, e.g. wish or must, are highly relevant for call control policies and are considered as fuzzy modalities. Finally a class of temporal modalities, containing items like 'in the future', 'periodically' or 'now', is defined. Interactions can occur within each of these modality groups, but modalities can also be used by conflict resolution approaches to identify which policy should be given precedence.

The action block simply contains one or more instructions to be executed when the policy is applied. These instructions are typically actions as provided by the target system. In a call processing system, they can be actions such as "forward call", "originate call" or "contact". Note that at this level we are concerned with abstract actions which are mapped by the policy server to messages in the underlying communications layer.

A trigger event block describes when the policy should be applied: if it exists the policy should only be considered when the specified trigger occurs. The omission of a trigger block means that the rule is always to be considered, and we would refer to such a rule as goal.

Conditions restrict the applicability of a policy rule further. Typical conditions are equalities or inequalities on parameters associated with the call. There is a large set of these parameters, and others can be readily added. Examples are: caller (a user), call content (email, video, language), media (fixed, mobile, high speed), call type (emergency, long distance, intracompany), cost (of the call) or topic (project x, weekend plans). Other conditions are based on the context and attributes of users such as location (my office, at customer site), identity, role) (Mary, service representative) , and capabilities (Java expert, German speaker).

As it is impractical to require the user to provide the relevant information when establishing a call, most of the required information will be inferred from the context. For example, roles may be defined in a company organisation chart, the location can be established from the user's diary or mobile home location register. Or better, "mobile devices have the promise to provide context-sensing capabilities. They will know their location – often a critical indicator to the user's tasks." [4]. So for example, if you are in your boss's office, it is probably an important meeting and you do not wish to be disturbed.

User Interfaces. The presented language is quite expressive. However, we do not expect the end user to define policies directly in the policy description language. This would be unrealistic, as we expect lay users to be able to define their policies. We have developed a simple wizard that aids the user in creating and handling policies. The current interface is web-based, so users can access it from anywhere, with voice based interfaces being considered. A typical end-user can create new policies or edit, (de)activate and delete existing policies.

The whole process takes place in natural language and the gaps are filled by selecting values from the context (e.g. time from clocks and users or domains from hierarchies). Furthermore, the approach is language-independent so that in principle the user can work in his native language. Once a policy is finished, the user submits the policy to the policy server.

Upload of Policies. A policy server provides an interface via a TCP/IP socket to receive changes to policies (with appropriate authorisation). The user interface layer connects to this socket to submit the gathered information and to receive any feedback on success or failure through this connection. In the absence of the feature interaction problem, the policy store is updated to reflect the new policy. However, as conflicts are rather likely, a consistency check is performed on the set of policies applicable to the user before updating policies.

Interactions. An architecture that does not give rise for conflicts appears to be possible only at the cost of reducing the expressiveness of the policy language to trivial levels. We introduce a guided design process that automatically checks policies for the presence of conflict and presents any detected problems to the user, together with suggested resolutions.

When a policy is uploaded, it is checked against other policies from the same user, but also against policies that the user might be subject to (e.g. due to her role in the enterprise). Here we check for static interactions, i.e. those that are inherent to the policies. This suggests the use of offline detection methods and filtering techniques. Any inconsistencies detected need to be reported to the user. The resolution mechanism is either a redesign of the policy base or provisioning of information to guide online approaches.

Some methods considered most appropriate for the policy context are Anise [12] (predefined or user-defined operators allow to build progressively more complex features while avoiding certain forms of interaction) and Zave and Jackson's [14] Pipe and Filter approach. Also, Dahl and Najm [3] (occurrence of common gates in two processes) and Thomas [11] (guarded choices), where the occurrence of non-determinism highlights the potential for interactions are suitable. Note that these methods are applied at policy definition time, so execution times are less of an issue (as long as they stay within reason).

In fact we are not restricted to static interactions: we can also detect the potential for conflict. That is, we can filter cases where an interaction might occur depending on contextual data. A suitable approach might be derived from the work of Kolberg et al. [5].

Consistency could be checked at the user's device rather than the policy server. However, performing the check in the policy server has the major advantage that in addition to the user's policies, those from the same domain are accessible and can be considered. Furthermore, the user side of the implementation is kept light-weight, which is important for less powerful

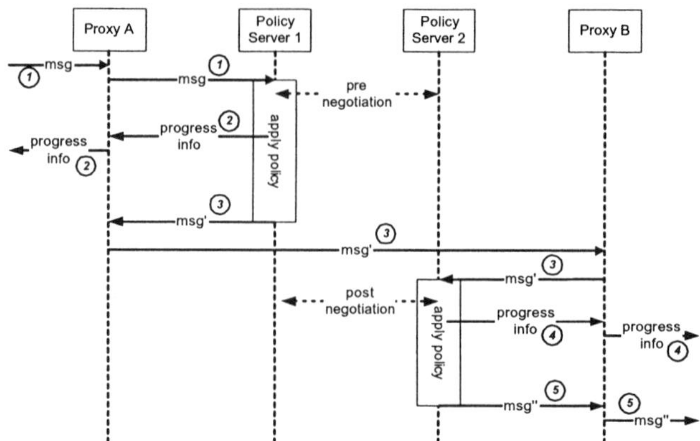

Figure 2: Policy Enforcement Process

end-devices such as mobile phones or PDAs.

5 The Call Process

In the previous section we have discussed how users can define and upload their policies to the policy server. Now, we discuss how policies are enforced to achieve the goals they describe.

Applying Policies to a Call. When attempting to setup a call from A to B, A's end device will generate a message that is sent via a number of switching points to B's end device. Assuming that policy servers are associated with switching points, every switching point along the call path can sent the message to the policy server. Routing is suspended until the policy server allows continuation. In SIP, we can intercept, modify, fork and block messages by using SIP CGI.

The policy server processes any messages that it receives in a four stages: (1) all policies applicable to the source or target of the message are retrieved. (2) The policies are filtered, removing those where the trigger event does not match the received event and those where additional conditions are not satisfied by the current context. (3) The remaining policies are analysed for conflict and these are resolved. (4) The outgoing message(s) are produced. If there are no applicable policies, the outgoing message is simply the incoming message. If one or more policies apply, the required changes are made to the original message. Alternatively, new or additional messages can be created. The example in Figure 2 shows "progress info" messages that are generated by the policy server. Also, if a policy requires forking of a call (e.g. "always include my boss in calls to lawyers") the respective messages to setup the extra call leg need to be created.

Routing is resumed as normal, with the next switching point again forwarding the message to a policy server for the application of policies. Figure 2 shows an example of this process for a call between two parties where policies are enforced in one direction. Note that the figure shows two message exchanges between the policy servers, labelled pre- and post-negotiation.

More Interactions. While a call is actually taking place more interactions can arise, either forced by context information or between policies of the different parties involved. Any detection mechanism incorporated in the enforcement part of the policy server needs to be able to detect and resolve such conflicts. The introduction of policy support must also not create unreasonable delays in call setups. However, if users are aware that complex steps are needed to resolve sophisticated policies, they should learn to live with the delays – and probably are happy to do so when the outcome is productive for them. It is for this reason that intermediate information messages are produced by the policy application process.

Online and hybrid feature interaction approaches are a possible solution. We believe that in a policy context the available information is sufficient to resolve conflicts. The underlying architecture provides protocols that are rich enough to facilitate exchange of a wide range of information. There are essentially two classes of run-time approaches. One is based around the idea of negotiating agents, the other around a (central) feature manager.

The feature manager in [7] detects interactions by recognising that different features are activated and wish to control the call. The resolution mechanism for this approach [8] is based on general rules describing desired and undesired behaviour. In negotiation approaches, features communicate with each other to achieve their respective goals [13]. Buhr et al. [2] use a blackboard technique for the negotiation, thus introducing a central entity.

Feature manager approaches lend themselves to the policy architecture, as their main requirement is that the feature manager is located in the call path. This is naturally the case with policy servers. However, feature manager approaches so far have suffered from a lack of information to resolve interactions (though some progress has been reported in [8]). Negotiation approaches can fill this gap. Their current handicap is the sophisticated exchange of information required; however this is addressed by allowing communication channels between policy servers. We foresee a combination of the two techniques to exploit their individual benefits.

Two forms of negotiation are practical in the policy architecture: pre- and post-negotiation. In the former case, the policy server contacts the policy server at the remote end and negotiates call setup details to explore a mutually acceptable call setup. The communications layer is then instructed to setup the call accordingly. In post-negotiation, the policies are applied while the call is being setup, thus potentially leading to unrecoverable problems.

6   Conclusion and Further Work

Evaluation. We have considered how policies can be used in the context of call control, especially how they can be seen as the next generation of features. The policy architecture allows calls to be controlled by policies. Each policy might make use of the context of a user. This allows context-oriented call routing, but goes far beyond routing by allowing for availability and presence of users to be expressed. Therefore, we can achieve truly non-intrusive communications that enable users to achieve their goals. Policies can be easily defined and changed by the end-user via a number of interfaces (e.g. Web or voice interfaces).

We have suggested some offline feature interaction techniques for policies to be checked for consistency when they are designed. They can be applied to new policies as well as to existing ones within the same domain. However, calls will eventually cross domains; then online techniques to detect and resolve any conflicts will be required.

Future Work. A prototype environment to create and enforce policies has been developed on top of a SIP architecture. Thus the proposed architecture has been implemented. The methods for detecting and resolving policy conflict identified in this paper need to be implemented in the prototype such that empirical data on their suitability can be gathered. These methods require some more formal understanding of the policy language as well as the communications layers to be used (the mapping of low-level messages to policy events).

In the future we would like to test the upper layers on top of other communications layers. For example H.323 and PBX are planned. Further development of additional user interfaces should strengthen the prototype. Another research area is the automatic gathering of context details, which is interesting in itself but beyond the scope of our work.

Acknowledgements

This work has been supported by EPSRC (Engineering and Physical Sciences Research Council) under grant GR/R31263 and Mitel Networks. We thank all people who contributed to the discussion of policies in the context of call control. Particular thanks are due to Tom Gray, Evan Magill and Mario Kolberg.

References

[1] M. Barbuceanu, T. Gray, and S. Mankovski. How to make your agents fulfil their obligations. *Proceedings of the 3rd Int Conf on the Practical Applications of Agents and Multi-Agent Systems*, 1998.

[2] R. J. A. Buhr, D. Amyot, M. Elammari, D. Quesnel, T. Gray, and S. Mankovski. Feature-interaction visualization and resolution in an agent environment. In K. Kimbler and L. G. Bouma, editors, *Feature Interaction in Telecommunications Networks V*, pages 135–149. IOS Press (Amsterdam), 1998.

[3] O. C. Dahl and E. Najm. Specification and detection of IN service interference using LOTOS. *Proc. Formal Description Techniques VI*, pages 53–70, 1994.

[4] A. Fano and A. Gersham. The future of business services in the age of ubiquitous computing. *Communications of the ACM*, 45(12):83–87, 2002.

[5] M. Kolberg and E. H. Magill. A pragmatic approach to service interaction filtering between call control services. *Computer Networks*, 38(5):591–602, 2002.

[6] E. Lupu and M. Sloman. Conflicts in policy based distributed systems management. *IEEE Transactions on Software Engineering*, 25(6), November/December 1999.

[7] D. Marples and E. H. Magill. The use of rollback to prevent incorrect operation of features in intelligent network based systems. In K. Kimbler and L. G. Bouma, editors, *Feature Interaction in Telecommunications Networks V*, pages 115–134. IOS Press (Amsterdam), 1998.

[8] S. Reiff-Marganiec. *Runtime Resolution of Feature Interactions in Evolving Telecommunications Systems*. PhD thesis, University of Glasgow, Department of Computer Science, Glasgow (UK), May 2002.

[9] S. Reiff-Marganiec and K. J. Turner. Use of logic to describe enhanced communications services. In D. Peled and M. Vardi, editors, *Formal Techniques for Networked and Distributed Systems (FORTE 2002)*, pages 130–145. Springer Verlag, November 2002.

[10] M. W. A. Steen and J. Derrick. Formalising ODP Enterprise Policies. In *3rd International Enterprise Distributed Object Computing Conference*. IEEE Publishing, September 1999.

[11] M. Thomas. Modelling and analysing user views of telecommunications services. In P. Dini, R. Boutaba, and L. Logrippo, editors, *Feature Interaction in Telecommunications Networks IV*, pages 168–182. IOS Press (Amsterdam), 1997.

[12] K. J. Turner. Realising architectural feature descriptions using LOTOS. *Networks and Distributed Systems*, 12(2):145–187, 2000.

[13] H. Velthuijsen. Distributed artificial intelligence for runtime feature interaction resolution. *Computer*, 26(8):48–55, 1993.

[14] P. Zave and M. Jackson. New feature interactions in mobile and multimedia telecommunication services. In M. Calder and E. Magill, editors, *Feature Interaction in Telecommunications and Software Systems VI*, pages 51–66. IOS Press (Amsterdam), 2000.

eSERL: Feature Interaction Management in Parlay/OSA Using Composition Constraints and Configuration Rules

Alessandro DE MARCO, Ferhat KHENDEK
Department of Electrical and Computer Engineering
Concordia University
1455 de Maisonneuve Blvd. W., Montreal, QC, Canada
{a_demarc, khendek}@ece.concordia.ca

Abstract. SERL is a language and framework for managing the triggering and execution of services in a single-user, single-network-component (SUSC) environment. We propose enhancements to SERL, dubbed eSERL, to allow for personalized customization of services by end-users who do not have expert knowledge of the services and of the environment, while guaranteeing, to a certain degree, that unwanted feature interactions will be avoided. SERL allows for such customization of service usage, but it does not consider the issue of providing the guarantee. Our approach involves validation of user-defined service configurations, specified in the form of rules, against constraints for service composition and inter-working imposed by experts.

1. Introduction

Next-generation telecommunication networks will provide new and enhanced capabilities and enabling technologies for application-layer service development. This emerging infrastructure is often referred to as a *multimedia service network*, or simply, a *service network*. It combines the advantages of the IP and cellular phone technologies in a converged framework. In addition, it offers access to network capabilities and services through Open Service Architecture (OSA) gateways. 3GPP [1] OSA is based on the Parlay suite of high-level and standardized APIs [2]. Parlay/OSA enables the provisioning of services independently of the underlying network technologies, and facilitates more rapid service development [3, 4]. The "open" paradigm allows for new players in the service network architecture, new streams of revenue for network operators, and new business opportunities for 3rd party service providers or developers with innovative services to offer. Alas, it also present a number of technological challenges that must be dealt with before such a model can be realized, not least of which is the Feature Interaction Problem.

The Feature Interaction Problem [5] exists in current telephony systems, but it is expected that the problem will be severely aggravated in next-generation systems due to the openness and distributed nature of the architectures, and the new types of services that will be developed [6]. Another factor to consider in that context is that non-expert users will be provided with means to customize service behaviour to an unprecedented degree, or even develop and deploy their own services. Consequently, powerful mediators or control mechanisms must be implemented, possibly at more than one point in the architecture, in order to avoid, or detect and resolve unwanted, erroneous, or malicious service network usage.

In this paper, we describe a mediation system in a Parlay/OSA context, which provides customizable, yet controlled end-user access to service network capabilities and usage of high-level services without imposing unwarranted restrictions that would obviate the benefits of the functionality offered. The system is based on the Service Execution Rule Language and Framework (SERL) [7, 8, 9]. We propose and describe extensions to the original framework. Our contributions reported in this paper involve language extensions to support the definition of service composition constraints by experts, and a validation scheme to ensure that user-defined service configurations do not violate these constraints.

This paper is organized as follows. In Section 2, we briefly introduce SERL. In Section 3, we describe our approach. We then explain our validation scheme in Section 4. Finally, in Section 5 we conclude by hinting at potential applications and our future work. Notice that we often use the terms *feature* and *service* interchangeably. Similarly, we often use the term *call* to refer to the concept of a *session*, which is more general than a typical telephone call today.

2. SERL: Service Execution Rule Language

SERL [7, 8, 9] is a language and a framework for managing the triggering and execution of services. It is based on *condition-action* rules, and a processing model involving interception of events, matching of event conditions to rule triggering properties, and then application of matched rules. SERL was originally developed for SIP, where deployment of rule processing engines implementing the model is more amenable to Proxy Servers, but not unimaginable in User Agents. The XML-based language is flexible enough to support other technologies, such as Parlay/OSA, and even certain heterogeneous environments.

SERL rules are grouped into Rule Modules, each with one owner. The owner of a Rule Module is typically a subscriber to services affected by the rules in the module. Services are classified into groups according to their behavior, and each group is assigned a Processing-Point identifier. Processing Points refer to the points in the call-signaling timeline where services within a certain group may be invoked. Services are triggered in response to events flowing downstream (i.e. requests) or upstream (i.e. responses) through a node. Events usually originate from the network, or from services. When triggering rules are encoded, they take on the same Processing-Point identifiers as the services they relate to. A rule-search algorithm, set to run upon *the* occurrence of events, takes into consideration Processing-Point identifiers, priority of rules, as well as the relevant event information (a.k.a. event context) when searching for matching rule conditions. Rule actions may involve delaying, overriding, canceling or generating events.

In addition to managing the execution and triggering of services, SERL may access system capabilities like a database, a Presence server, a Location server, Web services, etc. Examples demonstrating such functionality are not provided in the Internet-Drafts, however such functionality is implied. It is the responsibility of the developer of the SERL engine to build-in adaptors for communication with such network capabilities or distributed services. SERL is not a language to describe the behavior of services, nor is it intended to be used to detect feature interactions. Rather, it is a mechanism to allow for the application of feature interaction resolution policies in a SUSC [5] context. It is assumed that potential interactions are known *a priori*, and knowledge about how to resolve interactions is encoded in rules. Ordering through Processing-Points and priorities are the only means available to enforce resolutions. For instance, when several rules are matched, SERL does not define a resolution policy other than a simple ordering scheme based on priorities defined *a priori*.

3. Managing FI in a Parlay/OSA context with eSERL

In this section, we discuss our architecture in a Parlay/OSA context, and our enhancements to SERL to allow for personalized customization of services by end-users who do not have expert knowledge of the services and the environment, while guaranteeing, to a certain degree, the absence of unwanted feature interactions. We assume that any user may attempt to configure any service they subscribe to, or compose and inter-work several of them for added-value. By validating user-defined configurations against constraints imposed by experts, we are able to provide a guarantee that user requirements will be met.

3.1. Rationale and Related Work

Our goal is to develop a generic framework for service composition in which we allow for personalization, yet avoid unwanted feature interactions. We wish to impose no specific application-domain, even though we have focused on a Parlay/OSA context initially. We also aim to provide for a clear relationship between user-defined requirements and services or capabilities available in the network. We believe that by achieving the latter goal, we will facilitate the development of feasible billing and pricing strategies. Our vision is that the next-generation "killer app" is probably not going to be a single service, but the capability for an individual user to compose and personalize a multitude of services according to his own requirements.

With regard to related work, we are aware of two initiatives, namely CPL [10] and ACCENT/PDL [11]. CPL enables end-user service creation and guarantees the absence of feature interactions due to the nature of the language. On the other hand, it is not flexible enough to support a behavior that is not within the realm of Call Processing. ACCENT/PDL is a newer initiative, which is more flexible. End-users specify policies for service behaviors in a constrained natural language format. The goal is to allow comparison of policies online and/or offline in attempts to detect policy conflicts. However, ACCENT/PDL seems to lack a clear mapping between policies and features or network capabilities.

3.2. Overall Architecture

Our approach requires an architecture employing one or more Feature Interaction Managers (FIM), which act as mediators controlling the triggering and execution of features. The behavior of a FIM in our context differs from the commonly understood behavior(s) documented in [12]. Each of our FIMs "virtually" composes services according to user-defined requirements (i.e. rules). In other words, each FIM manages the events that affect the behavior of any number of deployed services, and this may lead to the overall appearance of composed service behavior from the user point of view. SERL is a technology that essentially implements a SUSC subset of the generalized model, and therefore we have selected it as the starting point for our project. Moreover, since Parlay/OSA defines an architecture with clearly specified points of control (i.e. application servers, service capability servers), we intuitively position the FIM(s) in a Parlay/OSA framework as shown in Figure 1.

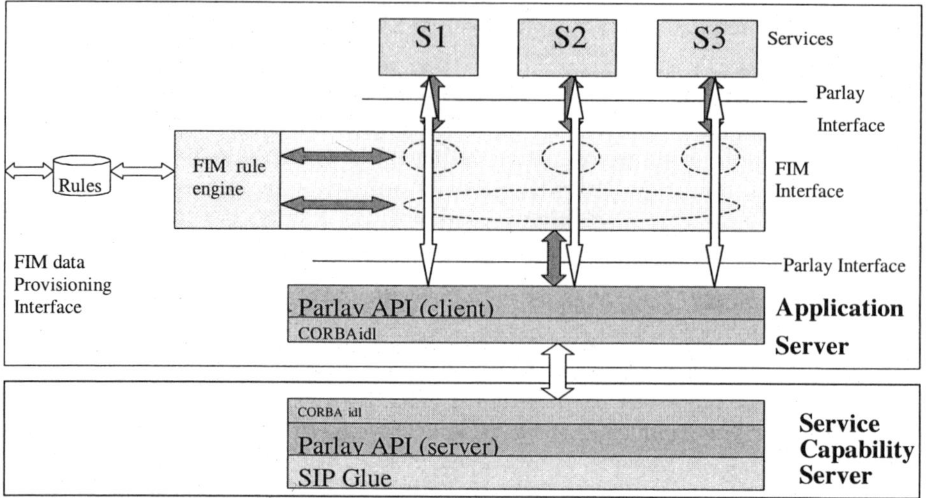

Figure 1. FIM for one Application Server

We take advantage of the standardized Parlay/OSA APIs, and the Half-Object Plus Protocol design pattern [13]. The FIM is inserted between services and the actual Parlay/OSA client side interface implementation. The FIM offers a virtual implementation of the Parlay/OSA APIs to services, and it uses the actual APIs provided by the Application Server and the Service Capability Server to implement the interface offered to services. The idea is to enable transparency of FIM behavior from the service point-of-view. A service developer should be able to develop a Parlay/OSA service and run it on an Application Server, whether a FIM is present or not. With the FIM positioned as such, it is able to intercept all events to or from services that are deployed on the Application Server.

3.3. eSERL: Enhanced SERL

We require each FIM in the architecture to implement our enhanced version of SERL (eSERL). As a primary enhancement, we define two types of SERL Rule Modules: Composition Constraint Rule Modules, and Configuration Rule Modules. Composition Constraints represent expert knowledge about how services may inter-work or be composed. Configuration Rules are user-defined and relate to instances of *acceptable compositions* of services (i.e. not in violation of constraints) along with personalized service data.

3.4. Composition Constraints

Since we are considering a SUSC context, it is reasonable to suggest that all possible compositions of services in the system are known *a priori*, which allows offline analysis of services to detect potential interactions. Moreover, an expert may use his or her knowledge, based on experience, to facilitate detection. The analysis procedure used by experts to detect interactions is out of the scope of this paper. Following analysis, the expert will have the required information to define constraints.

We highlight the fact that the set of possible compositions of services in the system may be reduced simply because we are dealing with SERL, involving service invocation according to

Processing Points and priorities. Intuitively, this may eliminate some potentially disruptive feature interactions. We define such constraints imposed by SERL as *implicit constraints* because they are intrinsic to the framework.

Composition Constraint Rule Modules express *explicit constraints*. To write these rules, an expert must consider the behavior of the services, their input/output data, and implicit constraints of SERL. Explicit constraints extend the service inter-working and composition protocol defined by SERL on a per system basis (i.e. single network component). We require that when services are deployed in the system, service providers will include *deployment descriptors* for each service. Contained within a deployment descriptor are items of information about the service that will be needed for feature interaction detection. If source code for services being deployed is available, then deployment descriptors may not be necessary.

Explicit constraints are classified into three types: *order-preserving*, *data selection*, and *mutual exclusion*. In the absence of a constraint, services may be invoked in parallel. We have extended the SERL language to be able to express these types of constraints and to be able to describe the service objects that will be invoked.

Composition Constraints usually express constraints for service composition and inter-working pair-wise, but triples, quads, etc., are allowed. This does not imply that services (assumed to be atomic) cannot be interleaved, so long as all constraints are satisfied for the duration of an event-flow leading to service invocations at the node, referred to a Cascaded-Chain in [7].

3.5. Configuration Rules

Configuration Rules specify conditions and actions to carry out when conditions are satisfied. They extend the run-time behavior of the call processing system. As such, a Configuration Rule Module may be seen as a meta-service or more abstract service layer. Existing language constructs allow end-users to write Configuration Rules, and as such, these rule sets would be compatible with SERL processing nodes not implementing the enhancements that we propose. On the other hand, we must constrain the SERL language in order to allow for validation. Hence, generic SERL Rules Modules would not be compatible with a SERL processing node expecting more specialized Configuration Rule Modules.

A user may have more than one Configuration Rule Module defined; however for simplicity, we allow for only one *activated* module at any given time. Also, we assume that the services considered in a module are actually in service while the module is active. Otherwise, it would not be eligible for activation.

3.6. Modified Feature Grouping Criteria

The final enhancement to SERL that we have devised is a modification to the criteria for determining the service groups and Processing Point identifiers. As defined, SERL provisions for this type of enhancement. Our modifications are summarized as follows:

- For services that may add, delete, or modify any part of the event context (e.g. SIP message, Parlay/OSA event), and are mostly related to routing, we assign Processing Point 1/-1. Such services may modify source, destination, or intermediary nodes identified in the event context, for instance.

- For services that may add, delete, or modify any part of the event context except routing information, and are mostly related to screening, we assign Processing Point 2/-2. Such services require read-only access to routing information, and are only allowed to block a call based on certain screening criteria. They may not re-route calls.

- For services that may add, delete, or modify any part of the event context except routing information, and are mostly related to the event context payload, we assign Processing Point 3/-3. Such services require read-only access to routing information, and may only modify event context payload (e.g. SIP message body).

- For services that may not add, delete, or modify any part of the event context, we assign Processing Point 4/-4. Such services require read-only access to event context.

4. Validation of Configuration Rules

Our approach for Feature Interaction Management hinges on the concept of validating user-defined Configuration Rules. The purpose of validation is to guarantee, to a certain degree, that user requirements (expressed as rules) can and will be satisfied.

4.1. Determining Acceptable Compositions

The set of acceptable compositions of services deployed in a system is generated automatically from Composition Constraints. Once acceptable compositions are calculated, they are stored in the system for future reference. A basic algorithm to achieve this involves enumeration and elimination of possibilities.

We assume that the set of Composition Constraints is complete. However, in theory, this assumption cannot be guaranteed. We rely on experts to know about possible problematic interactions between services and consider them when defining constraints. As more problematic interactions are discovered though service usage over an extended period of time, Composition Constraints will be updated, and acceptable compositions recalculated. However, this process of updating constraints may invalidate previously valid configurations. Solutions to this type of non-monotonic extension of the system will be explored in future work.

Inconsistency of Composition Constraints, an important issue, may, in the worst case, lead to a set of acceptable compositions where each composition is a service on its own. In such a scenario, an expert would intuitively try to detect inconsistencies and relax constraints when possible.

4.2. Validation of Configuration Rules

The algorithm to validate a user's Configuration Rule Module essentially tries to determine all potential service compositions from the user's set of rules, and then makes sure that each composition is acceptable. The issue requires us to examine the possibility of having separate rules with triggering conditional expressions satisfied simultaneously, thus causing their actions to be composed. In other words, if it is possible for a single event occurrence to satisfy conditions of two or more rules, we say that the rules *overlap* and assume that the services affected by the actions constitute a composition.

This problem has been studied in a different context in [14] where actions have well-known semantics (e.g. discard or forward packets). In our case the actions are unknown, but constrained by the Composition Constraints. If a set of actions is not forbidden by the Composition Constraints, we say that these actions will lead to a non-conflicting service composition.

Determining whether conditional expressions overlap has a solution in polynomial time as long as the set of possible values for variables is discrete, finite, and ordered as shown in [14]. Due to the usage of SERL language constructs in a Parlay/OSA context, this holds, but we also define a general principle.

> *General Principle:* Two rules are said to be overlapping, unless
> (a) conditional expressions have at least one common dimension, AND
> (b) at least one common variable in the common dimensions, AND
> (c) non-overlapping values for the common variables.

Notice that dimensions can be seen as Parlay/OSA APIs, or other discrete, finite, and ordered quantities.

5. Conclusion

Our approach for FI management when composing and personalizing services in a SUSC context is quite general. We expect to be able to apply it in any domain where the set of possible values for event variables (in one or more dimensions) is discrete, finite, and ordered. We have a clear relationship between user-defined rules and services or network capabilities. From a commercial point of view, a user may, for instance, subscribe to three traditional services and a Service Composition Manager (SCM) based on our approach. The user would then define rules to manage the three services as a composed set rather than individually, and deploy them into the SCM. This will lead to overall service behaviour tailored to user requirements, thus creating value that would potentially be much greater than the sum of the values created by the individual, independently configured services. Guaranteeing that a user's requirement can and will be met is of utmost importance in order to ensure customer satisfaction and justify the SCM subscription cost.

We view our contribution as a foundation for our future work towards a solution in a multi-user, distributed environment. This involves new architectures, enhanced algorithms, and accompanying tools. Moreover, we will explore Activation Rules for activating particular Configuration Rule Modules based on user-defined requirements. We are also planning to develop a framework to enable 3rd party packaging of services and theme-based rule writing wizards. Users would purchase or subscribe to such packages, for instance a "Family Package", "Driver's Package", or "World Traveller's Package". We believe that by having 3rd parties with intermediate-level expertise write theme-based wizards, users will be able to specify requirements in a controlled manner, and greatly improve the rate of "first-try" Configuration Rule Module acceptance.

Acknowledgements

We would like to acknowledge the participation and contributions of Roch Glitho, André Poulin, and Kindy Sylla from Ericsson Research Canada, during the initial phase of this project. We also acknowledge funding from les Fonds NATEQ (Québec) and Concordia University.

References

[1] Third Generation Partnership Project at http://www.3GPP.org

[2] Parlay API Specifications at http://www.parlay.org

[3] Ard-Jan Moerdijk and Lucas Klostermann, "Opening The Networks With Parlay / OSA APIs: Standards and Aspects Behind The APIs", at http://www.parlay.org/specs/library.

[4] Roch H. Glitho and Kindy Sylla, "Developing Portable Applications for Internet Telephony: An Overview of Parlay and a Case Study on its Use in SIP Networks", submitted to IEEE Network Magazine, 2002.

[5] E. J. Cameron, N. D. Griffeth, Y.-J. Lin, M. E. Nilson, W. K.Schnure and H. Velthuijsen, "A Feature Interaction Benchmark for IN and Beyond", in *Feature Interactions in Telecommunications Systems*, IOS Press, Amsterdam, pp. 1-23, 1994.

[6] J. Lennox and H. Schulzrinne, "Feature Interaction in Internet Telephony", 6th Workshop on Feature Interactions in Telecom and Software Systems, Scotland, June 2000.

[7] R.W. Steenfeldt and H. Smith, "SIP Service Execution Rule Language: Framework and Requirements", (work in progress), IETF Internet-Draft: draft-steenfeldt-sip-serl-fwr-00.txt, May 7, 2001.

[8] H. Smith and R.W. Steenfeldt, "SERL Examples with SIP and SDP", (work in progress), IETF Internet-Draft: draft-smith-serl-ex-00.txt, May 7, 2001.

[9] R.W. Steenfeldt and H. Smith, "Service Execution Rule Language (SERL 1.0) for SIP", (work in progress), IETF Internet-Draft: draft-steenfeldt-sip-serl-00.txt, May 21, 2001.

[10] J. Lennox, and H. Schulzrinne, "Call Processing Language Framework and Requirements", IETF RFC, http://www.ietf.org/rfc/rfc2824.txt.

[11] ACCENT Project at http://www.cs.stir.ac.uk/~kjt/research/accent.html

[12] M. Calder, E. Magill, M. Kolberg, and S. Reiff-Marganiec. "Feature Interaction: A Critical Review and Considered Forecast." *Computer Networks*, Volume 41/1, pp. 115-141, North-Holland. January 2003.

[13] Gerard Meszaros, "Half Object Plus Protocol", in *Pattern Languages of Program Design*, Vol. 1, James O. Coplien, Douglas C. Schmidt, eds. Addison-Wesley, 1995.

[14] Dong Wang, Ruibing Hao, David Lee. "Fault Detection in Rule-Based Software Systems", Concordia Prestigious Workshop on Communication Software Engineering, Montréal, Canada, Sept. 2001. Extended version to appear in the International Journal of Information and Software Technology, Elsevier, 2003.

Foundations

Feature Interactions in Telecommunications
and Software Systems VII
D. Amyot and L. Logrippo (Eds.)
IOS Press, 2003

Ideal Address Translation:
Principles, Properties, and Applications

Pamela ZAVE

AT&T Laboratories—Research, Florham Park, New Jersey, USA
`pamela@research.att.com`

Abstract. Address translation causes a wide variety of interactions among telecommunication features. Ideal address translation is a simple and intuitive set of rules for organizing features so that their interactions can be managed successfully. It is based on a classification of interactions and on principles that balance conflicting goals and reduce ambiguity. It has provable properties, is modular (explicit cooperation among features is not required), and supports extensibility (adding new features does not require changing old features).

1   The problem of feature interactions caused by address translation

In every telecommunication protocol, requests for communication carry at least two addresses: a *source address* indicating which object is making the request, and a *target address* indicating which object's participation is being requested. *Address translation* is a function performed by some features; it consists of modifying a request for communication by changing its source address, target address, or both.

Address translation is a very common feature function (any kind of "call forwarding" is translation of the target address). It also causes a wide variety of feature interactions. Between these two factors, address translation causes a huge number of feature interactions.

Examples of bad feature interaction due to address translation are easy to find. In one example, a customer calls a sales group. The call is forwarded to a sales representative; since the sales representative is not available, his voice mail offers to take a message. It would be much better for the failure to re-activate the group feature to find another representative. Because of forwarding, both group and personal features are invoked, and they interact badly.

In another example [5], two people with addresses *user1@host1* and *user2@host2* correspond by electronic mail. Since *user2@host2* wishes to remain anonymous in this correspondence, he is known to *user1@host1* as *anon2@remailer*, and the anonymous remailer retargets electronic mail for *anon2@remailer* to *user2@host2*.

However, *user2@host2* also has an autoresponse feature set to notify his correspondents that he is on vacation. When electronic mail arrives with source address *user1@host1* and target address *user2@host2*, it immediately generates a response with source address *user2@host2* and target address *user1@host1*. When *user1@host1* receives the response, he learns the identity of his anonymous correspondent. Thus the autoresponse feature undermines the purpose of anonymous remailing.

The goal of this work is full management of the feature interactions caused by address translation. This entails both preventing all the bad feature interactions, and enabling all the good ones.

Furthermore, management should be completely modular and extensible, in the following sense. To interact properly, features do not need to know about each other or to cooperate explicitly. If a system of features is well-behaved (all its feature interactions are properly managed), and if new features or other objects are added that are themselves well-behaved, then the resulting system is guaranteed to be well-behaved. Modularity and extensibility are important because new features are always being added to existing systems, and it is impractical to change the old features so that they interact well with new ones.

There are many reasons why this goal is difficult to achieve, beyond the obvious ones that modularity and extensibility are difficult to achieve in any software system. In particular:

- It is difficult to predict all the ways that features will interact.
- It is difficult to evaluate potential interactions. This is especially true because there are often real-life scenarios in which the same features should interact in different ways.
- There are some inherent conflicts among the desires of various stakeholders, so that no solution will appear perfect to everyone.

2 Outline of a solution

The first component of a proposed solution is a formal model of telecommunications (Section 3). The model is *partial* in the sense that its purpose is to formalize the effects of address translation, and it omits all aspects of telecommunication behavior not directly related to address translation. The model is currently *general* enough to describe the majority of telecommunication services (and will be extended to cover the missing ones). The model circumscribes the problem, and provides a foundation for rigorous discussion of it.

The second component of a proposed solution is a classification of feature interactions caused by address translation, along with principles for evaluating these interactions as desirable or undesirable (Section 4). The principles balance conflicting desires and design criteria, so that all can be satisfied to a reasonable degree.

The classification and principles are based on experience. Although there is no guarantee that either is complete, both are sufficient to cover hundreds of examples.

The third and central component of a proposed solution is the concept of *ideal address translation* (Section 5). This is a form of address translation intended to capture intuitively what address translation could and should be. It constrains *how* address translation is performed, without constraining significantly *what* address translation can accomplish.

Ideal address translation is based on *address categories*. Address categories reduce the usual ambiguity about why features are functioning, and on whose behalf. It then becomes easier to apply the principles from Section 4, and to understand how features should interact.

Ideal address translation is designed for management of feature interactions, modularity, and extensibility. A feature set in which all features obey the constraints automatically satisfies desirable properties derived from the principles. Its organization allows easy imposition of additional, more specific interaction policies. Features do not need to cooperate explicitly, and adding new features or other objects never requires changing old ones.

The remaining sections contain very brief discussions of exceptions to the constraints (Section 6), enforcement of the constraints (Section 7), and related work (Section 8). The entire paper is written in terms of closed networks. The results will be adapted to open networks in future work.

3   A partial formal model of telecommunications

3.1   *Informal description*

All telecommunication protocols support requests for communication. A request carries at least two addresses: a *source address* indicating which object is making the request, and a *target address* indicating which object's participation is being requested. There is some kind of response to each request.

All telecommunication services are set up by *request chains*. A request chain is a chain consisting of protocol requests and modules, where each module is an *interface module* or a *feature module*. An interface module provides an interface to a telecommunication device. A feature module provides feature functions.

Figure 1 shows an example of a request chain. The device with address *s1* is requesting communication with address *t2*. Its interface module *initiated* the request chain with those addresses. The network router routed the first request to the *source feature module of s1*, containing those features applicable to a chain whose source is *s1*.

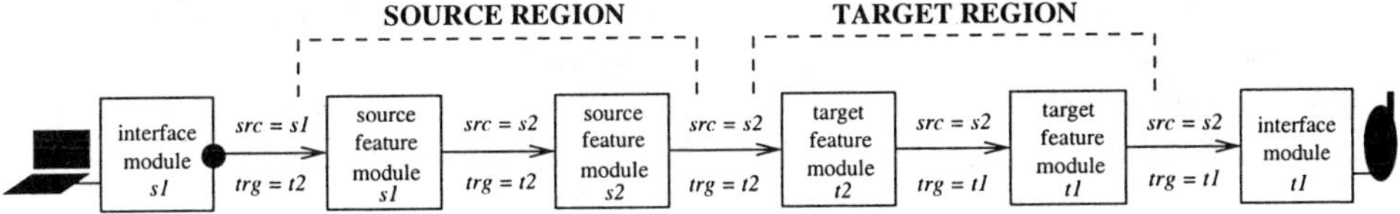

Figure 1: A request chain.

A module *continues* a request chain by making an outgoing request that corresponds to an incoming request it has already received. The source module of *s1* continued the chain, in doing so changing the source address from *s1* to *s2*. As a result of this change, the network router routed the outgoing request to the source feature module of *s2*.

The source feature module of *s2* also continued the chain. With no additional source modules to route to, the network router routed its outgoing request to the *target feature module of t2*. It contains those features applicable to a chain whose target is *t2*.

The target feature module of *t2* continued the chain, first changing the target address from *t2* to *t1*. The chain was routed to the target feature module of *t1*, and then continued to the interface module of *t1*.

Every feature module in Figure 1 is optional. The *source region* of a chain contains all the source feature modules, while the *target region* contains all its target feature modules.

3.2   *Formal definition*

Addresses are globally unique. Each address can be associated with a source feature module, a target feature module, both or neither. The important property here is association of an address with features, not module identity. The same module can serve as both the source and target feature modules of an address; there can also be many interchangeable instances of an address's feature module.

A request for communication has three mandatory fields. The *source* and *target* fields contain addresses. The *mode* field has three possible values: `source`, `target`, or `end`.

A request chain can be initiated by an interface module or a feature module. In all cases, *source* must be the address of the module, and *mode* must be `source`. If a chain is initiated by a feature module, the initiating feature module assumes the role of an interface module within that chain.

An incoming request can be continued by a feature module that has received it.[1] The module may change *source, target,* both, or neither. The outgoing request corresponds to the incoming request in the following sense:

- If the incoming *mode* is `source`, then if the feature module has changed *source* the outgoing *mode* is `source`. Otherwise the outgoing *mode* is `target`.
- If the incoming *mode* is `target`, then if the feature module has changed *target* the outgoing *mode* is `target`. Otherwise the outgoing *mode* is `end`.

If the module has changed an address while continuing a request, it has performed the function of *address translation*.

A network router processes a request by executing the following abstract program:

```
if (mode==source) then
      if (source has source module m) then route to module m
      else {mode:=target; restart routing}
else if (mode==target) then
      if (target has target module m) then route to module m
      else {mode:=end; restart routing}
else (mode==end)
      route to interface module of target
```

This formal model does not constrain any aspect of protocol or feature behavior other than what has already been mentioned. For example, it does not constrain media transmission, mid-call signaling, or disconnection behavior. Any part of any request chain can be torn down at any time. The behavior of feature modules can combine request chains to form overlapping and nonlinear patterns.

Some systems have a *network module* that all request chains must include. A network module can perform functions such as billing, or simply offer feature functions that anyone can use. The network module fits between source and target regions. It is trivial to add a `network` mode to the routing algorithm, so that all chains must pass through a network module if there is one.

3.3 Nonlinear examples

Figure 2 shows a source feature module that provides a three-way calling feature. It continues the same incoming request from its subscriber twice, once to each of the other two parties. As it maintains both continuations simultaneously, the source feature module and its incoming request belong to two distinct request chains.

---

[1] More precisely, the chain or chains to which the incoming request belongs can be continued.

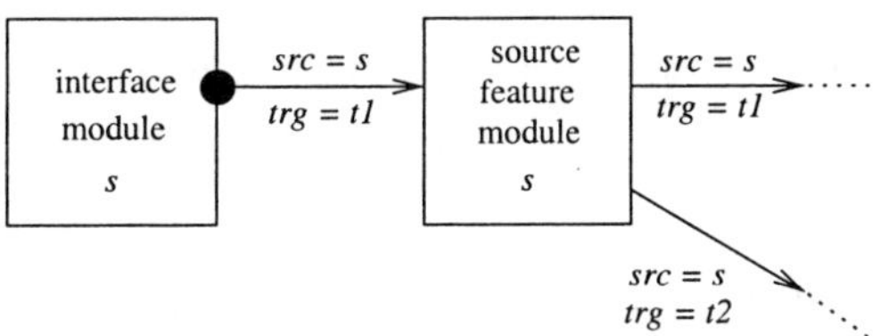

Figure 2: A feature module with Three-Way Calling.

Request chains for a broadcast would look the same, except that there would be many more of them. The main difference between a broadcast and a conference is that, in a broadcast, media flows in only one direction.

Figure 3 shows a feature module that provides a call-waiting feature. Here the same module serves as both the source and target feature module of x. The incoming request from z arrived after the outgoing request to y was continued by the feature module. The feature module, knowing that the interface module of x is busy, does not continue the request from z, but rather sends x a signal that a call is waiting.

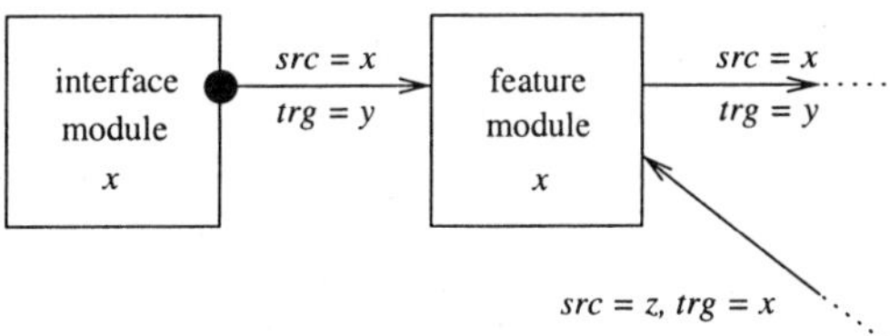

Figure 3: A feature module with Call Waiting.

3.4   Correspondence with real telecommunication protocols

Distributed Feature Composition (DFC) is a formally defined modular architecture for description of telecommunication services [7]. The formal model presented here is a simplification of the DFC routing algorithm. Routing in DFC is finer-grained because the features associated with one address can be decomposed into many independent modules.

The formal model here omits the "reverse" routing capability in the current version of DFC [8]. This capability is used, for example, by a mid-call move function in a source feature module. The purpose of the function is to allow the user to move to a different device without interrupting his conversation. When the module emits a request targeted to the new device, the "reverse" capability ensures that it is routed only to the target feature module of the device address, regardless of the source address. Reverse routing will be added to ideal address translation in future work.

The formal model describes a narrow aspect of the behavior of telecommunication protocols and features, concerning only addressing, routing, and the invocation of feature modules. When only this narrow aspect is considered, every telecommunication protocol conforms to this formal model, although sometimes in a disguised, degenerate, or restricted form.

As an example, Figure 4 shows request chains for electronic mail using SMTP and related protocols. The example is like the electronic-mail example in Section 1, except that no anonymity has been introduced.

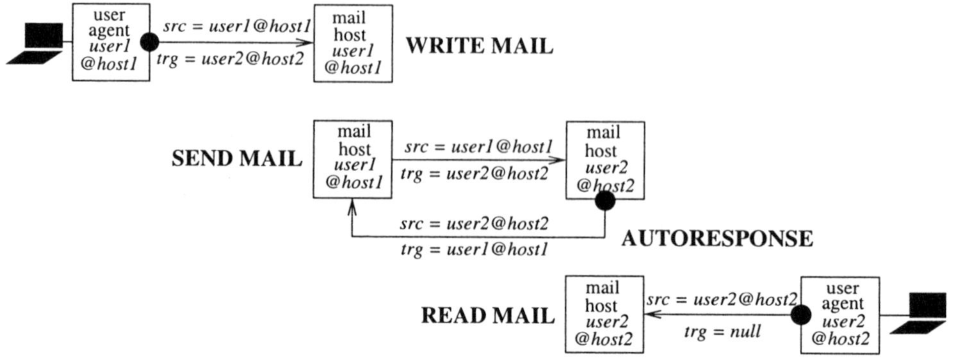

Figure 4: Electronic mail. The three snapshots arise during three different phases in the life of a message.

The telecommunication device is a user agent running on a personal computer. "Mail host *user1@host1*" is both the source and target feature module of *user1@host1*, and "mail host *user2@host2*" is both the source and target feature module of *user2@host2*.

To write mail, the user agent initiates a request chain which is first routed to its source feature module. This feature module buffers the mail, so that the initial part of the request chain can be torn down as soon as the mail has been transmitted to the buffer.

Electronic mail is sent by continuing the original request chain from the source feature module of the source address to the target feature module of the target address. A mail host acting as a target feature module only continues an incoming request if it is forwarding the mail. If it is not forwarding the mail, it buffers it until the addressee chooses to read it.

The target feature module of *user2@host2* includes an autoresponse feature which is currently enabled. When the module receives a request for *user2@host2*, the feature initiates a new chain with the source and target fields of the incoming request reversed. The feature module is acting automatically on behalf of its owner, and the new chain is treated exactly as if it had been initiated by the owner of *user2@host2*.

To read mail, a user agent initiates a request chain with a *null* target address.[2] Like all other request chains initiated by the user agent, this is routed to its source feature module. The chain goes no further, as its only purpose was to connect the agent with its mail host.

4 Principles for feature interaction

4.1 Ownership

Each address has one or more *owners* who are responsible for it and have rights concerning it. An owner is usually a person.

---

[2]The *null* value is a distinguished value of type *address*. It is often found in the source region of a request chain, either because the chain is not intended to extend beyond the source region, or because a source feature module will interact with the caller to produce a real target address.

With respect to knowledge, there is no point in distinguishing between a feature module and any owner of its address. If a secret is revealed to a feature module, the feature module can store it as data, and the data can be examined by any owner. If an owner knows a secret such as a password, he can insert it into the code or data of his feature module, so that the feature module can use it.

4.2   Address-translation functions

In telecommunications, the fundamental addresses are the addresses of telecommunication devices. However, addresses are used to identify many things besides devices.

A *group* address identifies a group of things, such as the departments of an institution, or the people of a work team. *Representation* is a feature function that translates a group target address to the address of an appropriate representative of the group. *Affiliation* is a feature function that changes the source address of a request to the address of a group. This allows a representative access to the source features and data of the group.

A *mobile* address identifies a mobile object such as a person. *Location* is a feature function that translates a mobile target address such as a personal address to the address of a device where the person is located. *Positioning* is a function that changes the source address of a request from a device address to a personal address. This allows the person access to his personal features and data from any device.

A *role* address identifies a role that can be played by another object such as a group, person, or device. *Assumption* is a feature function that changes the source address of a request to a role address, thus allowing the caller to assume that role. *Resolution* is a function that translates a target role address to the address of the object playing the role. Roles are assumed as identities, intended to reveal or conceal. In Section 1, *anon2@remailer* is a role address.

Often these concepts are combined in the meaning of an address. For example, the address of a physicians' office identifies a group of people. It also serves as an identity (role) that is more recognizable to patients than a physician's home address, and that a physician would rather give to patients.

Because the concepts of group addresses, mobile addresses, and role addresses are neither exhaustive nor mutually exclusive, we must aggregate them and all other non-device addresses into the general category of *abstract addresses*. Abstraction is a relative concept. An anonymous address is more abstract than a personal address that it conceals. A group address is more abstract than the address of any representative of the group. All non-device addresses are more abstract than device addresses.

4.3   Identification

People and feature modules use the addresses that they know to *identify* the parties with whom they are communicating. A feature that performs address translation interacts with other features by affecting the identification information they receive in requests.

As discussed above, privacy is a motive for address translation. When it comes to privacy, there is always an inherent conflict of interests between those who wish to know and those who wish to conceal. It seems that this conflict can be resolved fairly with the following two principles:

- *Privacy:* A person should be able to conceal a more private address that he owns behind a more public address that he owns.
- *Authenticity:* A person should not be able to pose as an owner of an address he does not own.

A private address is a more concrete address, while a public address is a more abstract address.

Achieving authenticity often requires *authentication,* another feature function related to address translation. An authentication function demands a password, voiceprint, or some other proof that a user is an owner of a particular address.

Authentication might be needed in many different feature modules, as shown in Figure 5. In this chain, *d1* and *d2* are device addresses. The source feature module of *d1* assumes the source role of *r1*, while the target feature module of *r2* resolves that role to *d2*.

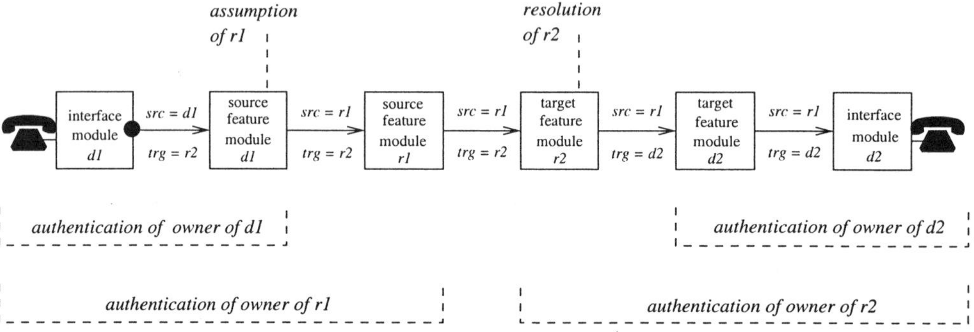

Figure 5: Many different feature modules may perform authentication.

In the absence of physical protection, anyone might walk up to device *d1* and start using it. If this is not acceptable, the source feature module of *d1* must authenticate that the user is authorized to use the device, which (in the simple authorization model used here) means that the user is an owner of *d1*. The brackets indicate that authentication is a dialogue between the feature module and the device/user on its left.

The assumption of source role *r1* can only be performed by a feature module of some address other than *r1*, so anyone can program his source feature module to assume the identity of anyone else! This serious problem is solved by putting authentication that the caller is an owner of *r1* into the source feature module of *r1*, as shown in the figure. This is secure because, once the source address has been changed to *r1*, the routing algorithm *must* route the request chain to the source feature module of *r1*.

Target feature modules can also authenticate that the person who answers the telephone is the expected person. The figure shows authentication of both *r2* and *d2*, through dialogues with the device/user on the right.

4.4 Contact

People and feature modules use the addresses that they know to *contact* the parties with whom they wish to communicate. A feature that performs address translation interacts with other features by affecting the contact information they receive in requests.

Concerning contact, there is a conflict of interests between the desire of owners of abstract target addresses to translate them to concrete addresses very freely and dynamically, and the desire of callers for reproducibility—if they call the same sales address twice, they want to get the same sales agent both times. In this conflict, owners of abstract addresses always win, and callers always lose. The only way for the caller to get back to the same agent is to ask for a personal address, and to call the personal address rather than the group address.

The important principle for contact is:

- *Reversibility:* A target-region module or callee should be able to call the source address of a request chain and thereby target the entity that initiated it.

The principle must be worded carefully because the source address may be abstract. For example, if the chain was initiated on behalf of a group, then the group is the initiating entity, rather than the group representative who actually picked up a telephone.

4.5 Invocation

Through the routing algorithm, the addresses in a request chain determine which feature modules are in the chain. A feature that performs address translation interacts with other features by affecting which feature modules are invoked by appearing in the chain.

Address translation causes the invocation of multiple source feature modules and multiple target feature modules. The worst case is having an unbounded number of them, because of an address-translation loop. The need for the following principle is obvious:

- *Boundedness:* The number of source feature modules in a source region should be bounded, and the number of target feature modules in a target region should be bounded.

A more subtle issue is that address translation can cause the invocation of several feature modules, in the same region, with competing features. This is often desirable, but it requires some coordination so that competing features interact properly.

Features compete with each other to handle the same situation. In the target region, for example, two features may both be treatments for unavailability of the target. In the source region, two features may both handle the situation that a user has initiated a call, but has not provided a valid target address. Such features assist "dialing" by allowing the caller to select from a list, identify a target through speech, or enter a short code instead of a real address. Even features with unrelated functions may compete with each other, however, if both attempt to use a unique resource—such as the voice channel to a user—at the same time.

There seems to be only one coordination mechanism for competing features that is feasible in a modular, extensible system with distributed authority. Competing features must share triggering signals. Triggering signals are propagated along the signaling paths created by request chains. A triggering signal acts as a token, so that when a feature module receives a triggering signal, it has permission to execute any feature triggered by the signal. The module itself decides when, and if, to propagate the triggering signal. When it propagates the signal, it is passing the permission token to other modules.

Competing features usually do share a triggering signal. Two unavailability treatments are both triggered by the protocol's failed-request signal. Two dialing features are both triggered by the request signal. Even unrelated features that compete for the same resource can both be triggered by the signal that initiates the phase in which they are active. For example, there are numerous target features that respond to a request by first playing an announcement on the voice channel to the caller ("Your call is very important . . .").

If competing features do not already share a triggering signal, it may be necessary to force them to. For instance, many features are treatments for a *no-answer* condition. Since most telecommunication protocols do not have a *no-answer* signal, a feature triggers itself by setting a timer for some locally determined no-answer interval. This makes it very difficult to coordinate separate no-answer treatments—even if the intervals are coordinated, race conditions may disrupt any intended prioritization. The straightforward solution to this problem is to have one timer that generates a *no-answer* signal, which is shared among features in the same way that other failure signals are.

Obviously, the coordination mechanism gives higher priority to features that are closer to the source of the shared triggering signal. Less obviously, this mechanism provides feature composition as well as coordination. Consider, for example, two dialing features. One is a voice-dialing feature using speech recognition to translate the caller's utterance to a text string. The other uses a personal directory to translate speed codes to addresses. If voice dialing is triggered first, the caller can utter a speed code. When voice dialing propagates the triggering request signal, it can include the recognized speed code in the request as if it had been dialed. If the personal directory feature is then triggered, it can translate the speed code to a real address.

In most cases, it works best for the feature modules of more concrete addresses to have priority over more abstract feature modules. This is true of the composition example above. The voice-dialing feature is a device feature, and therefore more concrete than a personal directory. This default priority is achieved by the following principle:

- *Monotonicity:* In a region, the feature modules of more concrete addresses should be closer to the outer end of the region than feature modules of more abstract addresses.

In the source region, the outer end is the source end; in the target region, the outer end is the target end. Because most triggering signals originate in the devices at the outer ends of request chains, these principles ensure that they will reach concrete features before they reach abstract features.

Occasionally a different priority order is needed, as the sales-group example in Section 1 shows. A group address is more abstract than a personal address, but group features should have priority over personal features in handling the unavailability of a person. A group consists of several people, so the quickest and most effective treatment for the unavailability of one person is to find another person.

In the distributed, modular setting of a request chain, an abstract feature module such as a group module in a target region cannot preempt a more concrete feature module in the target region. It can only ask the more concrete feature modules to relinquish their priority, and hope that they will cooperate.

Even in a cooperative setting, the principles of monotonicity and orientation are needed to maintain extensibility. The point is that a feature must never assume that other features are present or absent in the chain. A concrete feature module must not relinquish its function unless it receives a signal from a more abstract feature module asking it to do so. An abstract feature module must be able to send this signal to all the feature modules in the region more concrete than itself. All feature modules in the target region must know that the more abstract feature modules are upstream of them, and the more concrete feature modules are downstream of them, which is what monotonicity provides.

5 Ideal address translation

5.1 Address categories

In a system with ideal address translation, there is a finite set of *address categories*. Each address belongs to exactly one category. The set of address categories is partially ordered by abstraction. An address can have multiple meanings and multiple associations with other addresses, but they must all be compatible with respect to the abstraction order.

For an example of a violation that is easy to fall into, consider a new mobility service offered to office workers. Each worker already has an office telephone number, which is printed on his business card. So a worker subscribes to the mobility service by forwarding his office telephone number to his new mobile telecommunication address.

Now the office telephone number is a public role address, known to all, which is resolved to the more private mobile address. In this sense the office telephone number is more abstract than the mobile address. At the same time, the office telephone number represents a device which is sometimes the location of the mobile address. In this sense the mobile address is more abstract than the office telephone number. In ideal address translation, these two relationships between office telephone numbers and mobile addresses are incompatible.

Categorizing and ordering addresses tends to make it very clear what they identify. The resultant lack of ambiguity is very helpful in designing appropriate feature behavior and understanding feature interactions. It also has other beneficial side-effects, such as making presence information [3], which is necessarily based on addresses, more meaningful.

5.2 Constraints

A system with ideal address translation adheres to the following constraints.

The first constraint supports reversibility and other goals. If a target feature module changes the source address of a request chain, it is replacing source information with something else, for some other purpose. With true source information gone, the chain cannot be reversed.

- *Constraint 1:* A target feature module in a request chain does not change the source address of the chain.

Constraint 1 has no counterpart in the source region because it would be too restrictive (not because it would be useless). One of the most common functions of source feature modules is creating and modifying target addresses.

The second constraint supports privacy, boundedness, and monotonicity. It recognizes that the true purpose of address translation is to change to a different level of abstraction, as all the translation functions in Section 4.2 do. The constraint forces an orderly progression through the abstraction order. The constraint is symmetric across regions and has two parts, one for each of the two regions.

- *Constraint 2s:* If a source feature module in a request chain translates the source address, the new source address is more abstract than the old one.
- *Constraint 2t:* If a target feature module in a request chain translates the target address, the new target address is more concrete than the old one.

If the only places that addresses appeared on a signaling channel were in the *source* and *target* fields of requests, Constraint 2 would be the last constraint. Addresses also appear, however, in other signals and other fields of request signals. Constraint 3 extends the organization imposed by Constraint 2 to these additional signals.

- *Constraint 3s:* A source feature module in a request chain does not transmit downstream, in the role of a source of this chain, any address more concrete than its own.
- *Constraint 3t:* A target feature module in a request chain does not transmit upstream, in the role of a target of this chain, any address more concrete than its own.

The words *in the role of a source/target of this chain* are intended to include any address transmitted along the signaling path of a request chain for the purpose of providing auxiliary information about the routing of that particular chain. This kind of auxiliary signaling is quite common (in this paper, it can be found in Figure 7). The words are intended to exclude addresses transmitted for purposes not directly related to the particular chain. For example, a user might employ the signaling path to update his personal data. An address transmitted only so that it can be put in a database is not directly related to the particular chain along which it is transmitted.

5.3   Properties

The purpose of the constraints is to support the goals of privacy, authenticity, reversibility, boundedness, and monotonicity. Specifically, the constraints guarantee the properties stated in this section. Proofs can be found in the full version of the paper [11]. Systems with these properties are extensible, as discussed at the end of the section.

Constraints 2s and 2t are by far the most important ones. They guarantee the following boundedness and monotonicity properties.

- *Source Boundedness Property:* The number of source feature modules in a request chain is less than or equal to the depth of the abstraction order on address categories.
- *Target Boundedness Property:* The number of target feature modules in a request chain is less than or equal to the depth of the abstraction order on address categories.
- *Source Monotonicity Property:* If $m1$ and $m2$ are source feature modules in a request chain, and $m1$ precedes $m2$, then the address of $m2$ is more abstract than the address of $m1$.
- *Target Monotonicity Property:* If $m2$ and $m1$ are target feature modules in a request chain, and $m2$ precedes $m1$, then the address of $m2$ is more abstract than the address of $m1$.

The goals of privacy, authenticity, and reversibility are harder to achieve—features are too powerful for these goals to be feasible in any absolute form. Rather, we must settle for achieving them in restricted, but still useful, ways.

The authenticity of an address s cannot be guaranteed unless a feature module of s has an authentication function, so there are authenticity schemas rather than properties.

- *Source Authenticity Schema:* If s fills a source address field in the target region of a request chain, and if s has a source feature module m with unconditional authentication, then an owner of s is either present at the initiating device, or owns the address of a module preceding m in the chain.
- *Target Authenticity Schema:* If t fills a target address field in the target region of a request chain, and if t has a target feature module m with unconditional authentication, and if the request chain has reached an interface module on the target side, then an owner of t is either present at the target device, or owns the address of a module succeeding m in the chain.

Any source authenticity property (instance of the schema) is provable from Constraint 1. The qualifier "in a target region of a request chain" is necessary because an unauthenticated source address occurs in the source region, in the request routed to the authenticating feature module. The owner of s might encode the authentication secret for s in a feature module; if that module is present, it can supply the secret without the actual presence of the owner.

Any target authenticity property is provable from the formal model alone. It is similar to source authenticity properties. The qualifier "the request chain has reached a target interface module" is necessary because target authentication cannot usually take place until the request chain has reached a user.

The privacy of an address is protected by concealing it with a more abstract address. Thus there are privacy schemas rather than properties.

- *Source Privacy Schema:* If the source feature module of $s1$ changes the source address of a request chain to $s2$, then $s1$ is not observable as a source address of the request chain (either in a source address field or playing a source role) downstream of the source feature module of $s2$.
- *Target Privacy Schema:* If the target feature module of $t2$ changes the target address of a request chain to $t1$, then $t1$ is not observable as a target address of the request chain (either in a target address field or playing a target role) upstream of the target feature module of $t2$, except possibly in the source region of the chain.

Any source privacy property is provable from Constraints 1, 2s/t, and 3s/t, while any target privacy property is provable from Constraints 2t and 3t. The target privacy schema has the qualifier "except possibly in the source region of the chain" not because a private target address might leak into the source region of the chain, but because it might appear there independently. The address $t1$ could appear as a target in a request in the source region, be changed to $t2$ by another source module, and then re-appear in the target region.

The principle of reversibility says that it should be possible to target the entity on whose behalf a request chain was initiated. The best identification of this entity is the most abstract source address in the chain.

- *Chain Reversibility Property:* There is no more abstract address filling a source address field in a request chain than the one that crosses the boundary into the target region, and it remains the same throughout the target region.

The chain reversibility property is provable from Constraints 1 and 2s/t.

If a system with any of these properties is extended with new addresses and feature modules that do not violate the constraints, then the same property will hold in the extended system. This is because no proof relies on the presence or absence of any address or feature, except for those addresses and features named in the proved property itself.

5.4   Example: The anonymous correspondent

Figure 6 shows how anonymous electronic mail and an autoresponse feature can interact well within the framework of ideal address translation. The anonymous address *anon2@remailer* is a role address, used to conceal the more concrete personal address *user2@host2*. In the top half of Figure 6, the target feature module of *anon2@remailer* resolves the role address to the personal address.

The target feature module of *user2@host2* contains the autoresponse feature, set to notify all correspondents that the owner of this address is on vacation. The autoresponse initiates a new chain, which is first routed to the source feature module of *user2@host2*. The source feature module contains an "address book" feature [5] in which *user1@host1* is marked as a

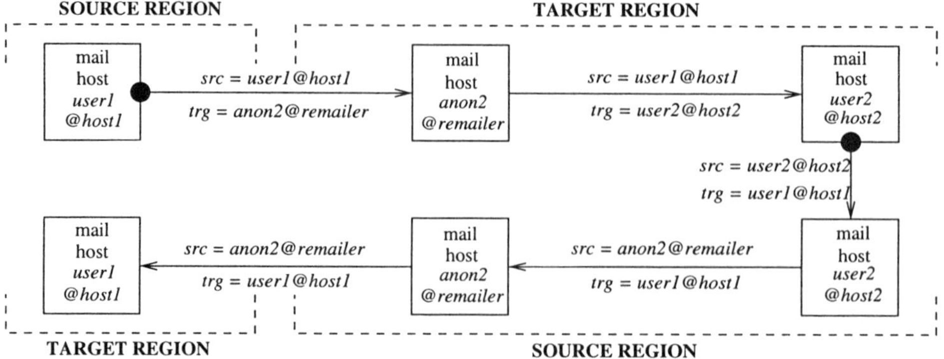

Figure 6: Anonymous electronic mail and autoresponse, in an ideal setting.

correspondent with whom anonymity must be preserved. Because the target of the chain is *user1@host1*, the source feature module performs assumption, changing the source address of the chain to *anon2@remailer*.

There can be a source feature module associated with *anon2@remailer*, to authenticate its use, if desired. The owner of *user2@host2* and *anon2@remailer* must encode his secret in the source modules of both addresses, so that the source module of *user2@host2* can send it automatically, and the source module of *anon2@remailer* can validate it automatically.

Because the source privacy property applies, *user2@host2* is not observable as a source of the autoresponse chain downstream of the source feature module of *anon2@remailer*. This is what the owner of *user2@host2* cares about most.

Obviously an autoresponse feature relies completely on the reversibility of the source address it receives. Because the chain reversibility property applies, autoresponse works.

Note that many forwarding features used in electronic mail change the source address to the address of the target feature module doing the forwarding [5], contrary to Constraint 1. If the target feature module of *anon2@remailer* behaved in this way, then it and the autoresponse feature would form an infinite routing loop.

Figure 6 is not realizable with the electronic mail available today. The mail host of *user2@host2* cannot send a message with an arbitrarily chosen source address such as *anon2-@remailer*. And if it could, there would be nothing to stop anyone else from using the same source address. This is because the protocol does not route to a subsequent source feature module, where authentication of the new source address can take place.

5.5   Example: The sales representative

Figure 7 shows three ways that the voice mail features of a sales group and of the members of the sales group can be coordinated successfully. Each snapshot is the target region of a request chain.

In the top snapshot, the caller has called the group address g. The feature module of g selects the sales representative with personal address p, and continues the request chain to that address. Both feature modules have voice mail as a failure treatment.

As explained in Section 4.5, group failure treatments should have priority over personal

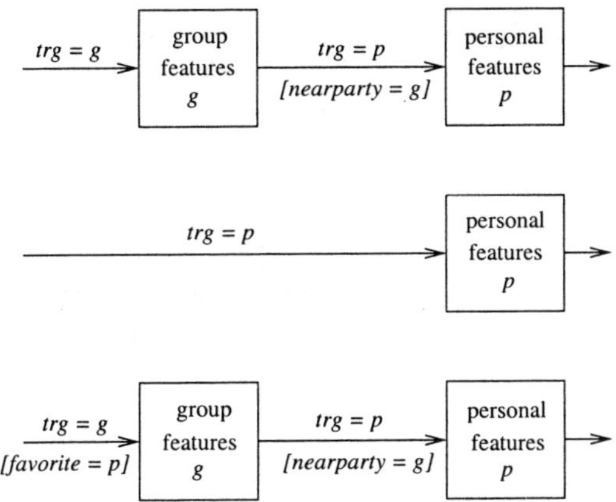

Figure 7: Three target regions containing the feature module of a sales representative.

voice mail, because this is really a call to the sales group. The feature module of g signals downstream that *nearparty* $= g$.[3] The feature module of p is programmed to relinquish all failure treatments in the presence of a feature module of g.

The field *nearparty* $= g$ should be propagated further downstream, announcing to the target device that the call was placed to g. If the device is a shared home telephone, and if the device displays this information, it will help family members know who should answer the call. Note that this auxiliary signaling does not violate Constraint 3t because it is sending an abstract address downstream, rather than sending a concrete address upstream.

The middle snapshot of Figure 7 simply reminds us that address p can be called directly, and that the target features of p will be unconditionally invoked.

The bottom snapshot of Figure 7 provides more complex behavior, combining the advantages of the top two. The caller's favorite sales representative is p. He calls g, including in the request a field indicating this preference. The feature module of g selects p as the sales representative; unlike the representation decision in the top snapshot, this decision is reproducible, so it has the advantage of the middle snapshot. If p is unavailable, however, p will relinquish treatment, and the unavailability will be treated by the feature module of g. Note that the auxiliary signal *favorite* $= p$ (which comes from a module in the source region) does not violate Constraint 3s because it concerns a target role, and does not violate Constraint 3t because the direction of propagation is downstream.

Figure 8 shows how the sales representative can use his home telephone. The source feature module of the device address h includes an assumption function, allowing the sales representative to assume his personal identity p.

The source feature module of h also includes a screening function, preventing certain kinds of outgoing calls. Screening applies to continuations of the request chain in which the source is unchanged, and not to continuations in which the source is changed. Thus the sales representative can make unscreened calls, while his children cannot.

The source feature module of p includes an authentication function, which is important

---

[3]A *nearparty* address is an address from the same region, here the target region. This is in contrast to an auxiliary *farparty* address, which here would be from the source region.

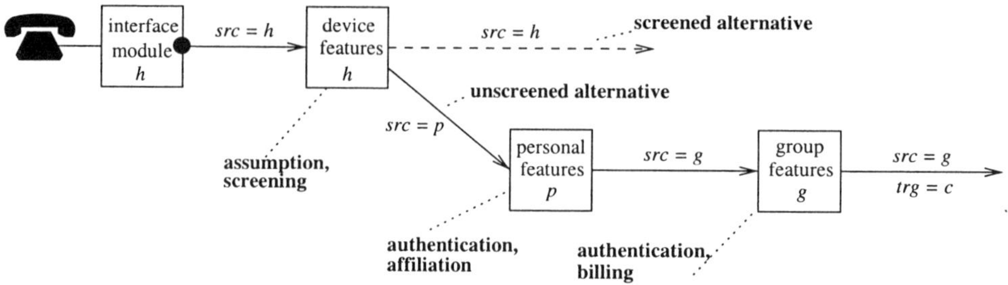

Figure 8: The sales representative makes work-related calls from a shared household telephone.

to prevent the children's misusing the telephone. It also has an affiliation function, so that the sales representative can make a call with source address g.

The source feature module of g also has an authentication function. After authenticating, it alters the billing so that work-related calls are billed to the company rather than the sales representative.

5.6 Deprecated functions

Most behaviors can be achieved, one way or another, within the confines of ideal address translation. Those that cannot be achieved are less useful or even harmful.

For example, it is not possible to have a target feature that blocks all calls from payphones: if some source feature changes the payphone address to a more abstract address, the payphone address is lost to all target features. It *is* possible, on the other hand, to have a target feature that blocks all payphone calls not identified as having an abstract source address. The latter feature seems far better, as it is not desirable to block calls from close friends just because they are calling from payphones.

It is always possible that a feature function will be required by a particular development context, even though it is deprecated by ideal address translation. Such a function must be treated as an exception.

6 Reasoning about exceptions

Address translation in real systems cannot always conform to the ideal. There will be exceptions to the constraints for the following reasons:

- There are legacy features and other objects that do not conform to the constraints, but cannot be abandoned.
- The infrastructure does not fully support the formal model of address translation, so that some necessary capabilities are unavailable.
- The constraints are not fully enforced.
- Some of the functions of ideal address translation are performed manually, so that pieces of it are missing from the system.

It is heartening that all of the constraint exceptions observed so far fall into one of these categories, because none of them indicates a fundamental flaw in the ideal.

An exception can be handled in two ways. We can supply more information about the system, and use it to prove that the system actually has all the desirable properties of ideal address translation, despite the superficial exception. Alternatively, we can diagnose exactly which desirable properties will not be preserved in the presence of the exception, and thus contain its bad effects.

The full version of the paper [11] contains several examples. In particular, one of them shows how to approximate the solution of Section 5.4 with electronic mail as it is today.

7 Enforcement of the constraints

Constraints 3s/t can be violated by a telecommunication protocol, regardless of the behavior of feature modules.

All the constraints can be violated by the behavior of feature modules. There are many interesting questions about how the contraints should be enforced on modules, among them:

- Should enforcement be at compile time or run time?
- How can the two be combined, so that compile-checked modules and legitimate exceptions are exempted from run-time checking?

The most difficult aspect of enforcement is that it requires global knowledge of address categories. The second most difficult aspect of enforcement is determining which signals and signal fields are transmitting addresses in the roles of sources or targets of their request chain.

Even with these problems unsolved, it is both realistic and useful to enforce the constraints on an "island" such as an IP PBX. All the addresses and modules implemented on the island should comply with ideal address translation. Addresses in the vast surrounding ocean belong to unknown categories, so no constraints apply to them.

8 Related work

Hall's study of 10 common electronic-mail features [5] revealed 26 undesirable feature interactions. Of these 26, 12 have nothing to do with address translation (they concern encryption and crude filtering based on domain names). All of the remaining 14 are predicted by the principles presented here, and would be eliminated by adherence to ideal address translation.

This work is based on a distributed, connection-oriented model of feature composition. Although PSTN central-office switches do not work this way, it clearly describes feature composition across network nodes, for both telephony and electronic mail. The example of DFC shows that it can apply to both intra- and inter-node feature composition in newer telecommunication architectures.

A new Mitel architecture for feature composition specifies features with rules. The rules are packaged within agents, agents communicate through a tuple-space blackboard, and many feature interactions take the form of competition among rules. With respect to addressing, the architecture is still in flux. In one version [1], the roles a person can play are simply the motivations for various rules in the person's agent. In another version [2], there are separate agents for devices, persons, and roles, and an address can identify a role as well as a person. Similarly, the Aphrodite architecture [10] has agents for devices, persons, roles, and related concepts.

Most research on feature interaction focuses on managing interactions at run time, in the design of the features, or in the design of the system architecture [6]. Ideal address translation can be enforced in any of these ways. It is unique in identifying global requirements and principles for telecommunications, and in focusing on how features should interact in order to satisfy them.

The disadvantages of this prescriptive approach, in comparison to behavior-neutral formal methods, are very clear. Many existing features do not comply. Some existing architectures and infrastructures do not comply, because of the source/target symmetry that is fundamental to ideal address translation. Some apparently reasonable behaviors are deprecated.

The advantages, on the other hand, justify the inconvenience. Compliant systems are modular, extensible, and guaranteed to have desirable properties. There are unambiguous, teachable guidelines for good feature design. These guidelines emphasize what function a feature is performing, for what purpose, and on whose behalf, and use the answers to those questions to organize appropriate interactions. Because escapes to a less-structured reality are accommodated, no feature behavior is prohibited absolutely.

Many of the issues of addressing and routing raised here are also issues in the Internet as a whole [9]. A form of organization discovered by contemplation of telecommunication services might very well be helpful in understanding other IP applications.

Acknowledgments

This work has benefited greatly from the contributions of Greg Bond, Eric Cheung, Bob Hall, Michael Jackson, Hal Purdy, and Chris Ramming.

References

[1] Magdi Amer *et al.* Feature-interaction resolution using fuzzy policies. In *Feature Interactions in Telecommunications and Software Systems VI*, pages 94-112. IOS Press, Amsterdam, 2000.

[2] D. Amyot *et al.* Feature description and feature interaction analysis with Use Case Maps and LOTOS. In *Feature Interactions in Telecommunications and Software Systems VI*, pages 274-289. IOS Press, Amsterdam, 2000.

[3] Russell Bennett and Jonathan Rosenberg. Integrating presence with multi-media communications. White paper, http://www.dynamicsoft.com.

[4] Gregory W. Bond *et al.* An open architecture for next-generation telecommunication service. Submitted for publication, 2002.

[5] Robert J. Hall. Feature interactions in electronic mail. In *Feature Interactions in Telecommunications and Software Systems VI*, pages 67-82. IOS Press, Amsterdam, 2000.

[6] Jonathan D. Hay and Joanne M. Atlee. Composing features and resolving interactions. In *Proceedings of the ACM SIGSOFT Eighth International Symposium on the Foundations of Software Engineering*, pages 110-119, 2000.

[7] Michael Jackson and Pamela Zave. Distributed feature composition: A virtual architecture for telecommunications services. *IEEE Transactions on Software Engineering* XXIV(10):831-847, October 1998.

[8] Michael Jackson and Pamela Zave. The DFC Manual. AT&T Research Technical Report, February 2003, http://www.research.att.com/info/pamela.

[9] E. Lear. What's in a name: Thoughts from the NSRG. IRTF Name Space Research Group, work in progress, 2002.

[10] Debbie Pinard. Enabling next-generation solutions for advanced business communications. Aphrodite Telecom Research, www.aphroditetelecom.com.

[11] Pamela Zave. Address translation in telecommunication features. In preparation.

*Feature Interactions in Telecommunications
and Software Systems VII*
D. Amyot and L. Logrippo (Eds.)
IOS Press, 2003

Feature Integration as Substitution

Dimitar P. GUELEV, Mark D. RYAN and Pierre Yves SCHOBBENS

School of Computer Science, Birmingham University, UK
E-mail: {D.P.Guelev, M.D.Ryan}@cs.bham.ac.uk
Institut d'Informatique, Facultés Universitaires de Namur, Belgium
E-mail: pys@info.fundp.ac.be

Abstract. In this paper we analyse the development of automated systems by means of adding features to a basic system. Our approach is to describe systems in temporal logic. We regard the process of integrating features as a transformation of those temporal logic descriptions. We use substitution of predicates as the basic means to achieve feature readiness of descriptions and define feature integration. We show that several description formalisms for features known from the literature can be fitted into the formal framework of our analysis, despite the fact that they have been initially motivated by different observations. Among these formalisms are Samborski's *stack service model* [12], the *feature construct* for SMV [10] and others. We argue that the way to address the verification problems which are specific to systems with features provided by our analysis is clear, convenient and based on classical and well established logical notions only. The logic we use in examples of description in this paper is the duration calculus with higher order quantifiers and iteration [15, 8, 1, 5].

Introduction

New models of automated systems, such as computerised systems, and new versions of software are typically obtained by making additions and small other changes to earlier ones. Such additions are known as *features* [2, 4]. Systems which have been built by adding features are more likely to have design faults than others, because features are typically designed separately from basic systems and from each other. Nevertheless, the method of obtaining better designs by integrating features is immensely more efficient that its known alternatives. That is why verification is important for such systems. Telecommunication systems are the greatest source of examples of systems with features. Sets of features, whose combined effect on the behaviour of a basic system is unconventional, are said to have an *interaction* in the literature on telecommunications. Identifying feature interactions is one of the most outstanding verification problems related to the development of systems by incorporating features.

In general, a feature F can be regarded as a mapping $F : S \rightarrow S$, where S is the class of systems in question, such that given a system $S \in S$ an (attempt to) integrate F into S results in a system $F(S)$. Less generally, but very commonly, a feature is a prefabricated item F, and the result of integrating F into a system S is then denoted by $S \oplus F$ where $\oplus$ stands for an operation of feature integration.

In this paper we are interested in systems' behaviour and therefore assume that the relevant properties of such behaviour are described in a temporal logic. We assume that there is a *modelling relation* $\models$ on $\mathcal{S} \times \mathcal{L}(\mathcal{S})$ where $\mathcal{L}(\mathcal{S})$ stands for the language of the chosen temporal logic based on the vocabulary which corresponds to the class of systems $\mathcal{S}$ such that

$$S \models \varphi$$

denotes that *the logic model(s) of all the potential behaviours of S satisfy φ*, and in particular,

$$S \oplus F \models \varphi$$

stands for *S with feature F added to it satisfies φ*. Using $\models$ and $\oplus$, some interesting statements about systems with features can be formalised:

Feature interaction

$S \oplus F_1 \oplus \ldots \oplus F_n \not\models \varphi$ — the "combined effort" of all the features breaks φ;

$S \oplus F_1 \oplus \ldots \oplus F_{i-1} \oplus F_{i+1} \oplus \ldots \oplus F_n \models \varphi, i = 1, \ldots, n$ — φ holds, as long as one of the interacting features is absent.

Implicit definition of feature composition

$(F_1 \oplus F_2) \equiv F$ iff $\forall S(S \oplus F \equiv (S \oplus F_1) \oplus F_2)$ where $S_1 \equiv S_2 \rightleftharpoons \forall \varphi(S_1 \models \varphi \Leftrightarrow S_2 \models \varphi)$.

Furthermore, we assume that the chosen temporal logic allows the representation of $S \models \varphi$ in the form $\models [\![S]\!] \Rightarrow \varphi$, where $[\![S]\!]$ stands for a logical formula which defines the class of the logic's models which represent behaviours of S.

In this paper we propose a representation for $\oplus$. We assume that a *feature ready* form $S_n \ldots S_1 B$ of $[\![S]\!]$ can be obtained in which the immutable base of S is denoted by formula B and the parts of S which can be affected by the integration of features are denoted by substitutions $S_1, \ldots, S_n$, S_i denoting parts of S which are mutable relative to the ith "stub" $S_{i-1} \ldots S_1 B$. We assume that a feature F is described as a sequence $F_1, \ldots, F_n$ of substitutions which represent the additions made upon its integration at the various levels of mutability in S, and the way it changes the roles of the default mutable parts of S, so that $[\![S \oplus F]\!]$ can be put down as $S_n F_n \ldots S_1 F_1 B$. Here n and the particular predicate letters subject to instantiation can vary, depending on the kind of systems and features in question.

Our approach was inspired by Samborski's *stack service model* [12] and builds on the semantics of the *feature construct* for SMV [10], given in [9]. We show that several description formalisms for features known from the literature can be fit into the formal framework we propose. Among these formalisms are Samborski's model [12] and the *feature construct* for SMV [10].

We choose the extension of the Duration Calculus (DC, [15, 6]) by iteration [1], a quantifier which binds state variables [8] and one which binds temporal variables with finite variability [5], as the logic for our examples in this paper. We include a concise definition of DC here only. The properties of DC that motivate our choice become manifest by the way we use it to describe systems' behaviour. We believe that other choices are compatible with our approach too.

1 Preliminaries on the Duration Calculus and Substitution of Predicate Letters

DC is a linear time first order interval-based temporal logic. It has one normal binary modality known as *chop*. A comprehensive survey on DC can be found in [6]. Extensions and variants of DC have been proposed and studied in a number of works, among which are [1, 14, 8, 13]. Since we use some of these extending constructs, and for the sake of self-containedness, we include a brief formal definition of DC as it appears in this paper.

Languages Along with the customary first order logic symbols, DC vocabularies include *state variables*. State variables P form *state expressions* S, which have the syntax:

$$S ::= \mathbf{0} \mid P \mid (S \Rightarrow S)$$

State expressions occur in formulas as part of *duration terms* $\int S$. The syntax of DC *terms* t and *formulas* φ extends that of first order logic by duration terms and formulas built using the modalities *chop* and *iteration*, denoted here by $(.;.)$ and $(.)^*$, respectively:

$$t ::= c \mid x \mid \textstyle\int S \mid f(t, \ldots, t)$$
$$\varphi ::= \bot \mid R(t, \ldots, t) \mid (\varphi \Rightarrow \varphi) \mid (\varphi; \varphi) \mid \varphi^* \mid \exists x \varphi$$

Constant, function and predicate symbols can be either *rigid* or *flexible* in DC. (The interpretations of rigid symbols are required not to depend on the reference interval.) Individual variables are rigid. State variables are flexible. The symbols 0, $+$, $=$ and $\leq$ with their customary roles are mandatory in DC vocabularies. In the BNF for formulas, x stands for either an individual variable, or a state variable, of a flexible constant. Flexible constants are also called *temporal variables* in DC.

Semantics The model of time in DC is the linearly ordered group of the reals, which is also the fixed domain of individuals of DC. Other models of time have been studied too. Domains are constant in DC in general. The set of the possible worlds in models for DC in the Kripke sense is $\{[\tau_1, \tau_2] : \tau_1, \tau_2 \in \mathbf{R}, \tau_1 \leq \tau_2\}$. We denote this set by $\mathbf{I}$. A DC *interpretation* I *of a DC language* $\mathbf{L}$ is a function on $\mathbf{L}$'s vocabulary. The types of the values of I for symbols of the various kinds are as follows:

$$
\begin{array}{ll}
I(x), I(c) \in \mathbf{R} & \text{for individual variables } x \text{ and rigid constants } c \\
I(c) : \mathbf{I} \to \mathbf{R} & \text{for flexible } c \\
I(f) : \mathbf{R}^n \to \mathbf{R},\; I(R) : \mathbf{R}^n \to \{0,1\} & \text{for } n\text{-ary rigid } f, R \\
I(f) : \mathbf{I} \times \mathbf{R}^n \to \mathbf{R},\; I(R) : \mathbf{R}^n \to \{0,1\} & \text{for } n\text{-ary flexible } f, R \\
I(P) : \mathbf{R} \to \{0,1\} & \text{for state variables } P
\end{array}
$$

$I(0)$, $I(+)$, $I(\leq)$ and $I(=)$ are required to be the corresponding components of $\langle \mathbf{R}, 0, +, \leq \rangle$ and equality on $\mathbf{R}$, respectively.

Interpretations $I(P)$ of state variables P are required to have the *finite variability property* which means that for any two $\tau_1, \tau_2 \in \mathbf{R}$, the set $\{\tau : I(P)(\tau) = 0$ and $\tau_1 \leq \tau \leq \tau_2\}$ is required to be either empty, or a finite union of intervals. This requirement corresponds to the assumption that observable states change only finitely often in bounded intervals of time.

Given an interpretation I, the value $I_\tau(S)$ of state expression S at time $\tau \in \mathbf{R}$, and the value $I_\sigma(t)$ of a term t at interval $\sigma \in \mathbf{I}$ are defined by the clauses:

$$
\begin{array}{lcl}
I_\tau(\mathbf{0}) & = & 0 \\
I_\tau(P) & = & I(P)(\tau) \\
I_\tau(S_1 \Rightarrow S_2) & = & \max\{1 - I_\tau(S_1), I_\tau(S_2)\}
\end{array}
$$

$$
\begin{array}{lcl}
I_\sigma(c) & = & I(c)(\sigma) \\
I_\sigma(x) & = & I(x) \\
I_\sigma(\textstyle\int S) & = & \displaystyle\int_{\min \sigma}^{\max \sigma} I_\tau(S)\, d\tau \\
I_\sigma(f(t_1, \ldots, t_n)) & = & I(f)(I_\sigma(t_1), \ldots, I_\sigma(t_n)) \text{ for rigid } f \\
I_\sigma(f(t_1, \ldots, t_n)) & = & I(f)(\sigma, I_\sigma(t_1), \ldots, I_\sigma(t_n)) \text{ for flexible } f
\end{array}
$$

The modelling relation $\models$ is defined on interpretations I of a given language $\mathbf{L}$, intervals $\sigma \in \mathbf{I}$ and formulas φ from $\mathbf{L}$ by the clauses:

$$I, \sigma \not\models \bot$$

$I, \sigma \models R(t_1, \ldots, t_n)$ iff $I(R)(\sigma, I_\sigma(t_1), \ldots, I_\sigma(t_n)) = 1$;

$I, \sigma \models \varphi \Rightarrow \psi$ iff either $I, \sigma \models \psi$ or $I, \sigma \not\models \varphi$;

$I, \sigma \models (\varphi; \psi)$ iff $I, \sigma_1 \models \varphi$ and $I, \sigma_2 \models \psi$ for some $\sigma_1, \sigma_2 \in \mathbf{I}$ such that $\sigma = \sigma_1 \cup \sigma_2$ and $\min \sigma_2 = \max \sigma_1$;

$I, \sigma \models \varphi^*$ iff there exists an ascending sequence $\tau_0, \ldots, \tau_n$ such that $\tau_0 = \min \sigma$, $\tau_n = \max \sigma$, and $I, [\tau_{i-1}, \tau_i] \models \varphi$, $i = 1, \ldots, n$;

$I, \sigma \models \exists x \varphi$ iff $J, \sigma \models \varphi$ for some J which is a x-variant of I.

Abbreviations The symbols $\top, \neg, \vee, \wedge, \Leftrightarrow, \forall, \neq, \geq, <$ and $>$ are used to abbreviate formulas and terms in the usual way. Infix notation is used wherever $+, =$ and $\leq$ occur. The following abbreviations are (more) DC-specific:

$$\mathbf{1} \rightleftharpoons \mathbf{0} \Rightarrow \mathbf{0} \quad \ell \rightleftharpoons \int 1 \quad \lceil S \rceil \rightleftharpoons \ell \neq 0 \wedge \int S = \ell \quad \Diamond \varphi \rightleftharpoons ((\top; \varphi); \top) \quad \Box \varphi \rightleftharpoons \neg \Diamond \neg \varphi$$

When omitting parentheses, we assume that $(.; .)$ has the *smallest* binding strength. We never omit the parentheses of $(.; .)$ itself. We also write $(\varphi; \psi; \chi)$ instead of $((\varphi; \psi); \chi)$, etc.

Substitution of predicate letters by predicates This is our main tool here. In this paper *variables* means individual, state or temporal variables. Let φ be a formula and $x_1, \ldots, x_n$ be free variables of φ. Let $\lambda x_1 \ldots x_n.\varphi$ be the predicate on $x_1, \ldots, x_n$ which φ defines. We do not require $FV(\varphi) \subseteq \{x_1, \ldots, x_n\}$, in order to enable the definition of parameterised families of predicates by single φs. Given an n-ary predicate letter P and a formula ψ, the *substitution* $[\lambda x_1 \ldots x_n.\varphi/P]\psi$ *of P by $\lambda x_1 \ldots x_n.\varphi$ is defined by the clauses:

$$\begin{aligned}
\theta \bot &\rightleftharpoons \bot \\
\theta R(t_1, \ldots, t_m) &\rightleftharpoons R(t_1, \ldots, t_m), \text{ if } R \neq P \\
\theta P(t_1, \ldots, t_n) &\rightleftharpoons [t_1/x_1, \ldots, t_n/x_n]\varphi \\
\theta(\psi_1 \Rightarrow \psi_2) &\rightleftharpoons \theta \psi_1 \Rightarrow \theta \psi_2 \\
\theta(\psi_1; \psi_2) &\rightleftharpoons (\theta \psi_1; \theta \psi_2) \\
\theta(\varphi^*) &\rightleftharpoons (\theta \varphi)^* \\
\theta \exists x \psi &\rightleftharpoons \exists y \theta[y/x]\psi \text{ where } y \notin FV(\varphi) \setminus \{x_1, \ldots, x_n\}
\end{aligned}$$

Simultaneous substitution $[\lambda x_1 \ldots x_{n_i}.\varphi_i/P_i : i \in I]$ of several predicate letters is defined similarly.

2   Hardware features: an introductory example

Let us show how our approach works when the integration of a feature amounts to connecting a piece of circuitry to a system, which itself is a circuit. Let the observable signals of our basic system be $x_1, \ldots, x_n$. Let $x_1, \ldots, x_n$ be also the names of the state variables which stand for these signals in the formula S describing the system. That is, an interpretation I for the signals $x_1, \ldots, x_n$ represents a behaviour of the considered basic system at interval σ iff $I, \sigma \models \llbracket S \rrbracket$. Naturally, $FV(\llbracket S \rrbracket) \subseteq \{x_1, \ldots, x_n\}$. There are extensive studies on the description of hardware and especially digital circuits by DC in this way. For instance, [11] proposes a set of derived DC constructs called *implementables* for the description of the basic temporal and causal relations between input and output signals of the simplest digital devices.

Let φ be a formula which represents the circuitry that comes with the feature F in a similar way. Let $FV(\varphi) = \{y_1, \ldots, y_m\}$. To describe the integration of the feature completely, we need to list the connections between the signals $x_1, \ldots, x_n$ of the basic system and the

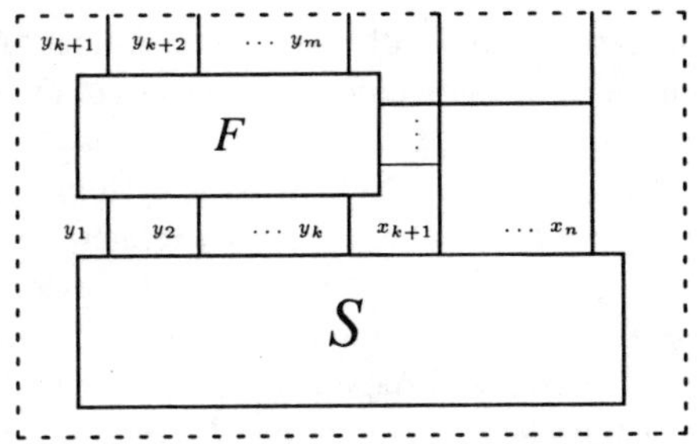

Figure 1: Feature F is integrated into system S by adding a circuit. S and $S \oplus F$, which is represented by the dashed rectangle, have the same observable signals.

signals of the new circuit $y_1, \ldots, y_m$. Some of these signals get identified, that is, connected by conductors, others remain accessible to the environment. Let the signals $x_1, \ldots, x_k$ of S be connected to the signals $y_1, \ldots, y_k$ of the feature. Let the system obtained after the integration interact with its environment through the remaining signals $x_{k+1}, \ldots, x_n, y_{k+1}, \ldots, y_m$ of its now two parts. Then its behaviour can be described by the formula

$$[\![S \oplus F]\!] = [x_1/y_{k+1}, \ldots, x_k/y_{2k}] \exists y_1 \ldots \exists y_k (\varphi \wedge [y_1/x_1, \ldots, y_k/x_k] [\![S]\!]) \tag{1}$$

We assume that $m = 2k$, that is, the number of signals that the feature contributes to the system is equal to the number of signals it hides from the environment, for the sake of simplicity. Then S can be written as

$$\underbrace{[\lambda x_1 \ldots x_n. [\![S]\!]/P]}_{S_1} \underbrace{P(x_1, \ldots, x_n)}_{B} \tag{2}$$

and $[\![S \oplus F]\!]$ can be written as

$$S_1[\lambda y_{k+1} \ldots y_m x_{k+1} \ldots x_n. \exists y_1 \ldots \exists y_k (\varphi \wedge P(y_1, \ldots, y_k, x_{k+1}, \ldots, x_n))/P]B \tag{3}$$

respectively, where P is an n-ary predicate letter. Since

$$[\lambda x_1 \ldots x_n. [\![S]\!]/P]P(y_1, \ldots, y_k, x_{k+1}, \ldots, x_n) \text{ is } [y_1/x_1, \ldots, y_k/x_k][\![S]\!],$$

the formulas (1) and (3) are equivalent. Besides, (3), which describes $S \oplus F$, is obtained by inserting a substitution between S_1 and B in the description (2) for S. The inserted substitution carries all the information relevant to the integration of the feature F. The formula $P(x_1, \ldots, x_n)$ on the right in both (2) and (3) determines the interface of the system with the environment, that is the list of its signals. The substitution $[\lambda x_1 \ldots x_n. [\![S]\!]/P]$ determines both the implementation of S as before the introduction of F and the behaviour of the circuit found in S upon the integration of F as part of $S \oplus F$.

Some observations can be made on the way the integration of a feature was described above.

The description $[\![S]\!]$ of S was rewritten into the form (2), as the instantiation of the system's interface B, by the system's actual behaviour S_1. This roughly corresponds to dismantling the system and preparing it for an upgrade. Upon dismantling the system, the identity between the implementation of its actual behaviour "inside" and its interface "outside" is lost, and the place holder P which is now explicitly available for substitution, defines the possible ways of revising this identity. The revision comes in the form of the intermediate substitution F which describes the feature to be integrated.

Here, F can fully control the extent to which the basic system's circuit affects the featured system's behaviour. For instance, F can choose to fully reinterpret the signals $x_1, \ldots, x_n$ of S; it can also reinterpret the signals $x_1, \ldots, x_k$ only, as we have done above.

The place holder P has at most one occurrence in the definition of the predicate on the left hand side of $/$ in F. This occurrence is positive and in the scope of existential quantifiers only. This is related to the feature being an addition of circuitry. Multiple occurrences of P would mean that whatever circuitry implements $\lambda x_1 \ldots x_n . [\![S]\!]$, should be multiplied upon the integration of F. This is unrealistic for hardware. Similarly, a negative occurrence or an occurrence in the scope of a universal quantifier may not correspond to any conceivable reassembly of the circuitry of S. In the next section we deal with software features, where this needs not be the case.

The description of the circuits of S and F as predicates on their signals is compatible with the approach to use *modules* and module *instances* as known from, e.g., SMV.

3 Software features

The huge variety of meanings that *executing programs* has in the various programming languages makes it very hard to search for universal and practically valuable formalisations to the integration of software features. In this section we introduce a simple imperative real time concurrent programming language similar to that in [3], in order to illustrate how software features' behaviour can be described by substitution. Programs in it are fixed sets of interleaving processes with shared variables. We assume that they interact with their environments by reading and writing signals like variables. We define the semantics of the language by a translation of programs into DC formulas.

3.1 A language for concurrent processes with shared variables

Programs $\mathbf{P}$ in our programming language are parallel compositions of sequential processes of the form

$$\mathbf{P} = P_1 \| \ldots \| P_n \tag{4}$$

which all have access to the same set of variables and input and output signals. We use x, y, ... to denote both variables and signals. The only difference is that signals may not occur on the wrong side of assignment statements. In the BNF for processes below e and c stand for arbitrary and boolean expressions respectively:

$$P ::= \texttt{skip} \mid x := e \mid \texttt{delay } e \mid \texttt{if } c \texttt{ then } P \texttt{ else } P \mid (P; P) \mid \texttt{while } c \texttt{ do } P$$

We denote the set of the variables occurring on the left side of assignment statements in process P by $Write(P)$. The DC language we use to describe the behaviour of programs contains a pair of temporal variables x and x' for every syntactically correct program variable x, the state variables $active_i$, and the flexible 0-ary predicate letters $[\![P]\!]_i$, $i < \omega$, for every syntactically correct process P.

The DC temporal variables x and x' denote the values of the program variable x in the beginning and in the end of a reference interval, respectively, $active_i$ holds at time τ iff P_i is active at time τ, and $[\![P]\!]_i$ hold at interval σ iff σ represents a complete run of a subprocess

P of the process P_i in (4). We consider only terminating behaviour here. Nonterminating behaviour can be handled similarly, using the extension of DC by *infinite intervals* [13].

For the rest of this section we fix a program $\mathbf{P}$ and $\models_{\mathbf{P}}$ to denote validity in the DC *theory of* $\mathbf{P}$. Obviously, for all program variables x

$$\models_{\mathbf{P}} \forall u \forall v ((x' = u; x = v) \Rightarrow u = v) \tag{5}$$

Consider the abbreviations:

$\mathsf{K}(X) \rightleftharpoons \bigwedge_{x \in X} \Box x' = x$, where X stands for a set of variables;

$A_i \rightleftharpoons \lceil active_i \rceil$;

$Sleep_i \rightleftharpoons \lceil \neg active_i \rceil \vee \ell = 0$.

$\mathsf{K}(X)$ describes that the variables X preserve their values within the reference interval. A_i describes that process P_i is active and $Sleep_i$ describes that P_i is inactive throughout the reference interval. Obviously, $\models_{\mathbf{P}} A_i \Rightarrow \Box(A_i \vee \ell = 0)$ for all i and $\models_{\mathbf{P}} \neg(A_i \wedge A_j)$ if $i \neq j$. A variable x can be updated only *within* an interval where some process is active:

$$\models_{\mathbf{P}} \Box \left(\neg \mathsf{K}(\{x\}) \Rightarrow \bigvee_{x \in Write(P_i)} \Diamond A_i \right)$$

Let t_e denote the time needed to evaluate expression e. Then the predicate letters $[\![P]\!]_i$ validate the formulas:

$\models_{\mathbf{P}} [\![\texttt{skip}]\!]_i \Leftrightarrow Sleep_i$

$\models_{\mathbf{P}} [\![x := e]\!]_i \Leftrightarrow (\ell = t_e \wedge x' = e \wedge \mathsf{K}(Write(P_i) \setminus \{x\}) \wedge A_i; Sleep_i)$

$\models_{\mathbf{P}} [\![\texttt{delay } e]\!]_i \Leftrightarrow \ell = \max(t_e, e) \wedge (A_i \wedge \mathsf{K}(Write(P_i)); Sleep_i)$

$\models_{\mathbf{P}} [\![\texttt{if } c \texttt{ then } P \texttt{ else } Q]\!]_i \Leftrightarrow \left(\begin{array}{l} (\ell = t_c \wedge c \wedge A_i \wedge \mathsf{K}(Write(P_i)); Sleep_i; [\![P]\!]_i) \vee \\ (\ell = t_c \wedge \neg c \wedge A_i \wedge \mathsf{K}(Write(P_i)); Sleep_i; [\![Q]\!]_i) \end{array} \right)$

$\models_{\mathbf{P}} [\![(P; Q)]\!]_i \Leftrightarrow ([\![P]\!]_i; [\![Q]\!]_i)$

$\models_{\mathbf{P}} [\![\texttt{while } c \texttt{ do } P]\!]_i \Leftrightarrow [\![\texttt{if } c \texttt{ then } (P; \texttt{while } c \texttt{ do } P)) \texttt{ else } \texttt{skip}]\!]_i$

Under these assumptions, an interpretation I and an interval σ represent a complete run of $\mathbf{P}$ iff

$$I, \sigma \models_{\mathbf{P}} \bigwedge_{i=1}^{n} [\![P_i]\!]_i \tag{6}$$

3.2 Adding features which are processes

Designing a software feature as a process to be run concurrently with the rest of the basic system's processes is intuitive, because the design can be started from an informal account of what conditions on the behaviour of a system should trigger action on behalf of a feature and how the feature should affect the working of the system. Besides, a process can easily be designed to monitor the behaviour of the processes it shares variables with.

Let F be a feature of this kind. Let P_f be the process to be added upon the integration of F and (4) stand for the basic system. Then the result $\mathbf{P} \oplus F$ of integration F into $\mathbf{P}$ can be represented as $\mathbf{P} \| P_f$. This suggests the following form of the basic system, which is feature-ready for features which are additional concurrently run processes:

$$\mathbf{P} = [\texttt{skip}/A] P_1 \| \dots P_n \| A$$

Now the substitution $[P_f \| A / A]$ can be used to describe the integration of F:

$$\mathbf{P} \| P_f = [\texttt{skip}/A][P_f \| A / A] P_1 \| \dots \| P_n \| A$$

Since the behaviour of **P** is described by means of a simple conjunction in (6), a similarly shaped feature-ready form of $[\![\mathbf{P}]\!]$ can be obtained, and (6) can be rewritten in the form:

$$I, \sigma \models_{\mathbf{P}} [Sleep_a/A] \left(\bigwedge_{i=1}^{n} [\![P_i]\!]_i \right) \wedge A \tag{7}$$

where $a \notin \{1, \ldots, n\}$, and then I, σ would represent a complete run of $\mathbf{P} \oplus F$ iff

$$I, \sigma \models_{\mathbf{P}} [Sleep_a/A][[\![P_f]\!]_f \wedge A/A] \left(\bigwedge_{i=1}^{n} [\![P_i]\!]_i \right) \wedge A \tag{8}$$

where $f \notin \{1, \ldots, n, a\}$.

3.3 Adding features which are procedures

Software features can also be implemented as modifications of subroutines such as drivers and operating system functions which are to be invoked through standard entry points. One way to demonstrate this in our programming language is to introduce named processes and allow occurrences of their names to stand for invocations of these subprocesses as procedures. Yet, in order to show how substitution can be used to describe the revision of such procedures upon the integration of a feature, we only extend the BNF for processes to allow place holders A to occur where subprocesses subject to revision is to be put:

$P ::= \texttt{skip} \mid \ldots \mid \texttt{while } c \texttt{ do } P \mid A$

Now a feature-ready program can be written in the form

$$\mathbf{P} = [D_1/A_1, \ldots, D_k/A_k]P_1(A_1, \ldots, A_k)\| \ldots \|P_n(A_1, \ldots, A_k) \tag{9}$$

where the parameter lists $(A_1, \ldots, A_k)$ indicate that the processes $P_1, \ldots, P_n$ may have been written with occurrences of the some of place holders $A_1, \ldots, A_k$, and $D_1, \ldots, D_k$ are the default implementations of the processes named $A_1, \ldots, A_k$, which are part of the basic system. Various kinds of substitutions can be inserted in (9) to change the working of $D_1, \ldots, D_k$ in various ways:

$[P_{f,i}/A_i]$ — The default procedure D_i gets replaced by one supplied by the feature. Further feature integration may be unable to change this again, because $P_{f,i}$ contains no place holders.

$[(P_{f,i}; A_i)/A_i]$ — D_i gets *chained*, that is, programmed to be run after some new code $P_{f,i}$. This is typical of exception handlers.

$\left[\texttt{if } c \; \begin{array}{l} \texttt{then } P_{f,i} \\ \texttt{else } A_i \end{array} /A_i \right]$ — The feature-supplied code replaces the default conditionally.

Just like in the case of adding processes by parallel composition, the role of place holders and the corresponding substitutions is preserved in the DC description of the behaviour of the considered systems.

3.4 The stack service model in terms of substitutions

Samborski's *stack service model* assumes that services in a distributed system are organised as a *network* of *stacks*, and process *tokens*, which represent user and system requests. A stack processes a token it receives by feeding it to its topmost service. A service may choose either to completely process a token itself, or resort to the service immediately below it in the stack by generating a token to be passed to this service. In this model feature integration is represented by the insertion of services into stacks. Different stacks in a network can have different services and have them appearing in different orders. This reflects the possibility for different users of the network to subscribe to different features, as known in the case of a telephone system, and prioritise the reaction of services in response to requests in different ways. In the stack service model, feature integration is done by inserting services in stacks.

The *state* of a stack of services is a tuple

$$\mathbf{S} = \langle n, L_u, L_s, \varepsilon, d, R_1 \cdot \ldots \cdot R_k, V \rangle$$

where n is an *identifier* for the stack in question, L_u and L_s are queues of incoming tokens originating from the stack's user and from the system, respectively, $\varepsilon \in \{0, 1\}$ denotes the state of a "door" which gives priority to user incoming tokens when open, and to system tokens when closed, d is the *coming down* token, $S = R_1 \cdot \ldots \cdot R_k$ is the *list of services* in $\mathbf{S}$, and V is the *valuation of the variables* of $\mathbf{S}$. If there is no coming down token d, δ is put in the place of d. Services R_i, $i = 1, \ldots, k$, can be regarded as functions which, given a token t, a state of the "door" ε and a valuation of the stack's variables V, return a tuple $\langle d, U, \gamma, W \rangle$, where d is a token to be passed to the next service in the stack, U is a set of tokens to be released in the network, and γ and W are the state of the "door" and the valuation of the stack's variables upon the completion of R. In case service R does not process token t, $R(t, \varepsilon, V) = \langle t, \emptyset, \varepsilon, V \rangle$.

The operational semantics of the stack service model is presented in detail in [12]. In this section we show how the feature integration in the sense of the stack service model can be described in terms of substitutions.

3.4.1   *Programming stacks in the language of shared variables*

A stack of services can be programmed as a process in the extension of our programming language by one process place holder A. Individual services R can be programmed in the form

$$P(R) = (C(R); \text{if } d \neq \delta \text{ then } A \text{ else skip}), \tag{10}$$

where $C(R)$ stands for some code which implements the relation $R(t, \varepsilon, V) = \langle d, U, \gamma, W \rangle$ by sampling and assigning ε, d and $x_1, \ldots, x_m$ appropriately, and releasing the tokens U into the network by means of, e.g., an atomic command $\texttt{putToken}\,(e)$. The designated occurrence of a place holder A is the only one allowed in (10). It enables the chaining of successive services $\ldots R_i \cdot R_{i+1} \ldots$ in a stack by means of a sequence of successive substitutions

$$\ldots [\![P(R_{i+1})]\!]_n/A][\![P(R_i)]\!]_n/A] \ldots$$

Here n is the identifier of the stack. Let the variables of a stack with identifier n, be $x_1^n, \ldots, x_m^n$. $P(R)$ can be programmed to use the same names $x_1, \ldots, x_m$ for the variables $x_1^n, \ldots, x_m^n$ of whatever stack R can be part of, and Id for the identifier of this stack. Let θ_n be the substitution $[n/Id, x_1^n/x_1, \ldots, x_m^n/x_m]$. Then $\theta_n P(R)$ will stand for the instance of $P(R)$ which can appear in the implementation of a stack with identifier n. We assume that the queues L_u and L_s, the door ε, the coming-down token d and an auxiliary variable t are among $x_1, \ldots, x_m$ for the sake of simplicity.

Let b' be an auxiliary process to acquire a coming-down token from the appropriate queue and transfer control to the topmost service of the stack:

$(\text{while } L_u = \emptyset \wedge L_s = \emptyset \text{ do skip};$
　$\text{if } \varepsilon \wedge L_u \neq \emptyset \text{ then } ($
　　　$d := head(L_u); L_u := tail(L_u); \varepsilon = 0$
　$) \text{ else } ($
　　　$d := head(L_s); L_u := tail(L_s)); A$
　$)$
$)$

Let b'' be an auxiliary process to feed tokens to L_u and L_s from the network. Let, given the stack identifier Id, the $TokensAvailable(Id)$ indicate whether the network contains tokens bound for this stack. Let $\text{getToken }(Id, t)$ assign t a Id-bound token from the network, if it contains any tokens bound for this stack. Then b'' can be programmed by means of the auxiliary boolean function $TokensAvailable$ and atomic command getToken as follows:

　$\text{if } TokensAvailable(Id) \text{ then } ($
　　　$\text{getToken }(Id, t);$
　　　$\text{if } UserToken(t) \text{ then } L_u := L_u * t \text{ else } L_s := L_s * t$
　$)$
　else skip

Let s_1 be an auxiliary process to output to the network whatever token the last service in the stack produces:

　$\text{if } d \neq \delta \text{ then } (\text{putToken }(d); d := \delta) \text{ else skip}$

Now, given that $R_1 \cdot \ldots \cdot R_k$ is the list of services in the stack, the behaviour of the stack can be represented by the formula

$$\theta_n[\![s_1]\!]_n/A][\![P(R_k)]\!]_n/A] \ldots [\![P(R_1)]\!]_n/A]([\![b']\!]_n^* \wedge [\![b'']\!]_{n'}^*) \, . \tag{11}$$

The occurrence of A in $P(R_i)$ guarantees that upon terminating R_i transfers control to R_{i+1} for $i = 1, \ldots, k - 1$, and R_k transfers control to s_1. Control gets transferred only if R_i terminates with $d \neq \delta$. The index n' in $[\![b'']\!]_{n'}$ here can be chosen to be e.g. $N + n + 1$, where N is the greatest value that a stack identifier can take, to ensure that the two processes corresponding to each stack in a system interleave correctly.

To describe the behaviour of several stacks running concurrently in the same network, one can take the conjunction of the formulas (11) with n ranging over the identifiers of the stacks in the system and $R_1^n \cdot \ldots \cdot R_{k_n}^n$, being the list of services of stack with identifier n.

4　The SMV feature construct in terms of substitutions

In this section we show that the feature construct introduced in [10] to the input language of SMV [7] fits into our scheme of representing feature integration as the insertion of substitutions. We first show how the description of a system in SMV can be translated into an appropriate DC formula, written using substitutions of predicate symbols. The predicate symbols subject to these substitutions and the predicates which substitute them stand for the names of the components of the modelled system which are subject to change by the integration of features and these components themselves, respectively. Finally, the integration of features as it can be described using the SMV feature construct is done by inserting appropriate substitutions in the original formula.

Assumptions about the initial SMV description and its form

For the sake of simplicity, we assume that the description of the basic, that is, feature-free, system consists of a single main module.[1] We assume that all the assignment statements in the considered main module are of one of the two forms:

```
next(⟨variable⟩) := ⟨expression⟩;
init(⟨variable⟩) := ⟨expression⟩;
```

and all variables occur in the left hand side of *exactly* one assignment of each of the two kinds, possibly with non-deterministic expressions on the right hand side. We assume that the expressions above are `case` expressions with disjoint guards and possibly nondeterministic unconditional subexpressions. (All SMV programs can be rewritten into this form.)

Under all the above restrictions, such an arbitrary single simple SMV module can be regarded as a set of variables X, a set of next value assignments and initial value assignments:

$$
\begin{array}{ll}
\texttt{next}(x) := \texttt{case} & \texttt{init}(x) := \texttt{case} \\
\quad g_{x,0} \ : \ v_{x,0}; & \quad h_{x,i} \ : \ w_{x,0}; \\
\quad \vdots & \quad \vdots \\
\quad g_{x,N_x-1} \ : \ v_{x,N_x-1}; & \quad h_{x,i} \ : \ w_{x,M_x-1}; \\
\texttt{esac} & \texttt{esac}
\end{array}
$$

Given x, let e_x and f_x stand for the expressions on the right side of the corresponding next and initial value assignments.

We deal with the `treat` and the `impose` components of the feature construct for SMV first here.

As in the previous sections, given the set X of the variables which occur in the considered system description, we assume that the vocabulary of our logical language contains the temporal variables named x and x' for every $x \in X$. We also assume that the syntax of the gs, vs, hs and ws is the same as that of the terms in the chosen logical language. Using this convention, a description of any finite initial subinterval of a behaviour of a system can be written in the form:

$$
\bigwedge_{x \in X} \left(\left(\ell = 1 \wedge \bigwedge_{i < N_x} \left(g_{x,i} \Rightarrow \bigvee_{v \in v_{x,i}} x' = v \right) \right)^* \wedge \left(\bigwedge_{i < M_x} \left(h_{x,i} \Rightarrow \bigvee_{w \in w_{x,i}} x = w; \top \right) \right) \right) \tag{12}
$$

Here, as above, we assume that the temporal variables x and x' from our logical vocabulary denote the initial and the final values of the corresponding SMV variable $x \in X$ in every reference interval, and therefore satisfy the axiom (5).

Making the description feature-ready

In order to enable the alteration of a system's description by means of substitution, we need to have symbols to subject to substitution at the places which we intend to allow the integration of features to affect. (This is so, because substitution is simplest to define if its target is an atomic entity.) The `impose` and `treat` statements in the SMV feature construct prescribe the alteration of expressions to be assigned to variables and the alteration of the meaning of variable occurrences in such expressions. This means that we need to be able to substitute these

---

[1] Of course, what comes below is hardly like what users of SMV have these days. Arguably, it only has comparable potential expressivity.

expressions and the occurrences of variables in them. Variable occurrences are represented by atomic entities in (12), but the possibly ambiguous right hand sides of SMV assignments are represented as flexible predicates on the variables x and x', which, given $x \in X$, stand for $\text{init}(x)$ and $\text{next}(x)$ in (12). To introduce symbols in their places, we rewrite (12) in the form:

$$\left[\begin{array}{l} \lambda u. \bigwedge_{i < N_x} (g_{x,i} \Rightarrow \bigvee_{v \in v_{x,i}} u = v) \quad / \quad E_x, \\ \lambda u. \bigwedge_{i < M_x} (h_{x,i} \Rightarrow \bigvee_{w \in w_{x,i}} u = w) \quad / \quad F_x \end{array} : x \in X \right] \underbrace{\bigwedge_{x \in X} ((\ell = 1 \wedge E_x(x'))^* \wedge (F_x(x); \top))}_{B}$$

$$(13)$$

Clearly, (12) and (13) evaluate to the same formula. Using the predicate letters E_x and F_x as place holders enables the presentation of practically arbitrary restrictions on the initial and subsequent values of the variables $x \in X$. We need this, because of the possibility to have nondeterministic assignments. Similarly, although variables' occurrences on the right hand side of assignment are atomic, it is not sufficient to use them immediately as place holders to be substituted by terms, because SMV expressions which occur in `treat` statements can be nondeterministic too, and therefore produce more than a single term in the chosen logical form of description. That is why we further partition (12). To do this, we assume that the temporal variables $\tilde{x}$, $x \in X$, are fresh wherever they occur, and we introduce the fresh unary predicate letters V_x, $x \in X$. Informally, $V_x(u)$ denotes that u equals the current value of the SMV variable x. We denote the set of SMV variables which occur in a given SMV expression e by $Var(e)$. Given that $v_1, \ldots, v_n$ is a finite set of SMV variables, we abbreviate the quantifier prefix $\exists \tilde{v}_1 \ldots \exists \tilde{v}_n$ by $\exists_{v \in \{v_1, \ldots, v_n\}} \tilde{v}$. Using this notation, we replace (13) by

$$\underbrace{[\lambda u.u = x / V_x : x \in X]}_{S_2} \underbrace{S_1' S_1''}_{S_1} B \qquad (14)$$

where S_1' and S_1'' denote

$$[\lambda u.\exists_{y \in Var(e_x)} \tilde{y}(\bigwedge_{y \in Var(e_x)} V_y(\tilde{y}) \wedge [\tilde{y}/y : y \in Var(e_x)] \bigwedge_{i < N_x} (g_{x,i} \Rightarrow \bigvee_{v \in v_{x,i}} u = v)) / E_x, : x \in X]$$

and

$$[\lambda u.\exists_{y \in Var(f_x)} \tilde{y}(\bigwedge_{y \in Var(f_x)} V_y(\tilde{y}) \wedge [\tilde{y}/y : y \in Var(f_x)] \bigwedge_{i < M_x} (h_{x,i} \Rightarrow \bigvee_{w \in w_{x,i}} u = w)) / F_x : x \in X]$$

respectively The advantage of this apparently longer form of description is that now the effect of integrating features with `treat` statements can be represented by substitutions on V_x, to be inserted between S_2 and S_1. Similarly, integrating features with `impose` statements can be represented by substitutions on E_x and F_x, to be inserted between S_1 and B. The precise form of these substitutions is explained below. Clearly, (13) is equivalent to (14). In the sequel we briefly denote (14) by $S_2 S_1 B$.

Describing the integration of a feature

Consider the SMV feature which, for the sake of simplicity, consists of the single `impose` statement targetting some variable $y \in X$

if c_i then impose next$(y) := e_i$;

and the single `treat` statement targetting some variable $z \in X$

if c_t then treat $z = e_t$;

Let e_i and e_t be of the same form as e_x and f_x, $x \in X$, above. Predicates like the ones in S_1' and S_1'' for e_x and f_x, $x \in X$, can be written to represent e_i and e_t too. Let us denote these unary predicates by $[\![e_i]\!]$ and $[\![e_t]\!]$ respectively, and assume, just as in the case of e_x and f_x, that they are defined in terms of $Var(e_i)$ and $Var(e_t)$, respectively. Then, given the description $S_2 S_1 B$ of the feature-free system considered above, the result of integrating this feature into this system can be described by

$$S_2(S_t S_1)|_{\{E_x, F_x : x \in X\}} S_i B \tag{15}$$

where $S|_A$ denotes the restriction of a substitution S to a set of symbols A, and

$$S_t = [\lambda u . \exists_{y \in Var(e_t, c_t)} \tilde{y} (\bigwedge_{y \in Var(e_t, c_t)} V_y(\tilde{y}) \wedge [\tilde{y}/y : y \in Var(e_t, c_t)] \left(\begin{array}{c} ([\![e_t]\!](u) \wedge c_t) \vee \\ (V_z(u) \wedge \neg c_t) \end{array} \right))/V_z]$$

$$S_i = [\lambda u . \exists_{y \in Var(e_i, c_i)} \tilde{y} (\bigwedge_{y \in Var(e_i, c_i)} V_y(\tilde{y}) \wedge [\tilde{y}/y : y \in Var(e_i, c_i)] \left(\begin{array}{c} ([\![e_i]\!](u) \wedge c_i) \vee \\ (E_y(u) \wedge \neg c_i) \end{array} \right))/E_y]$$

Integrating more than one feature

The integration of complex features can be regarded as the integration of a sequence of features of the simple form studied above. In case several features get integrated into a system in a sequence, the ones which get integrated later affect the ones which get integrated earlier, but not the other way around. The straightforward way to express this in terms of substitutions is as follows.

Let $F_1, \ldots, F_r$ be features, each consisting of either a single `treat` statement or a single `impose` statement. Let $S_{t,k}$ be the substitution which corresponds to the `treat` statement of the feature F_k in the way introduced above, in case F_k has a `treat` statement, or be the vacuous (identity) substitution otherwise, $k = 1, \ldots, r$. Let, similarly, $S_{i,k}$ represent the `impose` statement of F_k, if F_k has one. Then the subsequent integration of $F_1, \ldots, F_r$ into the system described by $S_2 S_1 B$ can be described by the formula

$$S_2 S_{t,r} \ldots S_{t,1} S_1 S_{i,1} \ldots S_{i,r} B$$

A simple way to motivate this approach is to notice that the composition of substitutions $S_t S_1 S_i$ in the description (15) of a featured system has the role played by S_1 alone in the description (14) of the corresponding basic system. This means that subsequent acts of integrating features $F_k, \ldots, F_r$, should correspond to inserting substitutions $S_{i,k}, \ldots, S_{i,r}$, which correspond to `impose` statements, and substitutions $S_{t,k}, \ldots, S_{t,r}$, which correspond to `treat` statements, on the right and on the left hand sides of $S_{t,k-1} \ldots S_{t,1} S_1 S_{i,1} \ldots S_{i,k-1}$, $\ldots, S_{t,r} \ldots S_{t,1} S_1 S_{i,1} \ldots S_{i,r}$, respectively. Yet this approach is incorrect, because the composite substitution which is obtained this way prescribes that *all* the `treat` statements affect the meaning of *all* the `impose` statements, disregarding the order of their integration. For instance, according to (14) the `treat` statement of the considered feature affects its `impose`, regardless to the intention of the feature's author, which might have been to exercise the effect of the `treat` statement on the basic system's description only.

The cause for this paradox is that a composition of substitutions of the kind of $S_t S_1 S_i$ from (14) can be safely regarded as equivalent to a single substitution of the kind of S_1 only when applied to a formula like B, because the symbols E_x, $x \in X$, which are subject to substitution in B are in the domain of both S_1 and $S_t S_1 S_i$. Unlike S_1, $S_t S_1 S_i$ substitutes symbols from among V_x, $x \in X$ too. That is why, in order to obtain a version of $S_t S_1 S_i$ which is strictly of the same form as S_1, we need to take the restriction of $S_t S_1 S_i$ to the set of symbols $\{E_x : x \in X\}$ instead of the entire $S_t S_1 S_i$. That is why we put the result (14) of integrating a single feature to a system in the form

$$S_2(S_t S_1)|_{\{E_x, F_x : x \in X\}} S_i B \ .$$

Now it can be safely assumed that $(S_t S_1)|_{\{E_x, F_x : x \in X\}} S_i$ plays the role of S_1 in the description of the result of integrating the considered feature into the considered system. Consequently, the integration of a sequence of single-statement features $F_1, \ldots, F_r$ can be described by substitutions in the form:

$$S_2(S_{t,r} \ldots (S_{t,1} S_1)|_{\{E_x, F_x : x \in X\}} S_{i,1} \ldots)|_{\{E_x, F_x : x \in X\}} \ldots S_{i,r} B$$

5   Verification of feature-ready descriptions and features

A Hoare triple $\{P\}\texttt{code}\{Q\}$ can be written in DC as

$$P \wedge [\![\texttt{code}]\!] \Rightarrow [\overline{x'}/\overline{x}]Q \ ,$$

where $\overline{x}$ and $\overline{x'}$ represent the vectors of the initial and the final values of program variables like above. That is why standard verification methods apply straighforwardly to the verification problems which appear in this paper. Here we only briefly mention some opportunities for verification which appear due to our choice of representation for systems and features.

Refinement of feature descriptions

Let the feature F_i be described by the substitutions $F_{i,n}, \ldots, F_{i,1}$ and let $S_n F_{i,n} \ldots S_k F_{i,k} = [\lambda x_1 \ldots x_{n_i} . \varphi_j^{S \oplus F_i}/A_j : j = 1, \ldots, m]$, $i = 1, 2$. Then, as long as all the occurrences of $A_1, \ldots, A_m$ in $S_{k-1} \ldots S_1 B$ are positive,

$$\models S_n F_{1,n} \ldots S_k F_{1,k} S_{k-1} \ldots S_1 B \Rightarrow \alpha$$

and

$$\models \varphi_j^{S \oplus F_2} \Rightarrow \varphi_j^{S \oplus F_1}, \ j = 1, \ldots, m, \tag{16}$$

imply $\models S_n F_{2,n} \ldots S_k F_{2,k} S_{k-1} \ldots S_1 B \Rightarrow \alpha$. The condition (16) can be used to define a relation of refinement between features of similar type.

Let us assume that no place holder occurs in more than one formula in a feature-ready description $S_n \ldots S_1 B$. This restriction entails a one-to-one correspondence between the place holders and the mutable parts of the considered system and allows us to associate requirements on mutable parts with place holders. Assume that all the place holders in the considered system have positive occurrences only. Let the place holder A occur in the substitution S_k. Let S_{k+1} be $[\ldots, \lambda x_1 \ldots x_m . \varphi_A/A, \ldots]$. Then a requirement ρ_A can be imposed on the mutable part instantiating A by putting

$$\models S_n \ldots S_{k+2} \varphi_A \Rightarrow \rho_A \ . \tag{17}$$

It is natural to assume that a mutable part of the system will satisfy its requirement, provided that its subparts satisfy their respective requirements. This means that (17) can be replaced by:

$$\models [\lambda x_1 \ldots x_m.\rho_C/C : C \in \mathbf{A}]\varphi_A \Rightarrow \rho_A \,. \tag{18}$$

where $\mathbf{A}$ stands for the set of all the place holders involved.

6 Integrating variables

Along with other things, integration of features can bring additional variables. In fact, the SMV feature construct enables the addition of variables in its full form, but we only focus on this feature construct's more specific elements in Section 4. Adding variables that are only handled individually requires no special care upon the integration of a feature. However, it is necessary to be able to vary the scope of operations which are to affect the values of collections of variables that are to be handled similarly, so that variables that features can contribute be included in the appropriate collections. Using $x = e$ to denote that variable x evaluates to e at a certain time is convenient as long as only fixed sets of variables are involved in every description. Revising collections of variables upon feature integration can be achieved, if equality $=$ is replaced by a flexible binary predicate M (for *Memory*) with its first place being of an also newly introduced sort VN for names for variables (or memory locations), and the second place being of the already known sort of variables' values.

Now unary predicates can be introduced to designate the collections of variables in question. For example, a feature-ready form of the property

$$M(v7, z) \Rightarrow M(v8, z) \wedge \ldots \wedge M(v15, z) \tag{19}$$

that enables the members of the conjunction to be changed is

$$\underbrace{[(\lambda v : VN).v = v8 \vee \ldots \vee v = v15/H]}_{S_1} \underbrace{(\forall v : VN)(M(v7, z) \wedge H(v) \Rightarrow M(v, z))}_{B}$$

Substitutions that revise the definitions of such unary predicates in feature descriptions can be used to define revisions of the scope of the respective operations upon feature integration. For example

$$S_1[(\lambda v : VN).H(v) \vee v = v16 \vee \ldots \vee v = v31/H]B$$

denotes that the value z of $v7$ is the same as that of 16 *more* variables, as compared to (19), where only 9 variables are said to have the same value.

Concluding remarks

We have shown that several formalisations for feature integration can be fit in the scheme of decomposing descriptions and recomposing them with the features included as substitutions of predicate letter place holders by (parameterised) predicates.

The feature integration construct for SMV [10] is the most complex case for the analysis we present, because of the diversity of ways in which substitution appears to implement the integration of a composite feature. Besides, the feature construct for SMV is the one where substitution appears in the most explicit form of all. Chaining of substitutions appears in the most general form in the case of the stack service model [12].

In our attempt to put all the cases in a unified framework we have found that there are readily available formal methods, such as DC, which have been developed for more general purposes, and can be part of our approach to features. This paper has been written with DC as the system of logic to illustrate the form of analysis we have pursued, because of the very small added cost on description in DC in the cases considered. However, we believe that our approach can be similarly successful with a variety of logical systems, if not for the conciseness of presentation, at least when description stages can be assisted by a tool, and with the benefit of better mechanizability of the verification stage.

The approach suggests that a system needs to be (1) described in a feature-ready form, (2) features should be described as revisions of *this* feature-ready form, and finally (3) interactions can be searched for at the various levels of mutability subdivision of the system with respect to the requirements that can be formulated in the feature-ready form. Search for interactions can be carried out in the form of verification of properties, which are either associated with the entire system, or with its various mutable parts, just like with feature constructs in general. The most decisive part of the job is, of course, to obtain the feature-ready description of the system. We believe that automating this cannot be done for the sole reasons of verifying, let alone for finding interactions. It should rather be part of the overall design process of the basic system. Being relevant to the partitioning of the work on the design and, possibly, of the implementation of a system, tool support for maintaining a feature-ready form should be part of a respective specification language. A conclusion that can be made here is that the capacity to describe feature integration is plausible not only for programming and specification languages meant for the specialist user, but also for intermediate languages that can possibly combine descriptions of diverse original forms that target diverse forms of implementation: very roughly speaking, both hardware and software. The gap between maintaining and obtaining a feature-ready form is a problem which the approach reveals as a very important one, if maximal automation is to be sought in all possible steps. The feature-readiness of a system can be decisive for its lifetime and creating feature-readiness is a development stage that developers are (ideally) well aware of and give it priority. Again, the flexibility in describing the existing practices of creating feature-readiness, rather than proposing particular architectures and then advocating the use of a corresponding system of techniques and tools, is the goal of this work. Further bringing the correspondences between such architectures to a level of precision can possibly contribute to understanding them and deciding how to enhance some of them or use them in parallel.

Acknowledgements

The authors are grateful to Alan Sexton, Dang Van Hung and Wang Yi for a number of discussions on the topic of this paper during its preparation.

References

[1] DANG VAN HUNG AND WANG JI. On The Design of Hybrid Control Systems Using Automata Models. *FST TCS 1996*, LCNS 1180, Springer, 1996, pp. 156-167.

[2] CALDER, M. AND E. MAGILL, *Feature Interactions in Telecommunications and Software Systems VI*, IOS Press, 2000.

[3] GUELEV. D. P. AND DANG VAN HUNG. Prefix and Projection onto State in Duration Calculus. *Electronic Notes in Theoretical Computer Science* 65, No 6, 2002, `http://www.elsevier.nl/locate/entcs/volume65.html`, 19 p.

[4] GILMORE, S. AND M. D. RYAN (EDS.), *Language Constructs for Describing Features*, Springer-Verlag, 2001.

[5] GUELEV, D. P. A Complete Fragment of Higher-Order Duration μ-Calculus. *Proceedings of FST TCS 2000*. LNCS 1974, Springer Verlag, 2000. Available as Technical Report 195 from http : //www.iist.unu.edu/newrh/III/1/page.html.

[6] HANSEN, M. R. AND ZHOU CHAOCHEN. Duration Calculus: Logical Foundations. *Formal Aspects of Computing*, 9, 1997, pp. 283-330.

[7] MCMILLAN, K. *SMV documentation postscript versions.* http : //www-cad.eecs.berkeley.edu/˜kenmcmil/psdoc.html. Accessed in February, 2002.

[8] PANDYA, P. K. Some extensions to Mean-Value Calculus: Expressiveness and Decidability. *Proceedings of CSL'95*. LNCS 1092, 1995, pp. 434-451.

[9] PLATH, M. C. AND M. D. RYAN. The Feature Construct for SMV: Semantics, In [2], pp. 129-144.

[10] PLATH, M. C. AND M. D. RYAN. Feature Integration Using a Feature Construct. *Science of Computer Programming*, 41(1), pp. 53-84, 2001. Available from http : //www.cs.bham.ac.uk/˜mdr/research/papers/index.html.

[11] RAVN, A. P. *Design of Embedded Real-Time Computing Systems*. Dr. techn. dissertation. Technical report 1995-170, Technical University of Denmark, 1995.

[12] SAMBORSKI, D. Stack service model. *Language Constructs for Describing Features*. In [4], pp. 177-196.

[13] WANG HANPIN AND XU QIWEN *Temporal Logics over Infinite Intervals*. Technical Report 158, UNU/IIST, P.O.Box 3058, Macau, May 1999. To appear in *Discrete Applied Mathematics*.

[14] ZHOU CHAOCHEN AND M. R. HANSEN. An Adequate First Order Interval Logic. *COMPOS'97*, LNCS 1536, Springer, 1998, pp. 584-608.

[15] ZHOU CHAOCHEN, C. A. R. HOARE AND A. P. RAVN. A Calculus of Durations. *Information Processing Letters*, 40(5), pp. 269-276, 1991.

Detection and Resolution
Methods II

*Feature Interactions in Telecommunications
and Software Systems VII*
D. Amyot and L. Logrippo (Eds.)
IOS Press, 2003

295

Hybrid Solutions to the Feature Interaction Problem

Muffy CALDER[1], Mario KOLBERG[2], Evan MAGILL[2], Dave MARPLES[3],
and Stephan REIFF-MARGANIEC[2]

[1]*University of Glasgow*
Computing Science
17 Lilybank Gardens
Glasgow G12 8RZ, UK
muffy@dcs.gla.ac.uk

[2]*University of Stirling*
Computing Science &
Mathematics
Stirling FK9 4LA, UK
{mko, ehm, srm}@cs.stir.ac.uk

[3]*Global Inventures Inc*
Bishop Ranch 2
2694 Bishop Drive, Suite 27
San Ramon, CA, 94583
dmarples@iee.org

Abstract. In this paper we assume a competitive marketplace where the features are developed by different enterprises, which cannot or will not exchange information. We present a classification of feature interaction in this setting and introduce an online technique which serves as a basis for the two novel *hybrid* approaches presented. These approaches are hybrid as they are neither strictly off-line nor on-line, but combine aspects of both. The two approaches address different kinds of interactions, and together they provide a complete solution by addressing interaction detection and resolution. We illustrate the techniques within the communication networks domain.

1 Introduction

The provision of communications and multi-media services in software is a major growth industry. Increasingly end-users and vendors find software solutions by combining software components, which may or may not be distributed across a network. Typically, such components are added incrementally, at various stages in the lifecycle. When deployed, each component functions well on its own. In an open market components will be provided by different vendors who will have limited ability to ensure that their products are compatible with those of other vendors. Indeed experience suggests that even within one enterprise new components may not be compatible with their existing products [4]. In call control systems components enhancing a basic service are referred to as *features*. Incompatibility between features is referred to as *feature interaction*.

It is important to note that an interaction between two features in call processing systems corresponds to a single *scenario*: which party is subscribing to which particular feature and who is calling whom. Consequently, a *single* pair of features may be involved in a number of *different* call scenarios which cause feature interactions.

As services proliferate with increasing market de-regulation, so too do the problems of feature interactions, particularly in the presence of legacy code and numerous third party vendors. In fact, feature interaction may jeopardise the goal of an open service provision within the telecommunications market and indeed any (distributed) software system employing a variety of software components from a variety of sources.

Within the context of call control software the terms interworking and incompatibility have well understood meanings. Features must *interwork* to share a (communications) resource. The features may interwork *explicitly* through an exchange of information with each other, or *implicitly* through this shared resource. In this second case, the features often have no knowledge that the other exists, however they may be aware of another controlling influence from the behaviour of the resource. Explicit interworking is possible in advanced session control architectures and was first proposed in TINA [14, 17], where interworking was both controlled and expected. Implicit interworking is common in traditional telephony and occurs simply because features attempt to share a resource.

When features interwork to share communication resources, they are *compatible* if the joint behaviour of the resource is acceptable. Typically this requires that the behaviour of the resource does not break the expectations of any controlling feature. Compatibility does *not* refer to simple coding errors, nor to the adherence of interfaces or protocols, but to the adequate behaviour of a resource under the joint control of interworking features.

A substantial body of work [11, 13, 5] exists on dealing with feature interactions. The work embraces both off-line and on-line approaches. Briefly, off-line approaches are those that are applicable at design-time whereas on-line approaches are applicable at run-time. The former being most useful at the early stages of the software lifecycle, the latter during testing and deployment [4]. The literature differentiates between the *detection* of interactions, and the *resolution* of interactions. Traditionally, there has been scarce interest in *avoidance* of interactions because of the enormous architectural changes required.

Off-line approaches are often based on the application of formal methods, and as such require considerable information of each individual software increment. Increasingly, as the market becomes more competitive, this information may be difficult to obtain. Also, as the number of features increases, the work in analysing pair-wise interactions increases with the square of the number of features. With a large number of features in an open market, this will quickly become untenable.

In contrast, on-line approaches simply carry out checks as required. Clearly there are computing resource issues, but the major issue is the paucity of feature information available to on-line approaches. This leads to an ability to *detect* interactions, but a poor ability to *resolve*. Note that off-line approaches at the early stages of development can resolve interactions by indicating how to re-design the features.

Strangely, little attention has been given to combining these approaches. Yet a *hybrid* approach offers the potential to develop a scalable on-line approach, informed by sufficient off-line information to allow not just detection, but also resolution. We suspect this lack of interest to be caused in part by the different academic background required by the two approaches. Another serious problem for hybrid approaches is the completely different representations of information employed in on-line and off-line approaches.

This paper describes two approaches that each combine aspects of on-line and off-line behaviour. The concept of such a hybrid approach was first suggested in [6]. Here, we report on results of a thorough investigation of those ideas, including implementation and experimentation. Preliminary results have been published throughout the investigation [15, 8, 7, 21]. In particular, we have tried to gain experience and exploit the advantages of both: a high level of abstraction in the approach presented in Section 4, and a lower level abstraction in the approach presented in Section 5. Both approaches are based on the same on-line technique, which is discussed in Section 3. The two approaches address different classes of interactions, and neither provides a complete solution. However, if the approaches are combined they provide a complete solution, as shown by the experimental evaluation (see section 6). The next section contains a discussion of four classes of interaction that occur.

2 Types of Interaction

We can identify four types of interaction based upon a model where an event triggers the system to perform an action. This action may itself trigger the system to perform further actions in a sequence of 'micro-steps'. When there are no further actions pending the system is said to have returned to a relaxed state. Systems of this type were described by Blom [2].

2.1 Shared Trigger Interactions

Shared trigger interactions (STIs) occur when more than one feature wishes to respond to a stimulus. It is possible to detect this type of interaction automatically simply by examining the set of responses from the features to a stimulus. If more than one feature responds, then an STI has occurred.

The "legendary" example of Call Waiting and Call Forwarding Busy is a typical case of an STI: both features are triggered by an incoming call to a busy line. This is the most dangerous type of feature interaction with respect to the stability of the system. There is usually very little protection built into the switch software against the effects that this type of feature interaction can cause and STIs are always considered to be detrimental. Typically an STI could result in a terminal being told to go into the busy state and the ringing out state at the same time; which one the terminal actually reaches being determined by the order in which the messages arrive, that is STIs usually appear as race conditions. It is not uncommon for this to happen in switching software and so it is extraordinarily important that STIs are prevented. Sections 3 and 5 discuss automatic detection and resolution of this type of interaction.

2.2 Sequential Action Interactions

Sequential action interactions (SAIs) occur when the responses generated by one feature cause another feature to be triggered. Thus, although all features do not respond to the same stimulus, from the perspective of the user all of the results occur as a direct result of their action. It is difficult for users to be able to distinguish between SAIs and STIs since the micro-steps inherent in an SAI are hidden from the user.

Call Forwarding Unconditional (CFU) causing Do Not Disturb (DND) to be triggered on a terminal would be a typical example of an SAI. Although SAIs may cause unexpected system operation, they do not usually lead to damage or errant behavior of the system overall because each of the individual features is operating in its 'standard' environment.

SAIs lead to new emergent behaviors of the system which may be desirable or undesirable depending upon the individual use case being considered. Two examples illustrate this point. The interaction between Call Forward Unconditional (CFU) and Call Waiting (CW), where a party that is forwarded to another terminal will receive CW treatment at the second terminal, is probably a desirable new behavior. The interaction between Hot Line (HL) and Terminating Call Screening (TCS), where a HL call from A to B will never be completed if A is on B's TCS list is probably not desirable.

Because some SAIs are beneficial, as in the first case above, it is not sufficient to say that when multiple features respond in the process of handling an original event then this is an interaction. Section 4 introduces an automatic qualification of SAIs to solve this problem.

2.3   Looping Interactions

Looping Interactions (LIs) occur when multiple features, or even multiple instances of a single feature, operate in concert to cause a cyclic sequence of events to occur. LIs are a special case of SAIs. This cycle can be detected since the call never reverts back to a relaxed state. The loop can be detected by simply detecting an event/response chain that is longer than would naturally occur in normal call processing. This is a slightly imperfect solution since it always allows the possibility that a perfectly legal, but long, sequence of events would lead to a false detection as a looping interaction.

2.4   Missed Trigger Interactions

Interactions where the presence of one feature in the system prevents the second feature from operating at all are termed Missed Trigger Interactions (MTIs). Although these interactions cannot be detected during runtime it should be possible, with an extension of the approach presented by Wakahara[26] and Kuisch et al.[18] to apply extended feature specification principles to provide additional information about the trigger and return points for the feature.

3   An On-line Transactional Approach

3.1   On-line Technique

An approach extending the earlier work of Homayoon and Singh[12], Schessel[24] and Cain[3] was developed in which features are represented as finite state automata which only perform state transitions in response to external events. In the absence of these external events they perform no action of their own. Even timers are treated as being external stimuli. Call processing events are offered to all features to see what action they would perform if they were allowed to progress. This is in the spirit of these earlier techniques, but allows application in environments where prior information about the features is not available.

This is an interesting approach but, unfortunately, once a feature has been triggered, the event has been consumed and the internal state of the feature has been updated. It would appear that explicit code must be added into every feature to introduce some sort of 'proposed' phase where its responses have been sent back to the system but they have not been accepted for commitment to the hardware. This would require extensive changes to the design of both current and future features.

An alternative to this significant modification to existing architectures is presented by allowing features to be rolled back using the UNIX *fork* mechanism. This allows for an exact copy of the operational feature to be created, with all of the process and state information of the parent. The triggering event can then be passed to the copy. There are now two copies of the feature in memory; an unmodified copy representing the state of the feature before receiving the event, and a modified copy which *has* received the event and responded.

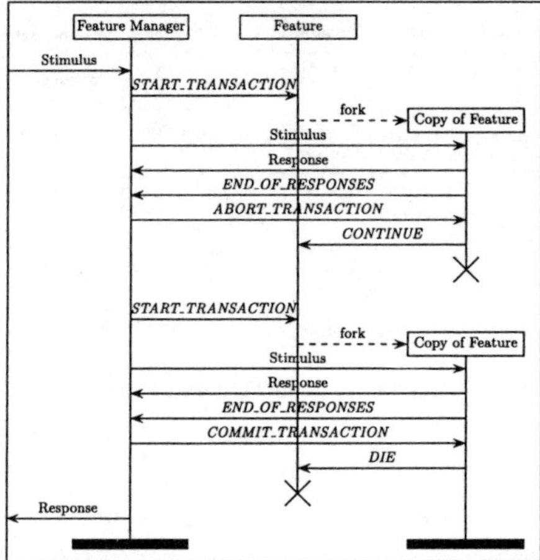

Figure 1: Feature Manipulation using Transactional messages

A feature manager which has performed this splitting process now has a choice: it can commit the responses to the event from the forked feature back to the hardware or it can throw the responses away. If it chooses the former the forked copy of the feature is allowed to continue and the unforked copy is deleted, if the latter then the forked copy is deleted and the feature is returned to its initial state. This is shown in Figure 1.

By using this approach, it is possible to effect rollback control over a feature [27]. Further, since this control is performed at the operating system level, no modification to the feature is required to allow its use. The approach shares similarities with software fault tolerance solutions [20] and error recovery techniques such as documented by Shin [25] and Campbell and Randell [10].

To implement this approach a feature is nested inside a feature controller. The feature is a conventional non-transactional state machine implementation providing call processing functionality and the feature controller is responsible for cocooning this to make it work correctly with the feature manager. We also refer to the controller as a cocoon. This is shown in Figure 2. The use of an all-enveloping cocoon makes it very straightforward to incorporate legacy code into the system since all of the functionality that the feature manager requires can be incorporated into the cocoon itself. Indeed, when using an operating system which supports shared libraries the functionality that the feature manager expects from the feature can be upgraded without even needing to re-compile the feature. This provides a very powerful

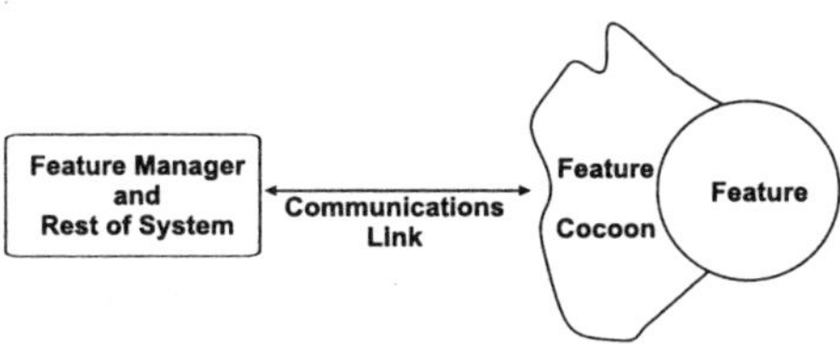

Figure 2: Arrangement of Feature and Cocooning Environment

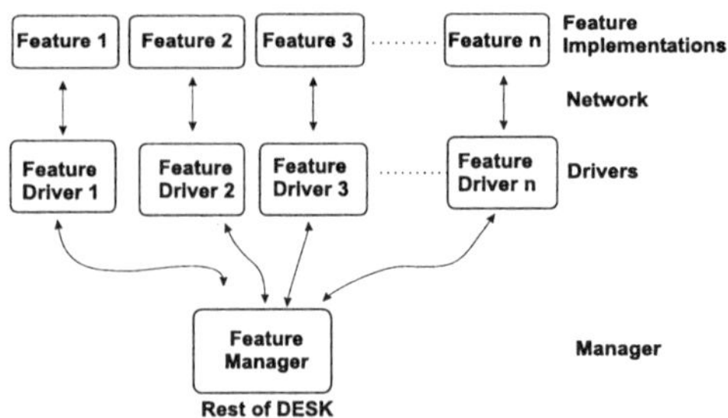

Figure 3: Relationship between the Feature Manager and the Features

upgrade capability.

This rollback technique has been coupled with a conventional feature manager to develop an approach which is capable of using the proposed responses to events to decide which features to allow to progress. The approach presents the stimuli to all of the features in parallel and uses the proposed responses back from the features to detect interaction between them. This can be done during runtime and no prior knowledge about the features is required.

3.2 System Operation

From the point of view of the algorithms that are employed, the structure of the system is as shown in Figure 3. The system consists of the switching and control system which provides stimuli to the feature manager. The feature manager is responsible for distributing these stimuli out to the features via the feature drivers and collecting the responses back again. It is within the feature manager that the feature interaction detection and resolution techniques are implemented. The feature manager has the capability to intercept, block and delay messages as required.

The experimental system is build on top of a switching platform created for the purpose of this work, known as DESK. Initially the system, comprising of the switch (DESK), the feature manager and the features, is in a stable state with no events or responses pending. An event then arrives from the switch, maybe caused by a user having taken one of the terminals off hook. The event is passed to all of the features in the system in parallel. The system has now moved into an unstable state as there are events and/or actions outstanding. All of the features respond to the event with the actions that they wish to perform. These actions are then broadcast back to the features and again they reply with further actions. This process continues until no more actions are sent back from the features and the system returns to a stable state or, alternatively, a loop is detected. In general, we refer to this process of sending out events to the features and capturing their responses as *farming* and in many ways it is analogous to Blom's[2] collection of *micro-steps* forming a *macro-step*. It also shares aspects in common with operating system recovery points as discussed by Rossi and Simone[23].

If in one of these micro-steps more than one feature responds to a stimulus a shared trigger interaction has occurred and requires resolution. If there is no pre-stored resolution available for this interaction the feature manager can try a number of different interaction resolutions to create a graph of all of the possible posterior system states. To achieve this the feature manager rolls all features back to the point just before the arrival of the interaction-causing event before attempting to trigger the responding features in any order (and also only some of those features). This will result in a graph showing all possible behaviours of the system.

Note that this approach is different to other feature manager based techniques in that *the interaction is detected based on the result of passing the triggering event to the features, rather than just on the fact that potentially contesting features are present in the system.* Manual (or in future automatic) study of this graph allows determination of the correct resolution for this interaction, which can then be selected, allowing the system to continue processing. More details about the design and operation of this system are available in [19].

3.3   Strengths and Weaknesses

This work has enabled the differentiation between, and automatic detection of STIs, SAIs and LIs at runtime. It does not require any modification to existing features and the processing overhead for using it is relatively small. Unfortunately, at the point at which this first phase of work was completed, only manual resolution techniques by means of table population were in operation and so further work was required to totally automate the process. Further, SAIs are simply detected by one feature triggering another, which can lead to both desired and undesired interactions. A qualification of SAIs provides a significant improvement.

Once a stable system state has been achieved the messages representing requests for changes to the state of the hardware are committed, the features themselves are committed to the hardware and the system relaxes to await the arrival of the next event.

3.4   Augmenting the On-line Approach with Off-line techniques: Two Hybrid Approaches

As discussed previously, the existing transactional approach is based upon the stimuli and responses of software components. These events are rather "low level", as one would expect for an on-line approach dealing with live implementations. In communication networks parlance these events can take the form of *off-hook* or *on-hook*, *start-ringing* and the like. Hence the detection mechanism is said to be based at the *technical* level.

In contrast, many of the off-line approaches adopt a *user-centric* view. This is often also a *connection-orientated* view. For example, user A is connected to user B, who forwards the call to user C. A user-centric view is natural as it is the *user* of the network that is affected by feature interaction, and it is the user who judges the efficacy of the network. It also allows a natural high level of abstraction that aids the application of formal methods. Abstractions including network details are prone to complexity issues. Hence for off-line techniques it can be appealing to work at a user-centric level.

We have developed two hybrid approaches, one is technical and the other user-centric. While both approaches originated in the off-line domain, the level of abstraction is very different. That is, the more technical approach (based on [7, 21]) is low-level and requires additional technical-level modelling, whereas the user-centric approach (based on [15]) is more abstract. While there is an overlap between the approaches, essentially the technical hybrid approach is applied to the problems associated with the resolution of Shared Trigger Interactions. The user-centric approach helps primarily with the qualification of Sequential Action Interactions.

As both approaches operate at run-time, they consider only features which are currently active in a call. Other features, which may be implemented, but are not activated or are not subscribed to by the parties involved in a call, are ignored. This helps to cut down the processing overhead incurred by the approaches. Additionally, the two approaches are able to detect interactions between features which have been subscribed to by different parties and consequently may be triggered at different ends of a call.

Furthermore, both approaches treat features as black boxes and assume no detailed knowledge about internal behaviour. This has a positive impact on the applicability of both approaches. Firstly, they can be applied in a live network, not just a captive run-time environment. Secondly, the approaches can be used in a competitive business environment where no detailed technical feature information will be available. And thirdly, the approaches are not limited to a particular network architecture. While in this paper the applicability of the approaches is demonstrated in an IN-like environment, it is believed that the approaches work regardless of whether the features are deployed on an Intelligent Network, on Parlay, or even in a Voice over IP environment using communication protocols such as SIP (Session Initiation Protocol) in combination with e.g. RTP (Real Time Protocol).

The next two sections consider each of the two approaches in detail. For comparison of detection and resolution capabilities, and efficacy with respect to both STIs and SAIs, we choose a single experimental set of features.

4 The User-centric Hybrid Approach

In this approach the selected off-line approach was originally developed as a filtering technique [15], although its detection of problematic scenarios is quite accurate. It has capabilities for detecting interactions which involve features subscribed to by different users; making the approach suitable for the detection of SAIs as these often involve features of multiple users.

The off-line approach is user-centric. In other words, it should detect that certain combinations of features change the behaviour as perceived by a user. This high level view was chosen in order to allow for simple algorithms with good run-time performance. Abstraction to the user level is acceptable, because empirical evidence suggests that all low level problems also have a manifestation at a higher, user perceived level.

The next section provides an overview of the approach and how the necessary information about features is provided. Section 4.2 discusses the interaction analysis in more detail.

4.1 Overview

The approach assumes call control features to be extensions of the Basic Call Service. Feature interactions are problems between different extensions to the Basic Call Service.

Treatments are an important aspect of this approach. Treatments are announcements or tones triggered by the network to handle certain conditions during a call, for example when a call is screened or blocked. Potentially, there may be multiple treatments involved in a call. For example, consider two features invoked during one call. One feature might connect a party to a busy treatment, whereas the other feature connects the (same or the) other party to a network unavailable treatment.

The behaviour of a feature is described in two parts: the triggering party and a connection type. The latter consists of two parts: the original connection to be set up before the feature is activated and the connection set up after the feature has been triggered.

An example should illustrate this. Call Forwarding Unconditional (CFU), which redirects all incoming calls to a predefined third user, can be described as shown in Fig. 4. Assume party A is the originator, B the terminator, and C the party where the call is redirected to. The behaviour has two parts, separated by a semicolon. In the first part, the notation TP.: X indicates that X is the triggering party, in this case it is B because CFU is triggered at the terminating end of a call. In the second part, notation $(X, Y) \to (U, V)$ indicates the connection type. (X, Y) is called the original connection and (U, V) is the connection after activating the feature. For each pair (A, B) we refer to A as the source and B as the destination.

$$\boxed{\text{TP.: B; } (A, B) \to (A, C)}$$

Figure 4: Description of Call Forwarding Unconditional

The call starts with A attempting to connect to B. However, because of CFU, A is connected to C instead. So the connection type is $(A, B) \to (A, C)$.

4.2   Interaction Analysis

Interaction cases are found by analysing pairs of features. Two feature descriptions are compared according to six rules. If a feature pair fulfills any of the six rules, then the pair is said to interact. We consider each of the rules in turn. Note that while the analysis is pairwise, features are still described in isolation.

Rule 4.1. *Single User - Dual Feature Control*

If both features have the same triggering party and either the original connections or the resulting connections are identical, the pair interacts. Note that, even if the features aim at setting up the same connection, the features may clash as they may be triggered simultaneously. An example is given in Example 4.1. The shaded portions indicate the identical parts of the two behaviour descriptions.

Example 4.1.
$$\boxed{\begin{array}{l} \textbf{CW: } \textit{TP.: B; } (A, B) \to (A, B) \\ \textbf{CFU: } \textit{TP.: B; } (A, B) \to (A, C) \end{array}}$$

Rule 4.2. *Connection Looping*

The original connection of one feature is identical to the connection after activating the second feature, and the original connection of the second feature is identical to the connection set up after triggering the first feature. As both features are trying to divert the call, a loop occurs (ref. Example 4.2). Again, the shaded portions indicate the identical connections.

Example 4.2.
$$\boxed{\begin{array}{l} \textbf{CFB: } \textit{TP.: B; } (A, B) \to (A, C) \\ \textbf{CFU: } \textit{TP.: C; } (A, C) \to (A, B) \end{array}}$$

Rule 4.3. *Diversion and Treatment*

Two features interact if one feature establishes a call (*not* to a treatment) and the resulting connection is the original one of a feature, which redirects the call to a treatment. The triggering parties of the features must be distinct. This rule captures the fact that the connection which one feature is trying to establish is prevented by the other (ref. Example 4.3).

<table>
<tr><td rowspan="2" align="right">Example 4.3.</td><td>CFB: TP.: C; (A, C) → (A, B)</td></tr>
<tr><td>OCS: TP.: A; (A, B) → (A, Treat)</td></tr>
</table>

Rule 4.4. *Reversing and Treatment*

This rule is very similar to the previous one, but differs as follows: One feature reverses the call and the resulting connection is equal to the original connection of the other feature which connects to a treatment. In contrast to the previous rule, the same users need to be involved in both original connections, although their roles (source, destination) may be swapped. Example 4.4 illustrates this case.

<table>
<tr><td rowspan="2" align="right">Example 4.4.</td><td>AR: TP.: B; (A, B) → (B, A)</td></tr>
<tr><td>OCS: TP.: B; (B, A) → (B, Treat)</td></tr>
</table>

Rule 4.5. *Diversion and Reversing*

One feature forwards a call which is subsequently reversed to the originator by the other feature. More precisely, one feature forwards the call (not to a treatment) and the resulting connection is the original one of the other feature which reverses the call. A common manifestation of the problem is the originator of the call is called by a party that was never contacted (see Example 4.5).

<table>
<tr><td rowspan="2" align="right">Example 4.5.</td><td>CFB: TP.: C; (A, C) → (A, B)</td></tr>
<tr><td>AR: TP.: B; (A, B) → (B, A)</td></tr>
</table>

Rule 4.6. *Reversing and Diversion*

This rule is closely related to rule 4.5, but the forwarding happens after reversing the call. Consequently, one feature reverses the call and the resulting connection is the original one of the other feature, which then forwards the call to a third party. Example 4.6 illustrates this rule. So, the "new" connection of the first feature is prevented by the second feature.

<table>
<tr><td rowspan="2" align="right">Example 4.6.</td><td>AR: TP.: B; (A, B) → (B, A)</td></tr>
<tr><td>CFB: TP.: A; (B, A) → (B, C)</td></tr>
</table>

4.3 Integration into the On-line Environment

As discussed in section 3 the call control software, i.e. the basic call and all features, are encapsulated by cocoons. Signals destined for a feature are therefore passed through the cocoon to the feature proper, whereas control signals destined for the cocoon, are not. These are processed inside the cocoon.

The underlying on-line detection approach detects sequential action interactions because the feature manager monitors the replies sent by the features to an initial event. If two or more features (other than the Basic Call) reply to an event, or to feature responses which have been fed back to features, then an interaction is detected. However, as described in Section 3 this is a very crude detection mechanism. To augment the approach, the feature descriptions, i.e. the connection type and the triggering party, are added to the cocoon. The descriptions are queried using special control messages. The features themselves are unaffected. Recall that the feature manager detects interactions by corresponding with the features but does not commit feature responses to the hardware before the feature interaction check has been carried out.

The addition of the user-centric hybrid approach is a qualification of SAIs. To qualify a detected SAI, i.e. to ascertain if this is a desirable or undesirable interaction, the feature manager prompts both features for their behaviour descriptions. On reception the analysis discussed in Section 4.2 is applied. The feature manager will log the encountered scenario and terminate the call.

For new features, the cocoon is provided on feature deployment. For legacy software the cocoon is provided at the time of introduction of the on-line approach. Hence it is possible to provide the feature descriptions on a *per feature basis*, at feature deployment, assuming the feature provider makes available such high-level information.

5   The Technical Hybrid Approach

We turn our attention to the second hybrid approach, based on a technical level off-line technique. This hybrid approach augments the on-line approach presented in section 3 with the ability to automatically select good (if not best) resolutions.

The advantage of this approach is again an independence from the detailed implementation of a feature, but rather than considering a high level user's view, the feature providers' understanding of contradictory messages is taken into account. This requires a common semantics of messages across different developers and network operators or at least a common classification of messages according to treatment type (e.g. forwarding feature). Such information would be expected, in any event, for minimal interworking.

5.1   Overview

The user centric approach assumes that features extend a basic call model. We make the same assumption here, but do not distinguish between interactions between features and between features and the basic call. The on-line feature manager allows to explore interacting features in different orders and combinations. Here we make precise the concepts of solutions and resolutions as basis for the technical hybrid approach.

For a given set of features, a *solution* is a trace of one or more of those features running concurrently. That is, it is an interleaving of messages generated by a subset of the features. The *solution space* is the set of all traces (i.e. all solutions), for all subsets of a given set of features. It should be noted that the solution space might contain many traces that lead to a violation of required properties (i.e. there might be traces that represent incorrect behaviour). A *resolution* is a trace in the solution space that does not violate any specified properties.

The solution space depends on the granularity of the interleaving. A coarse grained interleaving allows features to be run in any order but does not allow messages of features to be interleaved, (within one macro-step). In this case the solution tree grows exponentially with the number of interacting features. A fine grained interleaving allows individual messages to be interleaved, thus resulting in a solution space growing exponentially in both the number of features and the length of their responses. Note that only features that are actually triggered contribute to the solution space. A fine grained interleaving has been explored in [21, Chapter 5]. Here we assume a coarse grained interleaving, as supported by DESK.

5.2　Resolutions

The main contribution of this hybrid approach is the ability to *resolve* interactions. After the feature manager has constructed the solution space as described in section 3, the space is passed through a pruning and extraction process: that is all branches that lead to undesired behaviour are removed and only the one with the most preferable behaviour is retained.

The operations of pruning and extraction are based on rules determined (off-line) by domain experts. The purpose of the rules is to discriminate between bad and good solutions and also to describe the quality of resolutions.

The offline analysis by domain experts is conducted on the solution space gained by running the feature manager without the extraction and pruning processes. The result of the analysis is a set of rules which are used to prune the branch of the solution space: our ultimate goal is one single branch. The aim of these rules is to describe undesired behaviour in a domain specific, but scenario independent way. An example of such a rule is that two consecutive tones are undesirable. Rules which depend on domain knowledge and the underlying architecture are classified as *message dependent*. We have also identified *message independent* pruning rules. These are more general, as they do not refer to domain specific information. The rules acquired in the offline phase are supplied to the feature manager in order to automatically resolve detected interactions by pruning the solution space. The new solution space can be analysed again to obtain additional resolution rules, and the process iterates, as required.

5.3　Message Independent Rules – Extraction

Message independent rules are used for *extraction* of the best resolution. They are characterised by not requiring any information about the semantics of the messages. Thus, they can be applied to a solution space without any knowledge of the messages occurring therein, assuming that we are able to differentiate between messages from different features (i.e. we know which messages originated from the same feature).

Rule 5.1. *Duplicate subtrees sharing the same parent can be removed.*

The solution space may contain several duplicate branches. Consider an example: assume 3 features (f_1, f_2, f_3) two of which (f_1 and f_2) respond to the same trigger. A branch corresponding to this trigger with all three features active is identical to one with feature f_3 disabled. Furthermore any branch with two features active, one of which is f_3, is identical to the branch with just the other feature active.

Obviously a duplicate branch will not provide a new solution, so all duplicates can safely be removed. This rule is not required explicitly in the DESK environment, as the feature manager implemented there excludes features that do not respond from alternative orderings.

Rule 5.2. *Traces containing messages from the largest possible number of features are preferable (to those containing messages from a small no of features).*

A feature is *satisfied* by a trace when its intended behaviour is exhibited. So, if a feature f_1 responds with messages a_1, a_2 and a_3 in that order, every trace in the tree containing these messages in *that order* satisfies feature f_1. The construction of the solution space maintains the relative order of messages as intended by a feature. However, the messages can be interleaved arbitrarily with responses from other features. If feature f_2 responds with b_1 and

b_2 then some possible traces satisfying both f_1 and f_2 are $a_1a_2a_3b_1b_2$ and $a_1b_1b_2a_2a_3$, but $a_1b_2b_1a_3a_2$ is not acceptable.

While the fine grained interleaving presented in this section is the more general case and not applicable in DESK, the rule is still relevant: the main motivation for the approach is to satisfy as many features as possible.

Rule 5.3. *Traces satisfying features with the highest priority are preferable.*

A simple, but effective method of extraction is prioritising features. We do not possess information about features' identities per se, though we do know their relative position in the network. For example, we know that a message has been received from feature f_i, but we do not know the identity of feature f_i (i.e. whether f_i is *call waiting, three way calling* or any other feature). We refer to i as the feature's *connection number*.

A simple precedence scheme allows features with a low connection number to have higher priority. Clearly this scheme could be extended to a system of priorities in which each feature has an associated weight (features with the highest weight are preferable). Each trace has a weight relative to the weights of the features satisfied by that trace. The trace with the highest weight would be preferred. One could define any number of other priority schemes. This provides a mechanism for user preferences.

Rule 5.4. *If there are a number of "best" resolutions, choose one.*

A "best" resolution is not necessarily unique. Suppose that after applying all rules a tree with more than one trace remains. At this point we have found more than one resolution that we would classify as the best, but obviously the system can only commit to one trace. However, if all traces represent behaviour that from a qualitative point of view is indistinguishable, we can simply choose one.

Note that it would be preferable from the user's point of view if this is a deterministic choice. The user is not (and shall not be required to be) aware of the resolution process, so the presented behaviour should be consistent across a number of separate calls.

5.4 Message Dependent Rules – Pruning

Message dependent rules can be seen to be more powerful, but they also require more information. In order for message dependent rules to be useful, a semantic understanding of messages is required. This allows one to develop relations on messages, such as classes of treatments, or billing messages. This understanding enables us to build grammars describing good or bad behaviour. For example "a treatment following an onhook message is not useful unless there was an offhook in between". Message dependent rules are used for *pruning* the solution space.

While it might seem unreasonable to require a semantics of messages, especially in a setting where the internal behaviour of the features is unknown, a requirement for some knowledge about messages is practical because the message set in telephone switching systems is somewhat restricted and such understanding is required for interworking even in the absence of feature interaction.

Messages can be grouped into classes. These include the rather obvious classes such as "billing messages", "user messages" and "system messages". In addition, and more interesting, we can have classes like "announcements", "treatments", "tones", "hookevents". Classes of messages can be overlapping, for example "announcements" is a subclass of "treatments" and "tones" intersects with "treatments" (*ringtone* is a tone, but not a treatment, whereas *busytone* is both a tone and a treatment and *announce* is a treatment but not a tone).

We wish to express rules that describe sequences of messages and also the absence of a message. Often we wish to refer to whole classes in a simple way. Regular expressions are used to express these rules. Empirical evidence shows that regular expressions are sufficient to express any rule required. (We have found no evidence, yet, of the need for context free grammars.) This is not surprising, as messages generally involve call setup, manipulation and tear down. These do not usually require the user to count the occurrence of events, e.g. for an incoming call the third ring denotes a different meaning, but note, such behaviour is possible in the future.

In our case study based on DESK we find that only the following five message dependent rules (5.5.x) are required: (1) connecting a user to two different resources, (2) routing a call to two different locations, (3) routing a call away from a user and still changing that user's state, (4) routing a call away from a user and connecting the user to a resource, and (5) changing a user's state and connecting the same user to a resource. The actions described in these rules map directly onto the message set used by DESK: routing is performed by a *routing* message, a state change is caused by *move to state* and *connect to resource* connects the user to a resource.

A set of resolution rules is considered complete if all interactions can be resolved. However, there is no generic method of identifying a complete rule set (due to the diverse nature of features). This is a drawback, as an incomplete rule set can lead to a trace being identified as a resolution when the trace is not a resolution.

5.5   Integration into the On-Line Environment

The feature manager in the online approach described in section 3 produces a solution space which is passed to a (human) operator for resolution. Here, we pass the solution space to our automated pruning and extraction process.

The pruning and extraction process is essentially the application of the rules defined previously. Pruning is a function that operates on a solution space and removes all undesired traces by checking each trace against the message dependent rules. Thus pruning returns a solution space containing no traces with undesired properties. Extraction operates on a pruned solution space and removes traces according to rules 5.1 to 5.4. The main goal of the extraction process is to identify and extract the most desirable resolution from the possible desirable resolutions. In the run-time environment the two functions are simply added instead of the manual resolution process.

A major drawback of this approach is the complexity: a large number of branches must be constructed that will be discarded. We therefore have also implemented an "on-the-fly" approach. On-the-fly resolution works by trying to apply pruning to the current solution under construction. As soon as a behaviour is identified as undesired, the construction of the current solution is aborted. The branch constructed since the last choice is then removed and construction of another solution is attempted. This will greatly reduce the complexity if many features are active: bad solutions can be identified early in the construction. Extractions, **cannot** be performed "on-the-fly" as there is not enough information about the solution space available before the construction is completed. Obviously, the on-the-fly mechanism requires

	CFU	CW	CFB	OCS	TCS	VMS	AR	ACB	DND	HL
CFU		5.x	5.x, 4.2	4.3	5.x, 4.3	5.x, 4.3	5.x, 4.5, 4.6	5.x	5.x, 4.3	
CW			5.x		5.x	5.x	5.x	5.x	5.x	
CFB				4.3	5.x, 4.3	5.x, 4.3	5.x, 4.5, 4.6	5.x	5.x, 4.3	
OCS							4.4			4.1
TCS							5.x, 4.4	5.x	5.x	4.3
VMS							5.x, 4.4	5.x	5.x	4.3
AR								5.x	5.x, 4.4	5.x
ACB									5.x	
DND										4.3
HL										

Figure 5: Results of the case study

a deeper embedding into the feature manager, that is the detection process and resolution process are entwined rather than staged, so changes to either become more complicated.

6 Experimentation and Evaluation

The two hybrid approaches were applied to the same case study, for evaluation and comparison. In this section we give an overview of the case study and results.

For the case study ten common features [9, 16] were selected covering a wide spectrum of functionality. The selected features are: Call Forwarding Unconditional, Call Waiting, Call Forwarding on Busy, Originating Call Screening, Terminating Call Screening, Voice Mail System, Automatic Ringback (returns the call to the caller), Automated Callback (allows a caller to automatically recall a busy callee), Do Not Disturb and Hotline.

The case study was performed using a prototype based on the DESK switching system in conjunction with the two off-line techniques presented in Sections 4 and 5.

6.1 Results and Discussion

Between the ten selected features 49 interaction scenarios have been found. Note that a number of feature pairs exhibit interactions in several different scenarios, so the number of pairs of features involved is somewhat smaller. Fig. 5 provides details of the detected interactions.

Entries are of the form $A.B$. A indicates which approach has been used to detect the interaction and refers to the section number in this paper. Hence *4* refers to the detection of a Sequential Action Interaction using the user-centric approach presented in Section 4, and *5* refers to a Shared Trigger Interaction using the technical approach discussed in Section 5. B refers to a particular rule discussed in the paper. As the technical approach requires the application of multiple rules to resolve an interaction, the notation *5.x* has been chosen.

Of the 49 detected interactions, 28 are Shared Trigger Interactions. All of these could be *resolved* at run-time and the call progressed. With the remaining interactions, of Sequential Action type, the problem was detected and the call terminated.

As can be seen, there is only one case where an interaction was detected using rule 4.1. This is due to the fact that rule 4.1 detects interactions which are triggered by the same party. Very often these are in fact STIs. However, in this project, the technical hybrid approach was used instead to deal with STIs; this approach is preferable, as resolution allows for the call to continue. As a consequence, rule 4.1 is not heavily employed.

A more detailed study of the results of the case study shows that no scenarios which have not been highlighted by any approach contain feature interactions. In other words, no *false negatives* have been found in the case study. This point is fundamental for an interaction approach if its results are to be trusted and hence be useful. On the other hand, all scenarios which are highlighted, actually exhibit problematic behaviour. However, sometimes this is subjective to individual user expectations.

An example is the interaction between Call Forwarding and Voice Mail where a forwarded call is redirected to a voice mail treatment. Some users might see the Call Forwarding feature as a feature to increase their reachability, for instance when they are at a different location. While the voice mail feature does not restrict or affect the behaviour of the Call Forwarding directly, some subscribers to CFB might not expect a voice mail to answer their calls. Also the caller might be surprised to get a voice mail announcement from a party they did not try to reach. There are a few interactions detected which depend on user expectations and user preferences. This point is discussed further in Section 7.

For STIs, the best possible resolutions (based on our understanding of the features) have been found for all detected interactions. The patterns describing undesired behaviour as used by the pruning algorithm have been relatively obvious and only very few such patterns were required. Our understanding of the semantics of the existing messages made the formulation of the rules possible. The quality of the resolutions depends on the knowledge of the message semantics. However, even a general understanding of the messages is sufficient. For any system under consideration, it must be assumed that this general understanding exists, as otherwise enhancement is questionable even in the absence of the feature interaction problem.

Looping interactions are a special case of SAIs, and thus are handled by the user centric approach. Missed trigger interactions cannot be detected by the online technique. However could they be detected, both the user-centric and the technical approach can handle these.

7 Summary and Further Work

In this paper, two separate, complimentary offline techniques have been integrated with a run-time approach to form two hybrid approaches.

Both approaches improve the relatively weak detection (because it lacks any qualification) of SAIs and more importantly resolution capabililties of the run-time approach. The run-time approach detects interactions in two ways: either more than one feature responds to an event (Shared Trigger Interaction) or a feature perceives an event only as a consequence of the behaviour of another feature (Sequential Action Interaction). The two hybrid approaches are complimentary, as broadly each addresses one of the two interaction classes.

The first hybrid approach, the user-centric approach, refines the SAI detection mechanism through an analysis of high-level descriptions of the two features involved. The analysis, based on a set of rules, checks the potential consequences of the two "triggered" features and then classifies them as desirable or undesirable. The latter interactions are then excluded.

The technical approach, refines the STI resolution mechanism through the addition of rules reducing the number of potential solutions and then extract a best resolution. The rules which exclude potential solutions are based upon an agreed semantics of messages, whereas the rules to extract a best resolution are independent of message content.

The two hybrid approaches have been implemented and evaluated in a case study involving 10 features. 49 interactions were detected. Of those, 28 were STIs which were detected and resolved at run-time and the call was allowed to progress. The remaining interactions, all SAI, were detected and the call terminated.

For the first time, Shared Trigger Interactions can be automatically detected *and* resolved at run-time. Previously, only an automatic detection had been possible. Further we have implemented a novel on-line automatic detection and qualification of Sequential Action Interactions; previously, this was a manual task. As a consequence, the difference and the precise meaning of these two types of interactions have been clarified.

With the automatic detection and qualification of sequential action interactions some scenarios have been found which could not be categorised into desirable or undesirable interactions. These cases are dependent on the preferences of particular users. Solving these cases is beyond the work reported in this paper. With the increasing acceptance and hopefully deployment of new emerging architectures, such as SIP new opportunities to solve these cases arise. For example, the expression of caller preferences and the application of feature interaction approaches based on negotiation may be of great value. One such attempt, using policies to express user preferences can be found in [22]. Future work should study these technologies and exploit them in new interaction approaches.

A further valuable area of research is the application of feature interaction approaches into the wider and more general domain of component based systems [1].

Acknowledgements

The two hybrid approaches have been developed as a part of a joint EPSRC-funded project between the Universities of Glasgow and Stirling (EPSRC GR/M03429/01 and GR/M03573/02).

The original work by Dave on classes of interaction was the result of work carried out with the aid and funding of the Royal Commission for the Exhibition of 1851. We would also like to acknowledge that the technical hybrid approach was developed during Stephan's previous period of employment, at the University of Glasgow.

References

[1] L. Blair, T. Jones, and S. Reiff-Marganiec. A Feature Manager Approach to the Analysis of Component-Interactions. In B. Jacobs and A. Rensink, editors, *Formal Methods for Open Object-Based Distributed Systems V*. Kluwer Academic Publishers, 2002.

[2] J. Blom. Formalisation of requirements with emphasis on feature interaction detection. In *[11]*, pages 61–77, June 1997.

[3] M. Cain. Managing run-time interactions between call processing features. In *IEEE Communications Magazine*, pages 44–50, February 1992.

[4] M. Calder, M. Kolberg, E. H. Magill, and S. Reiff-Marganiec. Feature interaction: A critical review and considered forecast. *Computer Networks*, 41(1):115–141, Jan 2003.

[5] M. Calder and E. Magill, editors. *Feature Interactions in Telecommunications and Software Systems VI*. IOS Press (Amsterdam), May 2000.

[6] M. Calder, E. Magill, and D. Marples. A hybrid approach to software interworking problems: Managing interactions between legacy and evolving telecommunications software. *IEE Proceedings - Software*, 146(3):167–180, June 1999.

[7] M. Calder, E. Magill, S. Reiff-Marganiec, and V. Thayananthan. Theory and practice of enhancing a legacy software system. In Peter Henderson, editor, *Systems Engineering Business Process Change 2*. Springer Verlag, London, 2001.

[8] M. Calder and S. Reiff. Modelling legacy telecommunications switching systems for interaction analysis. In Peter Henderson, editor, *Systems Engineering Business Process Change*, pages 182–195. Springer Verlag, London, May 2000.

[9] E. J. Cameron, N. Griffeth, Y.-J. Lin, M. E. Nilson, W. Shnure, and H. Velthuijsen. Towards a Feature Interaction Benchmark for IN and Beyond. *IEEE Communications Magazine*, 31(3):64–69, March 1993.

[10] R. H. Campbell and B. Randell. Error recovery in asynchronous systems. *IEEE Trans. Software Engineering*, SE-12(8):811–826, 1986.

[11] P. Dini, R. Boutaba, and L. Logrippo, editors. *Feature Interactions in Telecommunication Networks IV*. IOS Press (Amsterdam), June 1997.

[12] S. Homayoon and H. Singh. Methods of addressing the interactions of intelligent network services with embedded switch services. *IEEE Communications Magazine*, page 42ff, December 1988.

[13] K. Kimbler and L. G. Bouma, editors. *Feature Interactions in Telecommunications and Software Systems V*. IOS Press (Amsterdam), September 1998.

[14] M. Kolberg and E. H. Magill. Service and feature interactions in TINA. In *[13]*, pages 78–84, 1998.

[15] M. Kolberg and E. H. Magill. A pragmatic approach to service interaction filtering between call control services. *Computer Networks: International Journal of Computer and Telecommunications Networking*, 38(5):591–602, 2002.

[16] M. Kolberg, E. H. Magill, D. Marples, and S. Reiff. Second feature interaction contest. In *[5]*, pages 293–310, May 2000.

[17] M. Kolberg, R. O. Sinnott, and E. H. Magill. Engineering of interworking tina-based telecommunications services. *Proceedings of IEEE Telecommunications Information Networking Architecture Conference*, April 1999. IEEE Press.

[18] E. Kuisch, R. Janmaat, H. Mulder, and I. Keesmaat. A practical approach towards service interaction. *IEEE Communications*, pages 24–31, aug 1993.

[19] D. Marples. *Detection and Resolution in of Feature Interactions in Telecommunications Systems at Runtime*. PhD Thesis, Communications Division, Department of Electrical and Electronic Engineering, University of Strathclyde, 2000.

[20] B. Randell. System structure for software fault tolerance. *IEEE Trans. Software Engineering*, SE-1(2):220–232, 1995.

[21] S. Reiff-Marganiec. *Runtime Resolution of Feature Interactions in Evolving Telecommunications Systems*. PhD Thesis, University of Glasgow, Glasgow (UK), 2002.

[22] S. Reiff-Marganiec and K. J. Turner. Use of logic to describe enhanced communications services. In *Formal Techniques for Networked and Distributed Systems – FORTE 2002, LNCS 2529*, pages 130–145, November 2002.

[23] G. P. Rossi and C. Simone. A multitasking operating system with explicit treatment of recovery points. *Microprocessing and Microprogramming 14*, pages 55–66, 1984.

[24] L. Schessel. Administrable feature interaction concept. *ISS'92*, 2:122–126, oct 1992.

[25] K. G. Shin. Evaluation of error recovery blocks used for cooperating processes. *IEEE Trans. Software Engineering*, SE-10(6):692–700, 1994.

[26] Y. Wakahara, M. Fujioka, H. Kikuta, H. Yagi, and S. Sakai. A method for detecting service interactions. *IEEE Communications*, pages 32–37, aug 1993.

[27] K-L. Wu. Rapid transaction-undo recovery using twin-page storage management. *IEEE Trans. Software Engineering*, 19(2):155–164, 1993.

*Feature Interactions in Telecommunications
and Software Systems VII*
D. Amyot and L. Logrippo (Eds.)
IOS Press, 2003

Mechanism for 3-way Feature Interactions Occurrence and a Detection System Based on the Mechanism

Shizuko KAWAUCHI and Tadashi OHTA

Faculty of Engineering, SOKA University

1-236, Tangi-cho, Hachioji-shi 192-8577 Japan

ohta@t.soka.ac.jp

Abstract. This paper proposes a mechanism for 3-way feature interactions occurrence and a system for detecting 3-way interactions based on the proposed mechanism. In this paper, by analyzing examples, the mechanism of 3-way interactions is clarified and a detection algorithm of 3-way interactions is proposed. The problem of terminal assignments for the detection of 3-way interactions is also discussed. To validate the algorithm, we applied the proposed algorithm to 120 combinations out of 12 services, and 82 3-way interactions were detected in the 42 combinations.

1. Introduction

This paper proposes a mechanism for 3-way feature interactions (abbreviated as 3-way interactions) occurrence and a system for detecting 3-way interactions based on the proposed mechanism. A service, which behaves normally, behaves differently when initiated with another service. This undesirable behavior is called a feature interaction.

Recently, as shown in AIN, JAIN [1], Parlay [2], and Active Network [3], telecommunication network architecture is changing to a new one whereby the third-party service providers can provide network services. This architecture enables multiple providers to provide services simultaneously in the same network. Consequently, it is inevitable that feature interactions between different providers' services occur, and this will cause serious problems to service deployment. Since feature interaction has been recognized as one of the possible bottlenecks for the development of software, there is a lot of research being carried out all over the world on detecting and solving the problems of feature interactions [4][5].

Among feature interactions, one, which does not occur between two services but occurs among three services, is called a 3-way interaction. Although the existence of 3-way interactions is predictable, concrete examples were only reported for the first time at FIW00 [6], held in May 2000 in Glasgow, UK. But the mechanism of 3-way interactions was still unknown.

In this paper, by analyzing examples, the mechanism of 3-way interactions is clarified and a detection algorithm of 3-way interactions is proposed. The problem of terminal assignments for the detection of 3-way interactions is also discussed.

To validate the algorithm, we applied the proposed algorithm to 120 combinations out of the following 12 services: Terminating Call Screening (TCS), Reverse Charge (RC), Call

Forwarding Variable (CFV), Call Forwarding Busy Line (CFB), Call Number Delivery (CND), Three Way Call (TWC), Universal Personal Telecommunication (UPT), Automatic Call Back (ACB), Automatic Re-Call (ARC), Third Party Charge (TPC), Call Waiting (CW), and Originating Call Screening (OCS). 82 3-way interactions were detected in the 42 combinations.

In section 2, a definition of a 3-way interaction and an example are shown. In section 3, following and analysis of the example, the mechanism of the 3-way interaction occurrence is proposed. In section 4, to confirm the proposed mechanism, the mechanism is applied to a combination of three services, TCS, CND and CFV. After discussing a detection algorithm for 3-way interactions in section 5, a problem of terminal assignment is discussed in section 6. In section 7, a detection system and the results of the experiment are discussed.

2. 3-way Interactions

2.1 Definition of a 3-way Interaction

Among feature interactions, one which does not occur between two services but occurs among three services is called a 3-way interaction (Figure 1).

There are two types of 3-way interactions: One: There are no feature interactions between 2 services. By providing a third service together with the two services simultaneously, feature interactions occur.

Two: Though there are feature interactions between two services, the interactions do not actually appear because some functionality, which causes interactions, is prevented from being executed because of one reason or another. By providing a third service together with the two services simultaneously, that which prevents the functionality is removed and the interactions appear.

This paper analyses the latter case.

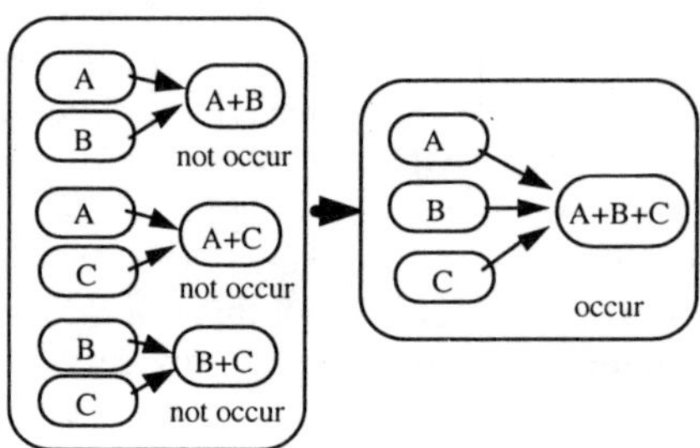

Figure 1: 3-way Interaction

2.2 An Example of a 3-way Interaction

A concrete example of a 3-way interaction among CFV, RC and TCS is introduced.

In CFV service, when a new call terminates to a terminal which has CFV service activated and has registered terminal C as a forwarded terminal (denoted by m-cfv(B,C)), the call is forwarded to terminal C and terminal A and C transit to talk state (denoted by talk(A,C) (Figure 2). In this case, a link between terminal A and B is charged for terminal A (denoted by chg(A,A,B)) and a link between terminal B and C is charged for terminal B (denoted by chg(B,B,C)).

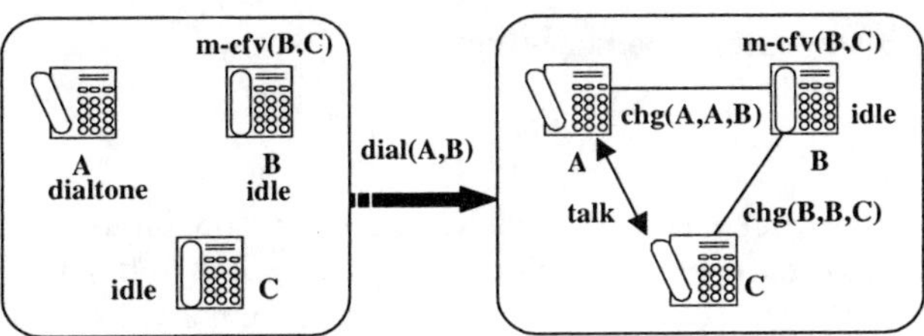

Figure 2: CFV Service

In RC service, when a call terminates to a terminal which has RC service activated (denoted by m-rc(B)), the bill for the call is charged to terminal B instead of terminal A (denoted by chg(B,A,B)) (Figure 3).

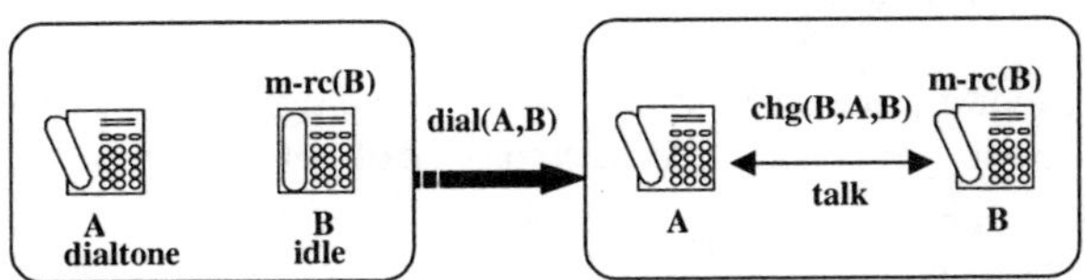

Figure 3: RC Service

In TCS service, when a call terminates to a terminal which has TCS service activated and has registered terminal A as a screened terminal (denoted by m-tcs(B,A)), the call connection between terminal A and B is rejected and, therefore, a link between both terminals should not be charged (Figure 4).

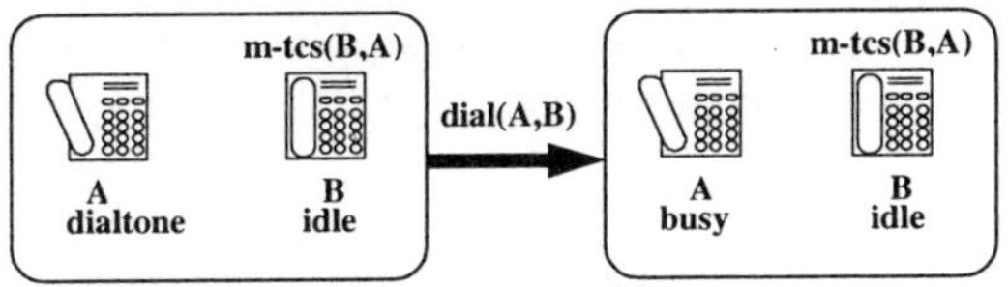

Figure 4: TCS Service

Suppose terminal B has these three services activated simultaneously and has registered terminal C as a forwarded terminal and terminal A as a screened terminal. When terminal A dials terminal B (denoted by dial(A,B)), the call is forwarded to terminal C and bills for a link between terminal A and B and for a link between terminal B and C are charged to terminal B, denoted by chg(B,A,B) and chg(B,B,C), respectively (Figure 5).

This means that though TCS service rejects a call from terminal A to terminal B the bill for a link between terminal A and B is charged to terminal B. This contradicts TCS service specifications. This feature interaction does not occur between two services, TCS and RC, TCS and CFV, and CFV and RC. So, this interaction is a 3-way interaction.

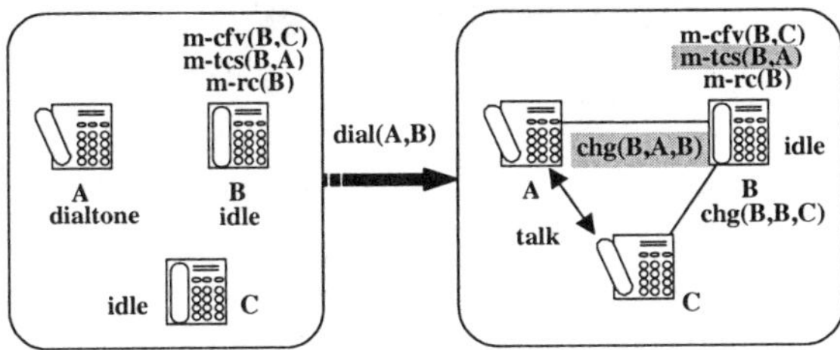

Figure 5: An Example for 3-way Interaction

3. Mechanism of 3-way interaction occurrence

3.1 Analysis

Analysis is done from three points of view: execution condition of services, potential interactions, and occurrence of interactions. Suppose terminal B has TCS service, RC service and CFV service activated and has registered terminal C as a forwarded terminal and terminal A as a screened terminal, respectively.

1) Execution conditions of a service

A condition that allows a service to execute is called an execution condition. Execution conditions for individual services are described as follows:

- An execution condition for TCS service is that a call terminates from terminal A to terminal B.
- An execution condition for CFV service is that a call terminates to terminal B.
- An execution condition for RC service is that a call connection between terminal A and terminal B is completed.

2) Specifications of services

From the interaction view point, the essence of service specifications is described as follows:

- TCS service rejects a call connection between terminal A and B, and therefore, a link between terminal A and B should not be charged.
- CFV service makes a call connection between terminal A and B and a link between terminal A and B is charged to terminal A.
- RC service charges a bill to terminal B instead of to terminal A when a call connection between terminal A and B is completed.

3) Potential interactions

A contradiction between specifications of two services is called a potential interaction (Figure 6).

- Since TCS service and CFV service are not compatible with a call connection and billing on a link between terminal A and B, there are two potential interactions between the two services.
- Since TCS service and RC service are similarly incompatible with respect to billing, there is a potential interaction between the two services.
- Since CFV service and RC service are similarly incompatible with respect to billing, there is a potential interaction between the two services.

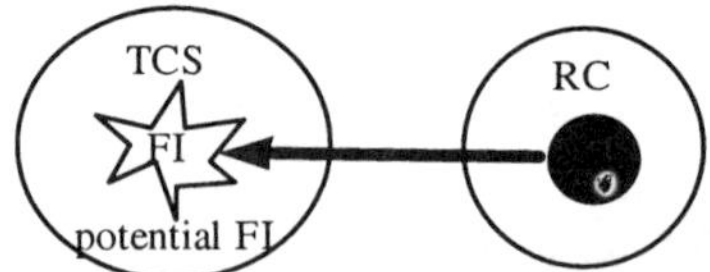

terminal B must pay for a call from terminal A

Figure 6: Potential Interaction between TCS and RC services

4) Possibility of 2-way interaction occurrence

- Suppose terminal A dials terminal B. Since execution conditions for TCS service and CFV service are satisfied, an interaction occurs: if CFV service is executed, a call connection between terminal A and B is established and a link between terminal A and

B is charged to terminal A. These interactions are 2-way interactions.
- Suppose terminal A dials terminal B. Since execution conditions for CFV service and RC service are satisfied, an interaction occurs: if RC service is executed, a link between terminal A and B is charged to terminal B. This interaction is a 2-way interaction.
- For TCS service and RC service, since TCS service rejects a call connection between terminal A and B, the execution condition for RC service is not satisfied, resulting in no interaction. That is, a 2-way interaction between TCS service and RC service, related to billing, does not occur (Figure 7).

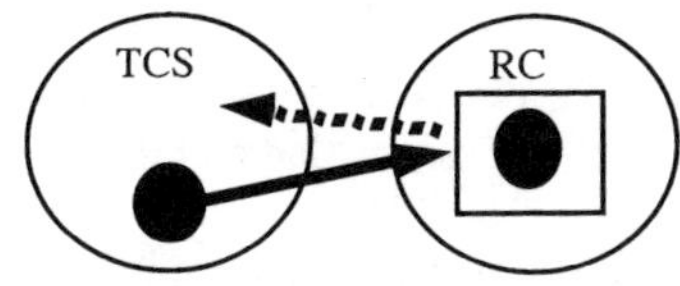

Figure 7: Prevention of an Execution Condition

5) Occurrence of a 3-way interaction
- Suppose three services are activated simultaneously. When terminal A dials terminal B, both services, CFV and TCS, are applicable. If CFV service is applied, the execution condition for RC service which is prevented by TCS service is satisfied. Consequently, an interaction between TCS service and RC service emerges: a link between terminal A and B is charged to terminal B. This interaction does not occur between just two services, TCS service and RC service.
Thus this interaction is a 3-way interaction.

3.2 Mechanism

By generalizing the example described in section 3.1, a mechanism for 3-way interaction occurrence can be described as follows:
1) There is a potential interaction between two services.
2) When two services are provided simultaneously, if an execution condition for one service that causes the potential interaction is not satisfied, a feature interaction does not actually occur even if there is the potential interaction.
3) When the third service, which creates the execution condition for the service that is prevented from execution, is applied, the service can be applied. Consequently, the potential interaction emerges. This feature interaction is a 3-way interaction.

4. Confirmation of Mechanism

To confirm the mechanism proposed in section 3.2, the proposed mechanism is applied to another combination of three services, TCS, CND and CFV.

4.1 Analysis

In CND service, when a call from terminal A is terminated to terminal B which has CND service activated (denoted by m-cnd(B)), a directory number of terminal A is displayed at terminal B.

Terminal C has CND service and TCS service activated and has registered terminal A as a screened terminal. Terminal B has CFV service activated and has registered terminal C as a forwarded terminal.

1) Execution conditions of a service
  - An execution condition for CND service is that terminal C is in calling state (denoted by calling(A,C)).
  - An execution condition for TCS service is that a call terminates to terminal C.
  - An execution condition for CFV service is that a call terminates to terminal B.

2) Specifications of services
  - CND service displays a directory number of terminal A (denoted by cnd(A)) in calling state.
  - TCS rejects a call from terminal A.
  - CFV service makes a call forward to terminal C.

3) Potential interactions
  - Since CND service and TCS service are incompatible with respect to displaying a directory number of originating terminal in calling state, there is a potential interaction between the two services.
  - Since TCS service and CFV service are incompatible with respect to a call connection between terminal A and C, there is a potential interaction between the two services.
  - Since CND service and CFV service are incompatible with respect to displaying a directory number of terminal A during a calling state, calling(A,B), there is a potential interaction between the two services.

4) Possibility of 2-way interaction occurrence
  - Suppose terminal A dials terminal B. Since an execution condition for CFV service is satisfied, the call is forwarded to terminal C and an execution condition for TCS service is satisfied. If CFV service is applied here, terminal C transits to calling state. This contradicts a specification of TCS. But, this interaction occurs without CND service. So, this interaction is a 2-way interaction.
  - Suppose terminal A dials to terminal C. Since an execution condition of TCS is satisfied, the call connection is rejected. So, a directory number of terminal A is not displayed. Since terminal C does not transit to calling state this is not an actual feature interaction.
  - Suppose terminal A dials terminal B. Since an execution condition for CFV service is satisfied, the call is forwarded to terminal C and a directory number of terminal A is displayed on terminal C. As for terminal B, since terminal B does not transit to calling state this is not an actual feature interaction.

5) Occurrence of a 3-way interaction
  - Suppose three services are activated simultaneously. Terminal A dials terminal B. Since an execution condition for CFV service is satisfied, the call is forwarded to terminal C and then, an execution condition for CND service is satisfied. Consequently, a directory number of terminal A is displayed. Thus, though terminal C rejects a call from terminal A, terminal C transits to calling state and a directory number of terminal A is displayed at terminal C (Figure 8). This interaction does not appear between any two services. So, this is a 3-way interaction.

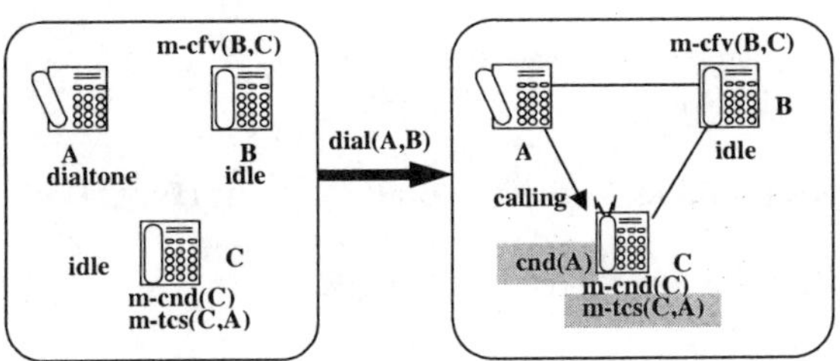

Figure 8: 3-way interaction for TCS, CND and CFV

4.2 Application of the Proposed Mechanism

The mechanism proposed in section 3.2 is applied to a combination of CND, TCS, and CFV services.

1) There is a potential interaction between CND and TCS services; displaying a directory number of terminal A on terminal C. There is a potential interaction between TCS and CFV services also; a call connection between terminal A and C. There is a potential interaction between CND and CFV services also; displaying a directory number of terminal A on terminal B.

2) When CND and TCS services are provided simultaneously, an execution condition for CND is not satisfied. Therefore the potential interaction between CND and TCS does not emerge. In the same manner, the potential interaction between CND and CFV does not emerge. When TCS and CFV services are provided simultaneously, both execution conditions for TCS and CFV are satisfied. Therefore the potential interaction between TCS and CFV emerges. So, this interaction is a 2-way interaction.

3) When the third service, CFV, is added to CND and TCS services and these three services are provided simultaneously, the execution condition of CND service is satisfied. Then, the potential interaction between CND and TCS services emerges; terminal C transits to calling state and a directory number of terminal A is displayed. This feature interaction is a 3-way interaction.

5. Detection Algorithm

5.1 Specification Description Language

The detection algorithm depends on service specification description languages. Therefore, before discussing the algorithm, a service specification description language used in the algorithm is explained. STR (State Transition Rule), which was developed at ATR (Advanced Telecommunication Technology Research Institute), is used in the proposed algorithm. A brief explanation, required to understand this paper, is provided. Please refer to [7][8] for details.

1) Language Specification

As is well known, telecommunication service specifications can be represented as state transition diagrams. So, a system state can be represented as a set of states of terminals or relationships between terminals connected to the system, called primitives. STR has the form of Pre-condition, event and Post-condition. It is a rule to define a condition for state transition and a system state change while the rule is applied. Pre-condition and Post-condition consist of primitives, respectively. An event is a trigger, which causes the state transition, e.g. a signal input to the node, and a trigger occurs in the node. A description example of STR is shown in Figure 9.

m-cfv(y,z),dialtone(x),idle(z)   dial(x,y):   m-cfv(y,z),calling(x,z)

Figure 9: An Example of STR

The example in Figure 9 is explained. Terminal y has CFV service activated and has registered terminal z as a forwarded terminal, denoted by m-cfv(y,z). A primitive, whose primitive name begins with m-, such as m-cfv(y,z), is called a service primitive. Terminal x is hearing dialtone and terminal z is in idle state, denoted by dialtone(x) and idle(z), respectively. If terminal x dials y, denoted by dial(x,y), the call is forwarded to terminal z and terminal x and z transit to calling state, denoted by calling(x,z). All arguments in primitives are described as variables, called terminal variables, so that a rule can be applied to any terminals.

[Application rule]: When an event occurs, a rule which has the same event and whose Pre-condition is included in the state of the system is applied.

[Precedent application rule]: If more than one rule is applicable, the rule whose Pre-condition covers all Pre-conditions of other rules is applied (precedent application rule).

[State change rule]: When the rule is applied, the state of the system changes as follows. A state corresponding to the Pre-condition of the applied rule is deleted from the current system state and a state corresponding to Post-condition of the applied rule is added (Figure 10). Here, a state corresponding Pre/Post-condition is obtained by replacing arguments in Pre/Post-condition with actual terminals when the rule is applied.

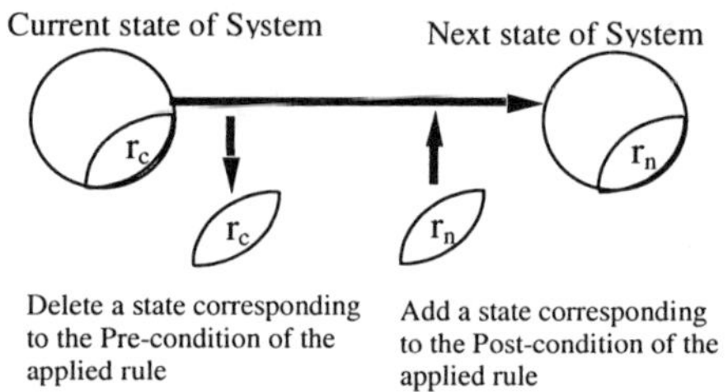

Figure 10: System State Change by Rule Application

5.2 Algorithm

Let a rule of service b, which creates a potential interaction between service a and service b in the mechanism described in 3.2, be rb. There are two ways for a rule of service a, ra, to prevent rb's execution. In one case, ra always takes precedence over rb (denoted by ra>>rb). In this case, even if a rule of service c, rc, can create an execution condition for rb, since ra>>rb, ra is always applied instead of rb. So, in this case, a 3-way interaction does not occur.

In the other case, in rules of service b, only rb2, which is another rule of service b, can create the execution condition for rb, and ra>>rb2. In this case, if rc creates the execution condition for rb, rb is applied resulting in causing a 3-way interaction. So, the detection algorithm based on the latter case is discussed.

The outline of the detection algorithm is described. The detection algorithm consists of four steps. In step 1, and step 2, identify if a potential interaction exists between two services. If there is no potential interaction, a 3-way interaction does not occur. In step 3, the potential interaction is checked if it actually emerges between two services or not. If it emerges, the potential interaction is a 2-way interaction, not a 3-way interaction. In step 4, rules of the third service are examined to establish if any rule can make the potential interaction emerge. If there is any rule which can make the potential interaction actually emerge, this potential interaction is a 3-way interaction. According to the outline of the detection algorithm, more detailed conditions are discussed to introduce detection algorithm.

step 1) Select a rule of service b, rb, which causes an interaction with specifications for service a. If rb can be selected, go to step 2. Otherwise, a 3-way interaction does not occur.

step 2) Select another rule of service b, rb2, which creates an execution condition for rb. The condition for selecting rb2 is that after rb2 is applied rb can be applied. Based on the Application rule and the State change rule for STR, described in section 5.1, the sufficient condition for rb to be applied after rb2's execution is that the Post-condition of rb2 includes the Pre-condition of rb. But, a service primitive for service b has existed. So, the Post-condition of rb2 need not include a service primitive for service b.

If rb2 can be selected, go to step 3. Otherwise, a 3-way interaction does not occur.

step 3) In this step, it is checked whether the potential interaction, obtained in step 1, actually emerges or not. In other word, select a rule of service a, ra, which can be applied in precedence over rb2 and does not create an execution condition for rb. According to the Precedent application rule for STR, described in section 5.1, the necessary condition for ra is that the Pre-condition of ra includes the Pre-condition of rb2 and the Post-condition of ra does not includes the Pre-condition of rb.

If ra can be selected, go to step 4. Otherwise, this potential interaction is a 2-way interaction, not a 3-way interaction.

step 4) Select a rule of service c, rc, which creates an execution condition for rb. The condition for rc is that after execution of rc, rb can be applied. This condition is the same condition for rb2, described in step 2. Therefore, the condition for selecting rc is that the Post-condition of rc includes the Pre-condition of rb. Thus, select a rule of service c whose Post-condition includes the Pre-condition of rb except for service primitive for service b.

If rc can be selected, this potential interaction is detected as a 3-way interaction. Otherwise, a 3-way interaction does not occur.

To simplify the explanation of the algorithm, in this paper, service constraints are considered as potential interactions. To offend service constraints for service a means as follows: A primitive set, which should not appear in any system states of service a, appears actually in some system state of service a, or a primitive set, which should appear in the specific system state of service a, does not actually appear in the specific system state [9]. The primitive set, which should not appear in any system state of service a, is called an inhibit primitive set for service a. The primitive set, which should appear in the specific system state of service a, is called an intention primitive set for service a. In this paper, in order to simplify, only inhibit primitive sets are considered. Assume a set of inhibit primitive sets has been given as knowledge beforehand [9][10].

Based on discussions mentioned above, the algorithm for detecting 3-way interactions can be described as follows:

step 1) Select a rule of service b, rb, whose Post-condition includes one of the inhibit primitive set for service a. If rb can be selected go to step 2. Otherwise, a 3-way interaction does not occur.

step 2) Select another rule of service b, rb2, whose Post-condition includes the Pre-condition of rb except for a service primitive for service b. If rb2 can be selected go to step 3. Otherwise, a 3-way interaction does not occur.

step 3) Select a rule of service a, ra, whose Pre-condition includes the Pre-condition of rb2 and whose Post-condition does not include the Pre-condition of rb. If ra can be selected go to step 4. Otherwise, this potential interaction emerges as a 2-way interaction, therefore it is not a 3-way interaction.

step 4) Select a rule of service c, rc, whose Post-condition includes the Pre-condition of rb except for service primitive for service b. If rc can be selected, the potential interaction is detected as a 3-way interaction. Otherwise, a 3-way interaction does not occur.

6. Terminal Assignment in Implementing the Detection System

Based on the detection algorithm described in section 5.2, a problem in implementing the detection system is discussed. One of the causes of complexity in detecting feature interactions is the number of ways of terminal assignments, called the number of terminal assignments hereafter. In 3-way interactions there are 4 related rules; an increase by 2 compared with 2-way interactions. As the number of terminal assignments increases, there is a possibility of an increase of computation time to detect feature interactions. This may cause a huge amount of computation time. Therefore, even if we implement the detection system based on the proposed detection algorithm, there is a possibility that it cannot actually be used because of the huge amount of computation time required to detect feature interactions. So we obtain the number of terminal assignments in the proposed detection algorithm, and evaluate whether the detection algorithm can actually be used or not.

6.1 Terminal Assignment

Terminals described in the specifications are written as terminal variables so that the specifications can be applied to any terminals. For detecting feature interactions, the real terminals should be assigned to terminal variables in the specification. Assigning real terminals to terminal variables is called terminal assignment. Even for the same specification, depending upon terminal assignments, feature interactions occur or do not occur. Therefore, to detect all feature interactions, all ways of terminal assignments should be considered. So the number of terminal assignments becomes huge. To solve this problem, the authors proposed a method for reducing the number of terminal assignments in detecting 2-way interactions [11]. For example, when the number of terminal variables in two rules is three, and the number of real terminals is six, the maximum number of terminal assignments is $(_6P_3)^2=14,400$. Here, $_mP_n$ means the number of ways of selecting n objects from m objects and arranging them in line; represented as the following formula:

$$_mP_n = m(m - 1)(m - 2)...(m - n + 1)$$

According to the proposed method, the maximum number of terminal assignments is only 33 by excluding terminal assignments which give equivalent states. But, since the number of related rules in 3-way feature interactions is four, when the number of terminal variables in each rule is three, the number of terminal assignments is 428,301 even if the proposed method is adopted. If this value is correct, it takes an enormous amount of computation time to detect 3-way interactions by the proposed algorithm. Therefore, we examine the number of terminal assignments in each step of the detection algorithm in detail.

6.2 The number of terminal assignments in each step

Using an example for detecting 3-way interactions between TCS, RC and CFV described in section 2.2, the number of terminal assignments is obtained based on the detection algorithm described in section 5.2.

First, step 1 is discussed. In step 1, the condition for selecting a rule is that the Post-condition of rb includes the constraint condition of service a. The constraint condition of TCS is that a link between terminal A and B should not be charged. This means that the primitive which represents charging on a link between terminal A and B (denoted by achg(-,A,B)) should not exist in any system state. Here, '-' represents an arbitrary terminal. Suppose that rb, which includes achg(x,y,x) in the Post-condition, exists in rules of service b. Real terminals are assigned to judge whether the condition for selecting a rule is satisfied. To satisfy the condition, achg(-,A,B) and achg(x,y,x) are the same primitive including arguments. Thus, x must be B and y must be A. Since there are no other terminal assignments required, the number of terminal assignments in step 1 is one.

Next, step 2 is discussed. In step 2, the condition for selecting a rule is that the Post-condition of rb2 includes the Pre-condition of rb. The Pre-condition of rb selected in setp1 is achg(A,A,B). Therefore, to satisfy the condition for selecting the rule, the Post-condition of rb2 must include at least achg(x,x,y). In addition, since achg(x,x,y) must be the same as achg(A,A,B), x must be B and y must be A in the same way as step 1. Since there are no other terminal assignments required, the number of terminal assignments in step 2 is one.

Next, step 3 is discussed. In step 3, the condition for selecting a rule is that the Pre-condition of ra includes the Pre-condition of rb2. The Pre-condition of rb2 selected in setp2 is {dialtone(A),idle(B)}. Therefore, to satisfy the condition for selecting the rule, the Pre-condition of ra must include at least {dialtone(x),idle(y)}. In addition, as {dialtone(x),idle(y)} must be the same primitive set as {dialtone(A),idle(B)}, x must be B and y must be A. Since there are no other terminal assignments required, the number of terminal assignments in step 3 is one.

Finally, step 4 is discussed. In step 4, the condition for selecting a rule is that the Post-condition of rc includes the Pre-condition of rb except for a service primitive for service b. The Pre-condition of rb selected in setp1 is achg(A,A,B). Therefore, to satisfy the condition for selecting the rule, the Post-condition of rc must include at least achg(x,x,y). The Post-condition of rc is {calling(x,y),achg(x,x,y),achg(y,y,z)}. Therefore, to satisfy the condition for selecting the rule, either achg(x,x,y) or achg(y,y,z) must be the same primitive as achg(A,A,B). Consequently, there are two ways of terminal assignments as follows: x=A, y=B, and z=C or y=A, z=B, and x=C. As there are no other terminal assignments required, the number of terminal assignments in step 4 is two.

For step 1, 2, and 3, mentioned above, there is only one pair of the same primitives as follows: a primitive in the constraint condition for service a and a primitive in the Post-condition of rb, a primitive in the Post-condition of rb and a primitive in the constraint condition for service a, a primitive in the Pre-condition of rb and a primitive in the Post-condition of rb2, and a primitive in the Pre-condition of rb2 and a primitive in the Pre-condition of ra, respectively. Therefore, the number of terminal assignments to satisfy the condition for selecting the rule in each step is only one. But, if the number of primitives, which has the same primitive name as those in the Pre-condition of rb and exist in the Post-condition of rc, is more than one, the number of terminal assignments to satisfy the condition for selecting the rule becomes more than one.

Next, the maximum number of terminal assignments is discussed. Now, suppose that the constraint condition of service a, the Pre-condition of rb, or rb2 are denoted by U, and the Post-condition of rb, rb2, and rc, or the Pre-condition of ra are denoted by V. The condition for selecting the rule in each step can be generally described by using U and V as follows: V includes U. Now, let the number of terminal variables used in U and the number of terminal variables used in V be n and m, respectively. As shown in the detection algorithm described in section 5.2, the terminal assignments in U have been already made. The same real terminals used in U should be used in V to satisfy the condition, V includes U. The number of ways to assign n real terminals, used in U, to m terminal variables used in V depends on the number of primitives in U and V which have the same primitive name. The maximum number of ways to assign n real terminals used in U to m terminal variables used in V is the same as the number of ways to choose n objects from m objects and arrange in line denoted by $_mP_n$. Even if more terminal assignments are done, V obtained by new terminal assignments is equivalent to one of V's which have already obtained; a set V obtained is a set equivalent to a set V given already. Therefore, the maximum number of terminal assignments in step i is given as $_{mi}P_{ni}$. Here, mi and ni are the number of terminal variables used in U and V, respectively. Now, let the number of terminals used in the constraint condition of service a, rb, rb2, ra, and rc be n1, n2, n3, n4, and n5, respectively. Consequently, each maximum number of terminal assignments at step 1-4 is given as $_{n2}P_{n1}$, $_{n3}P_{n2}$, $_{n4}P_{n3}$, and $_{n5}P_{n2}$, respectively. As the way of terminal assignments in each step is independent, the maximum number of terminal assignments to detect 3-way interactions is given as $_{n2}P_{n1}*_{n3}P_{n2}*_{n4}P_{n3}*_{n5}P_{n2}$.

Suppose, in step 3, both of the constraint conditions of service a and ra contain the same service primitive. In this case, since terminal assignment has been already made in step 1, the number of terminal assignments in step 3 decreases. Let the number of terminal variables in the service primitive be k. Then, the maximum number of ways for terminal assignments in step 3 is given as $_{n4-k}P_{n3-k}$.

Thus, the number of terminal assignments depends considerably on the number of the primitives, which have the same primitive name and different arguments, in U and V. So, we examined the actual number of primitives that have the same primitive name and different arguments.

6.3 Investigation of actual situation

To evaluate the number of terminal assignments, we investigated the actual number of primitives that are involved in U and V, and have the same primitive name and different arguments. Concretely, the distribution of the number of primitives, that are involved in Pre-conditions of each rule and have the same primitive name and different arguments, are investigated. Investigation results for 170 rules are shown in Table 1. More than 70 percent of the rules do not have the same primitive name. 20 percent of the rules have two primitives that have the same primitive names. 8 rules have more than three primitives that have the same primitive names. Next, the actual number of terminal assignments for 28 combinations of four rules (ra, rb, rb2, and rc) is discussed. The mean number of terminal assignments for one combination is 1.3. Thus, it can be expected that the total number of terminal assignments in detecting 3-way interactions isn't actually a huge number. Consequently, it was confirmed that the proposed detection algorithm can actually be used.

Table 1: Distribution of the Number of Primitives

the number of the same primitive name	the number of rules	distribution
don't have the same primitive name	127	74.7%
have two primitives that are the same primitive names	35	20.6%
have more than three primitives that are the same primitive names	8	4.7%
total	170	100%

7. Implementation of the Detection System and Discussion

7.1 The detection system

To evaluate the proposed algorithm described in section5.2, the detection system based on the detection algorithm was implemented. The software structure of the detection system is shown in Figure 11. In the detection system, first, inhibit primitive sets for service a are obtained (InhibitPrimGet function). Next, for one of the inhibit primitive sets, a rule which satisfies the condition for selecting rb in step 1 is selected (RbSelect function). For the rule selected as rb, a rule, which satisfies the condition for selecting rb2 in step 2, is selected (Rb2Select function). For the rule selected as rb2, a rule, which satisfies the condition for selecting ra in step 3, is selected (RaSelect function). A rules which satisfies the condition for selecting rc in step 4, is selected (RcSelect function). The procedure mentioned above is repeated for all inhibit primitives sets.

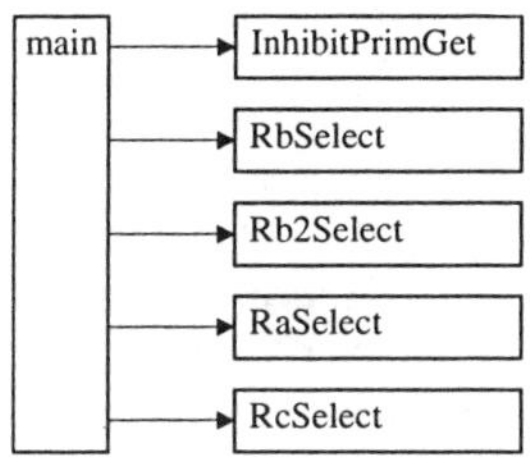

Figure 11: Software Structure of the Detection System

7.2 Discussion

The detection system, which was implemented based on the detection algorithm, was applied to the following 12 services: TCS, RC, CFV, CFB, CND, TWC, UPT, ACB, ARC, TPC, CW, and OCS.

As an example, an inhibit primitive set for CFV used as the constraint conditions is explained. Suppose terminal x has registered terminal y as a forwarded terminal. When terminal z dials terminal x, terminal x should not transit to calling state, denoted by calling(z,x). The primitive set {calling(z,x)} is the inhibit primitive set for CFV.

There are totally 10 inhibit primitive sets for 4 services, UPT, TCS, CFV, and OCS. For these 4 services, 3-way interactions detected are shown in Table 2. In Table 2, each service b has potential interactions which don't appear as 2-way interactions. As a result of applying the detection system, 82 3-way interactions were detected in the 42 combinations.

Table 2: 3-way Interactions Detected

(a) (c) \ (b)	UPT			TCS			CFV			OCS			total
	CND	RC	TPC	CND	RC	TPC	CND	RC	TPC	CND	RC	TPC	
CFB	0	0	0	1	2	2	0	0	0	1	2	2	10
CND	-	0	0	-	0	0	-	0	0	-	0	0	0
RC	0	-	0	0	-	0	0	-	0	0	-	-	0
UPT	-	-	-	1	2	2	0	0	0	1	2	2	10
ACB	2	0	0	2	2	2	2	0	0	2	2	2	16
TCS	0	0	0	-	-	-	0	0	0	0	-	-	0
TWC	0	0	0	0	0	0	0	0	0	0	0	0	0
CFV	0	0	0	1	2	2	-	-	-	1	2	2	10
CW	2	0	0	2	1	1	2	0	0	2	1	1	12
TPC	0	0	-	0	0	-	0	0	-	0	0	0	0
ARC	3	0	0	3	3	3	3	0	0	3	3	3	24
OCS	0	0	0	0	0	0	0	0	0	-	0	0	0
total	7	0	0	10	12	12	7	0	0	10	12	12	82

(a) service a　　(b) service b　　(c) service c

The numbers described in Table 2 represent the number of different scenarios where 3-way interactions occurred. That is, if a scenario, where a 3-way interaction occurs, is the same as another scenario, the 3-way interaction is not counted as the number of a new 3-way interaction. For example, suppose a feature interaction occurs when terminal A dials terminal B while terminal B is talking with terminal C. In this case, strictly there are two cases as follows: In one case, the call originated from terminal B and terminates to terminal C, in the other case, the call originated from terminal C and terminates to terminal B. If the 3-way interaction detected is related to only talking state, both cases are regarded as the same 3-way interaction. So, in this case, the number of detected 3-way interactions is one. Besides, even if the state of a terminal is different, if the terminal is not related to the 3-way interaction, this 3-way interaction is not counted as a new interaction.

All 3-way interactions, which were found during desk work for confirming the proposed mechanism, were detected. Moreover, we could detect 3-way interactions which had not been found during the desk work. As an example, a 3-way interaction among TCS, RC, and CW is explained. Suppose terminal B has these three services activated simultaneously and has registered terminal A as a screened terminal. When terminal A dials terminal B while terminal B is talking with terminal C, if CW is applied, terminal B transits to audible ringing state. At this moment, if terminal B makes onhook, terminal B transits to ringing state. Then, when terminal B makes offhook, a link between terminal A and B is charged to terminal B by RC. This means that though TCS service rejects a charge for the call from terminal A to terminal B, the bill for the call from terminal A to terminal B is charged to terminal B. Since this interaction does not occur between two services, this interaction is a 3-way interaction.

Considering discussions above, it is confirmed that the proposed detection algorithm for 3-way interactions is effective enough.

8. Conclusion

In this article, the mechanism for 3-way interactions occurrence and the detection system based on the mechanism were proposed. Then, the problem of terminal assignment, a problem in implementing the detection system based on the proposed detection algorithm for 3-way interactions, was described. The number of terminal assignments for each step of the detection algorithm was investigated. As a result, it was clarified that there was in fact no problem. Then a detection system for 3-way interactions was implemented and was applied to 12 services. As a result, 82 3-way interactions were detected in the 42 combinations. All 3-way interactions, which had been found during desk work for confirming the proposed mechanism, were detected. Moreover, we could detect 3-way interactions which had not been found during the desk work. Thus, it was confirmed that the proposed detection algorithm is effective enough.

In future work, other detection scenarios should be investigated. The algorithm to resolve the 3-way interactions detected should also be investigated.

References

[1] J.D.Keijzer et al., "JAIN: A New Approach to Services in Communication Networks," IEEE, Comm. Magazine, Vol.38, No.1, pp94-99, Jan. 2000.

[2] S. Moyer and A. Umar, The Impact Network Convergence on Telecommunications Software, IEEE, Comm. Magazine, Vol. 39, No. 1, pp78-84, Jan. 2001.

[3] K.L.Calvert et al., "Directions in Active Networks," IEEE Communications Magazine, Vol. 36, No. 10, pp72-78, Oct. 1998.

[4] N. Griffeth et al., "Feature Interaction Detection Contest of the Fifth International Workshop on Feature Interactions," The Interanational Journal of Computer and Telecommunications Networking, Computer Networks 32 (2000) pp.487-510, April 2000.

[5] N. Griffeth et al., "A Feature Interaction Benchmark for the First Feature Interaction Detection Contest," The Interanational Journal of Computer and Telecommunications Networking, Computer Networks 32 (2000) pp.389-418, April 2000.

[6] M. Kolberg et al., "Results of the Second Feature Interaction Contest," Proc. of FIW00, pp. 311-325, May 2000, IOS Press.

[7] Y. Hirakawa, et al., "Telecommunication Service De-scription Using State Transition Rules," Int.Workshop on Software Specification and Design, Oct. 1991.

[8] T. Yoneda and T. Ohta, "The declarative language STR," Language Constructs for Describing Features, pp.197-212, Aug. 2000, Springer.

[9] T. Yoneda and T. Ohta, "A Formal Approach for Definition and Detection of Feature Interactions," Proc. of FIW'98, pp.202-216, Sep. 1998, IOS Press.

[10] T. Yoneda and T. Ohta, "Automatic Elicitation of Knowledge for Detecting Feature Interactions in Telecommunication Services," IEICE transactions on information and systems, vol.E83-D No.4, pp.640-647, April 2000.

[11] T. Yoneda and T. Ohta, "Reduction of the number of Terminal Assignments for Detecting Feature Interactions in Telecommunication Services," Proc. of ICECCS, pp.202-209, Sep. 2000.

Author Index